VARIORUM COLLECTED STUDIES SERIES

Science, Philosophy and Religion in the Age of the Enlightenment

John Gascoigne

John Gascoigne

Science, Philosophy and Religion in the Age of the Enlightenment

British and Global Contexts

ASHGATE
VARIORUM

Published in the Variorum Collected Studies Series by

Ashgate Publishing Limited
Wey Court East
Union Road
Farnham, Surrey
GU9 7PT
England

Ashgate Publishing Company
Suite 420
101 Cherry Street
Burlington, VT 05401–4405
USA

Ashgate website: http://www.ashgate.com

ISBN 978–1–4094–0058–5

British Library Cataloguing in Publication Data

Gascoigne, John, Ph. D.
  Science, philosophy and religion in the age of the Enlightenment : British and global contexts.
  – (Variorum collected studies series ; 946)
  1. Enlightenment – Great Britain. 2. Great Britain – Intellectual life – 18th century. 3. Philosophy and religion – Great Britain – History – 18th century.
  4. Science – Great Britain – History – 18th century. 5. Newton, Isaac, Sir, 1642–1727 – Influence. 6. Culture diffusion – Europe – History – 18th century.
  7. Discoveries in geography, European – History – 18th century.
  I. Title II. Series
  940.2'5–dc22

  ISBN 978–1–4094–0058–5

Library of Congress Control Number: 2009942162

VARIORUM COLLECTED STUDIES SERIES CS946

CONTENTS

This volume contains x + 316 pages

PUBLISHER'S NOTE

The articles in this volume, as in all others in the Variorum Collected Studies Series, have not been given a new, continuous pagination. In order to avoid confusion, and to facilitate their use where these same studies have been referred to elsewhere, the original pagination has been maintained wherever possible.

Each article has been given a Roman number in order of appearance, as listed in the Contents. This number is repeated on each page and is quoted in the index entries.

INTRODUCTION

In an increasingly globalised and pluralistic world the core values of modernity have been exposed to ever more exacting scrutiny – something which has entailed rigorous examination of the Enlightenment, the fountainhead of modernity. Ever contentious, the Enlightenment has, of late, prompted both passionate apologetic as well as critical examination. For Stephen Bronner in his tellingly entitled, *Reclaiming the Enlightenment: Toward a Politics of Radical Engagement* (New York, 2004), the *philosophes* of the Enlightenment 'project the type of world that every decent person wishes to see' (p. 167). Similarly, for Jonathan Israel, in his weighty and formidably erudite *Enlightenment Contested: Philosophy, Modernity, and the Emancipation of Man 1670–1752* (Oxford, 2006), the values of the Enlightenment are 'above any possible alternative' (p. 869). In Israel's account the true Enlightenment, which the world neglects at its peril, is embodied in the Radical Enlightenment which challenged root and branch the values of a traditional order based on religion and inherited hierarchy. The propagators of this Radical Enlightenment were in conflict with those espousing the more pallid and compromised Moderate Enlightenment which sought to change the existing order by evolution rather than revolution. This conviction that the true, Radical Enlightenment meant a total reshaping of society – a continuing project basic to modernity – is accompanied by a strong commitment to the view that ideas have an intrinsic power which has been underestimated by the tendency of Enlightenment scholarship to focus on institutional factors in the spread of values.

The collection of essays in this present work stands at variance with such a revisionist view of the Enlightenment. The emphasis in this collection is on the evolutionary rather than the revolutionary character of the Enlightenment and its ability to change society by adaptation rather than demolition. There is little support, too, in these essays for a distinction between Moderate and Radical Enlightenments; rather there is a spectrum of views which merge into each other in a way that defies any such clear dichotomy. Radicals and conservatives engaged with each other and drew on each other's work – I, for example, traces the interplay backwards and forwards between Christian and unbeliever in the development of a view of history which downplayed the significance of the ancient Israelites as the Chosen People. The evolutionary potential of Enlightenment values is also the thread that runs through **II–V**. **II** shows how forms of Newtonian-based natural theology acted as one of the major vehicles for dissemination of Enlightenment-based values in Britain. This tradition was also sufficiently multi-layered to adapt to the changing wider social context and, with a view to discerning the longer trajectory of the Enlightenment, its modulations are traced up until Darwin. It is argued, however, that the fall of Newtonian natural theology cannot be explained by the impact of Darwin alone. Rather the increasingly effective separation of Church and State reduced the imperative to align religion and science, an instance of the way in which ideas draw much of their potency from their context, something which Israel's emphasis on the intrinsic potency of ideas underplays.

For ideas to have purchase on a society they have to be given forms which will have meaning to those whose lives are shaped by the institutions and practices of the world into which they have been born. In their different ways, **III**, **IV** and **V** all provide examples of such a view. **III** and **V** depict the way in which Newtonianism was diffused and institutionalised in the curriculum at Cambridge through an adaptation of existing academic forms and with the support of a confident belief of the consonance of theology and science. **IV** shows the way in which the attack on teleology, which Israel sees as such as a defining feature of the Radical Enlightenment (p. 812), did not altogether eradicate the hold of the traditional belief (long nurtured by Aristotelianism) that the True and the Good were one. In keeping with the evolutionary character of the Enlightenment's impact, however, a teleological understanding of the True, in the form of science, was modernised by exploring aspects of the Newtonian heritage. The ability of traditional institutions to remould themselves and to incorporate forms of knowledge central to the Enlightenment is a theme that runs throughout **V** with its detailed account of the changing forms of philosophical instruction and discussion in the teaching institutions and learned societies of the British Isles in the eighteenth century. The Scottish universities stand out particularly for their ability to reshape the strong philosophical heritage of scholasticism through rigorous attention to philosophical enquiry. They thus brought to bear Enlightenment canons of rationality on such perennial concerns as an understanding of the workings of the human mind or more contemporary issues such as the moral basis of civil society or the well-springs of modernity. Such a transformation is a potent example both of the way in which Enlightenment values could build on traditional forms in an evolutionary manner and the shaping role of institutions. In both respects this calls into question Israel's view of an Enlightenment which had as its true core a set of ideas fundamentally at odds with its ambient society – ideas, moreover, capable of achieving change by their intrinsic impact rather than institutional dissemination.

One important and relatively understudied aspect of the context which provided the breeding ground for the formulation and the dissemination of Enlightenment views was the European encounter with a wider world – hence the sub-title of this collection, 'British and Global Contexts'. **VI** demonstrates how important were Enlightenment values of science and the promotion of knowledge in justifying exploration and incipient imperialism, thus taking the place of religion in the defence of earlier, largely Iberian, expansion. Such an Enlightenment-based ideology had force even within the counsels of government – thus challenging the view that the true Enlightenment was fundamentally radical and even subversive in character. One of the most potent embodiments of such an Enlightenment-tinged determination to shed the light of reason and science on the dark corners of the globe was the use of maps. **VII** takes up this theme by examining the way in which this impulse was pursued by the politically conservative but Enlightenment-influenced Joseph Banks in different settings and levels from the highly local to the national and thence to the imperial. **VIII** takes up another aspect of Joseph Banks' merging of imperial expansion with the promotion of Enlightenment forms of science through his relations with the Göttingen professor, Johann Blumenbach. The material provided by Banks and others in turn provided Blumenbach with the basis of a form of racial classification – thus taking further that systematising spirit of figures such as Linnaeus with their Enlightenment-inspired attempt to render the world knowable and amenable to human improvement. But such forms of knowledge did not necessarily promote the values of universal tolerance that the Enlightenment's advocates now would attribute to it: Blumenbach argued that all races shared a common humanity but, as **IX** shows, such monogenesis was to be challenged by the polygenesis of other German proto-anthropologists such as Samuel von Sömmerring and George Forster. The analysis here of George Forster also demonstrates the permeability of moderate and radical views during the Enlightenment, with Forster's enthusiasm for the French Revolution drawing on ways of understanding society's development which had much in common with more conservative figures such as his father. In **X** the Enlightenment quest for classifactory order is illustrated in the Royal Society's experiences in pursuing forms of natural history which included ethnography in a world ever more open to European and, especially, British scrutiny. Such globalising discourses were to form an important part of the Enlightenment's attempts to construct a 'Science of Man'. As this collection suggests, the Enlightenment was too multi-faceted and too diverse to be readily contained within such clear boundaries as 'Radical' and 'Moderate'.

JOHN GASCOIGNE

University of New South Wales
August 2009

ACKNOWLEDGEMENTS

Grateful acknowledgement is made to the following publishers for permission to reprint the articles included in this collection, fuller information concerning relevant publications being given in the Contents: Springer, Dordrecht (for **I**); Cambridge University Press (**II**, **III** and **X**); the editors of *Enlightenment and Dissent* (**IV**); Schwabe AG, Basel (**V**); University of Hawai'i Press, Honolulu, HI (**VI**); Manchester University Press (**VII**); Wallstein Verlag, Göttingen (**VIII**); and Stanford University Press, Stanford, CA (**IX**).

For the images in **III** grateful acknowledgement is made to Cambridge University Library for figures 22, 24 and 27; Kevin Knox for figure 23; Archives and Special Collections, California Institute of Technology for figure 25; and The British Library Board for figure 26.

I

'THE WISDOM OF THE EGYPTIANS' AND THE SECULARISATION OF HISTORY IN THE AGE OF NEWTON

One of the basic tensions with which the proponents of the Scientific Revolution had continually to contend was to find some accord between the mechanical philosophy — which was found to provide every-increasing explanatory utility — and the concept of Providence, which underlay the religious foundations of their society. If, as the mechanical philosophy appeared at times to suggest, all was determined by particles in motion it was difficult to find a place for a guiding Providence impressing his will on Creation; or, at best, the Deity was consigned to a marginal role as a remote initiator who thereafter took little interest in terrestrial affairs. But in a society in which religion was inseparably intertwined with the political and social order it was essential to find some sort of a *modus vivendi* between the mechanical philosophy, which tended towards a form of determinism, and a conception of the role of Providence which still allowed for God's involvement in directing human affairs. That the scientific developments to which we give the name the Scientific Revolution continued to grow and prosper — despite such setbacks as Galileo's condemnation — is an indication that early modern European society was persuaded that such a reconciliation was possible.

The Scientific Revolution, then, involved not only a transformation in humankind's understanding of the physical universe but also a reappraisal of the theological conceptions which underlay so many facets of early modern society. As Funkenstein[1] has recently argued at length, the intellectual matrix in which the Scientific Revolution was formed was the result in large part of a long process of theological and philosophical debate, which had its roots in the Middle Ages, out of which developed an understanding of the role of Providence which allowed for the possibility of uniform physical laws. Such a conception of God's relations with the physical world was to underlie the mental universe of many of the major figures we associate with the Scientific Revolution — in particular Kepler, Boyle and Newton regarded the lawlike and mathematically predictable behaviour they discerned in nature as instances of the mind of the Creator at work in shaping and sustaining his creation. By the late seventeenth century, then, natural philosophers and, to an increasing extent, theologians, tended to emphasise the regularity and predictability of Providence's actions. In the language of the theologians ordinary (or general) Providence — God's direction of the universe as a whole through a set of uniform laws — received greater emphasis than extraordinary (or particular) Providences — occasions on which God intervened directly in human or terrestrial affairs. Or, to put the same intellectual transformation in the language of the natural philosophers, secondary causes — the instruments of God's will — tended to be given closer attention than the final causes which were concerned with God's ultimate purposes.

One further reflection of this same change in mentality was the growing tendency to regard the study of history, as well as the study of nature, as principally concerned with understanding those natural, secondary causes which were Providence's agents rather than with directly tracing God's role in human affairs. As in natural philosophy so, too, in history there was an increasing reluctance to explain events by involving God's direct intervention; while few historians (or natural philosophers) denied God's ultimate control of events Providence was more and more portrayed as working through the natural processes of history. Such an understanding of history — which we may, for convenience, describe as more secular (though, not necessarily, anti-religious) in character — was not altogether new in the seventeenth century. The Renaissance humanists had drawn on the example of classical historians to promote the study of history as a discipline with its own methods and concerns which could be distinguished from those of theology.[2] But, as Smith Fussner and Sypher have argued, such an approach to history may well have been strengthened by the increasing tendency of the Scientific Revolution to separate the study of events explicable in terms of secondary causes from the final causes which were the natural concern of the theologians.[3] Certainly the realisation of such an intellectual demarcation was a lifelong goal of Francis Bacon and was reflected in his historical as well as his scientific writings. It is the argument of this essay that such a secularised approach to historical issues had, by the late seventeenth century, become so pervasive that it coloured not only historical accounts of political events, where the influence of classical models and the Renaissance humanists was naturally strong, but also shaped the way in which the biblical record of the history of the Chosen People was understood. Such a reappraisal of the role of Providence in relation to so sensitive an area as the interpretation of Scripture was, moreover, to be closely linked with Newton and a number of his contemporaries who sought to defend true religion both from the superstitious and the unbelieving by showing how Providence worked primarily through secondary causes both in nature and in history.

The basic outlines of a form of Christian historical apologetics had been developed in the early Church in the context of the debates between Christian and pagan scholars. Some Church Fathers and, in particular, the influential Clement of Alexandria (d. 213 AD) had argued that classical learning, including that of the Greeks, had been largely derived from the Jews — a line of argument which, not surprisingly, had previously been advanced by Hellenistic Jews.[4] The writings of the supposed ancient Egyptian seer, Hermes Trismegistus, were used both by the Church Fathers and, subsequently, by Renaissance scholars such as Ficino to defend an extension of this position: that the Greeks not only derived their learning directly from the Jews but also indirectly through the Egyptians who, in turn, had been greatly influenced by the Jews.[5]

This ancient problem of the relations between the history of the Jews and other peoples of the ancient world became a matter of renewed concern after the Renaissance when the recovery of an increasing number of classical texts focussed attention on the problem of reconciling the biblical chronology with that which could be derived from pagan authors. Much scholarly labour was devoted to harmonising the biblical chronology with the classical histories of Greece and Rome and with the account by Greek historians of the Egyptians and other ancient civilizations of the Middle East. Newton's chronological writings formed part of this erudite tradition and, like many of his scholarly brethren in the field, he developed ingenious explanations to help to explain why the claims of Egypt to a past much older than that of the Jews could be dismissed — thus Newton argued that the supposed ancient pharoah, Sesostris, was but another name for Sesac, who posed no challenge to the biblical chronology since he was recorded in the Old Testament as invading Judea after the death of Solomon.[6]

Underlying the debate about such arcane matters of chronology was the desire by Christian scholars to defend a conception of world history which accorded primary importance to the Jews. In a society in which the biblical account still provided the basic framework for understanding history a defence of the historical primacy of the Jews as against the Egyptians, or other Gentile peoples, was also a means of defending an essentially providentialist conception of history in which God's relations with his Chosen People provided history with its most ancient and enduring theme. To further bolster this providentialist conception of history, with its premise that the biblical account of the Jews provided the key for understanding the history of the ancient world, a number of late seventeenth-century scholars restated the traditional argument that the Jews were not only the most ancient people but also that it was from them that the learning and even the languages of the ancient world were derived.

One of the lengthiest defences of this view was the work of Theophilus Gale, a dissenting divine, whose vast compilation of erudition, *The Court of the Gentiles* (1669—77) set out to demonstrate

That the wisest of the Heathens stole their choicest Notions and Contemplations both Philologic, and Philosophic, as wel Natural and Moral, as Divine, from the sacred Oracles.

One chapter of the work was even devoted to the theme 'Of the Traduction of al[l] Languages and Letters from the Hebrew'. The origin of this dissemination of Jewish learning to the Egyptians and hence to the Greeks Gale traced back to the period when Joseph and other partriarchs dwelled in Egypt for, before then, 'the Egyptians were no way famous for Wisdom, or Philosophie'. Much later, the Egyptians and the Greeks also benefited from the translation of the Hebrew Bible into Greek.[7] Among Gale's readers was the mathematician, John Wallis, who invoked Gale's authority when arguing that:

'Tis well known (to those conversant in such Studies) that much of the Heathen Learning (their Philosophy, Theology, and Mythology) was borrowed from the Jews, though much Disguised, and sometimes Ridiculed by them.[8]

Though the late seventeenth century saw the confident restatement of this ancient form of apologetics it was also in this period that a number of scholars began to question the traditional importance accorded to the Jews. And, though this scholarly re-evaluation was largely the work of those within the Christian fold, this tendency to downplay the significance of the Jews brought with it the need to reevaluate the highly providentialist understanding of world history which for so long had been associated with the view that the biblical account of the Chosen People was the lynch-pin for any understanding of the history of the ancient world. But such a reappraisal of world history did not necessarily mean abandoning a providentalist conception of history altogether — just as Christian exponents of the mechanical philosophy argued for the importance of understanding the way in which God used secondary causes to achieve his ends so, too, those engaged in developing a more critical understanding of the relations of the Jews with other peoples and cultures were frequently to argue that their work was an illustration of the way in which Providence could work through the normal processes of human history. The God of such historians, as of many of the mechanical philosophers, was one who intervened in terrestrial affairs only on rare occasions but one, nonetheless, whose guiding hand could be discerned in the overall pattern of events.

Such an understanding of both the workings of history and of nature can be seen in the writings of John Spencer, whose work did much to prompt a re-evaluation of the traditional conception of the relations between the Jews and their neighbours in the ancient world. While master of Corpus Christi College, Cambridge (a post he held from 1667 until his death in 1693), Spencer produced his pioneering work in the field of comparative religion, the massive *De Legibus Hebraeorum Earum Rationibus* . . . (1685). In it he explored in great detail and with formidable erudition the theme he had already ventured upon in his relatively brief, *Dissertatio De Urim and Thummim* (1669) — a discussion of the origin of Hebrew divinatory practices — namely the extent to which the Jews derived their religious practices from the Egyptians. Spencer's answer was that, although some of the Jewish laws were framed in opposition to Egyptians and other Gentile practices, none-theless

God tolerated and transferred not a few of the rites that were in use among the Pagans into his own law and worship; after he had corrected and reformed them.[9]

Spencer pointed out that the Egyptians of the age of Moses were a deeply conservative people who were unlikely to borrow rituals and customs from others — least of all from the Jews whom they despised. Moreover, the Greeks, who freely acknowledged their debt to the Egyptians, had nothing to say about the Jewish origins of Egyptian culture. Thus the close parallels between Jewish and Egyptian customs which Spencer minutely described in relation to a whole range of practices from the cult of worshipping God in a tabernacle to funeral customs could only mean that the Jews derived such practices from the Egyptians — just as the Egyptians had earlier been influenced by the customs of such neighbouring peoples as the Syrians.

Such a view ran counter to the commonly accepted Christian framework of world history in which the Jews were portrayed as the fountain head of the civilization of the ancient world. But in Spencer's view such a revision of the received wisdom did not entail any breach with orthodoxy. For it was Spencer's goal to show how God used the natural processes of history and, in particular, the cultural assimilation of Egyptian practices among the Jews to win over his Chosen People. Thus he stressed that the Jews did not come to a knowledge of the true God as a result of a single, dramatic, divine intervention in human affairs but rather that it was the result of a slow, often imperfect process of development. The ancient Jews, he wrote were 'rude, ignorant, obstinate' and very 'superstitious' so that

it was almost necessary that God should indulge them the use of some of the ancient [Gentile] rites, and accommodate his sanctions to their taste and capacity.

In particular God allowed the Jews to retain

very many of the Egyptian usages; especially those, that by their pompous appearance and shew were likely to please and take with the populace.[10]

Spencer, then, saw his work as contributing to an understanding of the way in which Providence could use secondary causes to influence history. True religion had to work through the natural limitations of human history — even the distinctive rituals of Christianity took some time to develop from their Jewish origins, a process that, by implication, Spencer parallels with the way in which Judaism gradually developed from earlier forms which owed much to the Gentile cultures of the Near East.[11] In Spencer's understanding of history, then, God does not achieve his goals 'immediately and as if through straight lines'[12] but rather is generally content to shape human affairs by the natural process of history with all its attendant delays and complications.

A similar conception of the way in which Providence works through secondary causes, rather than by means of direct intervention, can also be found in Spencer's work dealing with God's relations to the natural world. This theme he developed in a Cambridge sermon entitled *A Discourse Concerning Prodigies* . . . which was delivered in 1663 in the still tense aftermath of the Restoration when many feared that religious enthusiasts could undermine the restored order in Church and State.[13] Thus it was Spencer's purpose to show that the comet of that year, and indeed all such prodigies, should not be seen as harbingers of providential intervention. Such an aim, argued Spencer, was not only 'profitable to serve the just interest of the State' but also 'serve[d] the honour of Religion which the common reverence of Prodigies doth greatly trespass upon'.[14] For the argument which underlies Spencer's sermon is that God's will is manifest in the regular, law-like behaviour of the universe and that occasional spectacular events (such as comets) still conform to such divinely ordained laws; in short, he emphasizes the role of ordinary rather then extraordinary Providence. Thus he sees it as a matter of religious awe that one could discern

the faithfulness of Nature to its Original Laws of motion, the continuance *of all things as they were from the beginning of the Creation.*

God, he continues, with a metaphor natural to an author,

never saw it necessary (as upon maturer thoughts) to correct and amend any thing in this great Volume of the Creation since the first edition thereof.

Thus God's will is achieved not through the sort of direct intervention which some claimed to see in the case of comets but rather through his control of 'the several motions and mutual aspects of Secondary Agents, from the beginning of time to the end thereof'.[15] In support of his view that since comets and other such prodigies could be explained as the manifestations of a set of uniform cosmic laws they do not constitute a direct providential intervention in human affairs he invokes the authority of Kepler and Tycho Brahe. Spencer uses their work to demonstrate both that the traditional Aristotelian understanding of comets as a fiery 'exhalation' of the sub-lunar world was false and that the distance of comets from the earth undermined the view that terrestrial evil was the result of 'the malign aspects of Comets'.[16]

Though Spencer's basic aim was to 'assert *Prodigies* to be none of God's designed tokens' he was mindful, nonetheless, of the delicate intellectual balancing act that such a line of argument required of him as an Anglican opponent not only of enthusiasts but also of unbelievers. Spencer was well aware that while too great an emphasis on the role of direct providential intervention might lead to superstition and enthusiasm an exclusive concern with the natural means by which God's will is achieved might lead to infidelity. As Spencer himself put it:

As we must not loose [sic] our Philosophy in Religion, by a total neglect of second causes and turning superstitious; so neither must we loose our Religion in Philosophy, by dwelling on second Causes, till we quite forget the *First* and become profane.[17]

Spencer's solution was to emphasise the ultimate dependence of such secondary causes on the will of God — a point he made in the manner of a true mechanical philosopher by invoking the analogy between the world and a clock which

though it contain[s] very strong and powerful Springs of action within it self . . . yet these blind and decaying Powers must be managed and perpetually wound up by an Hand of Power and Counsel.[18]

This balance between the role of first and secondary causes was one which Spencer sought to achieve not only in his natural philosophy through his emphasis on the role of Providence in creating and sustaining a system of uniform laws but also in his historical writings by arguing that God used secondary causes — such as the cultural influence of the Egyptians — to help bring into being the form of religion which was to develop into the mature Judaism of God's Chosen People.

Predictably, however, Spencer's *De Legibus*, with its implied critique of the primary place traditionally accorded to the Jews in the history of the ancient world prompted considerable controversy. Spencer's view of Providence as employing the natural processes of human history to bring to fulfillment the Jewish religion prompted Hermann Witsius, a professor at Utrecht, to decry in his work, *Aegyptiaca* (1683), such an interpretation since

it is a dishonouring of GOD, who has the hearts of men in his power . . . to conceive of him as standing in need to the tricks of crafty politicians.[19]

Closer to home, John Edwards, another Cambridge don, also reacted sharply against Spencer's attempt to demonstrate the way in which God shaped history through the natural processes of human society. Edwards, a vigilant defender of Christian orthodoxy who was to be one of Locke's most pertinacious critics, devoted part of his modestly entitled *Polgpioikilos Sophia. A Compleat History or Survey of All the Dispensations and Methods of Religion, From the Beginning of the*

World to the Consummation of All Things ... (1699) to arguing that the way in which Spencer portrayed the Jews developing many of their religious rituals from forms derived from the Egyptians and other Gentile peoples made 'the True God most diligently and precisely tread in the steps of the false Gods and Idols'. Edwards then vigorously reasserted the traditional primacy accorded to the Jews by setting out to demonstrate

that many of the Pagan Rites and Customs in Religion (as well as in Secular Affairs) were borrow'd from the *Jews* and their Secular Usages

— something, he added, with a pointed reference to Spencer,

which is directly contrary to what this author asserts, *viz* that the Rites and Ceremonies injoyn'd by God himself to the *Jews* were of Pagan Extraction.[20]

In an earlier work, *Cometomantia. A Discourse of Comets* ... (1684) Edwards had attacked another aspect of Spencer's portrayal of the role of Providence: his argument that comets were not portents of calamity since God's will was made manifest in nature through uniform laws rather than divine intervention. Just as Edwards saw Spencer's conception of history as undermining the role of an active Providence so, too, he saw Spencer's form of natural philosophy as tending the same way. For Edwards 'A Christian Divine cannot have a better Text to preach Repentance from than a Comet' since a comet was an overt sign of the immediate and direct involvement of Providence in human affairs. God, he argued, 'speaks to and instructs Mankind' not only through Scripture but also 'by the course of Nature, and by Acts of Providence'. Edwards concedes that the Moderns had shown that some comets were above the moon but maintained that at least some others were sublunar, as Aristotle had taught, and could therefore influence human affairs.[21] Edwards' immediate target was probably Pierre Bayle whose famous pamphlets on comets — in which he, like Spencer, argued that the appearance of a comet had no supernatural significance — were published in 1682 and 1683,[22] just before Edwards' work. However, as a fellow Cambridge resident Edwards must have known Spencer's influential work on prodigies and there is little doubt that, along with other clerical adversaries such as Gassendi, Edwards had Spencer in mind when he castigated those among the clergy who undermined belief in comets as divine portents. Such writers, stormed Edwards:

carry on the Plot of Atheists and Epicures to root out the Notion of a God, to extirpate Providence, to debauch Mens Lives and Manners, and to blot out the Sense of another World.[23]

The geologist, physician and lay Christian apologist, John Woodward, expressed his reservations about Spencer's historical work more temperately than the irascible Edwards and acknowledged his 'infinite industry' but he, too, regarded Spencer's challenge to the commonly accepted framework of world history as a threat to orthodoxy. Though he acknowledged that it may have been unintentional Woodward viewed Spencer's work as tending in the same direction as Giordano Bruno 'and others of like libertine principles, who bear no good-will to Christianity' and who, in order to subvert it, 'extoll the Egyptians beyond measure, decry the Jews, and vilify that nation, their archives and laws, as meerly of Egyptian extract'.[24] Woodward's response in his posthumously published *Of the Wisdom of the Ancient Egyptians* was to undermine such an historical interpretation by a systematic denigration of Egyptian civilization and learning. The Egyptians, he wrote, 'were the most ostentatious, boasting people in the universe' and such much vaunted achievements as the pyramids were vastly overrated. Far from being the source of many of the Jewish rituals the Egyptian religion

was undoubtedly the wildest and most fantastic that the sun ever saw . . . They were above all other nations, so sunk in idolatory, that they seem to have known little if anything of God.

Indeed, Spencer's whole thesis needed to be turned on its head — it was the Egyptians that derived many of their rituals and laws from the Jews rather than vice versa.[25]

Along with Spencer, another of Woodward's targets in this same work was Thomas Burnet who, in his *Archaeologiae Philosophicae sive Doctrina Antiqua de Rerum Originibus* (1692), had, like Spencer, contrasted the civilization of the Egyptians with the backwardness of the Jews. Burnet had in fact gone so far as to suggest that the Mosaic cosmogony was (as Woodward summarises him)

drawn up . . . in imitation of those of their neighbours the Egyptians, Phoenicians and Chaldeans, all of whom the Jews want greatly to admire.[26]

Burnet's *Archaeologiae Philosophicae*, then, complemented his earlier, much better-known work, *The Sacred Theory of the Earth* (1681),[27] in which he had argued that the biblical account of the creation of the earth should largely be understood metaphorically. Burnet had then devoted the bulk of his *Sacred Theory* to demonstrating that the formation of the world could largely be explained in terms of natural forces even though the scriptural record could be used to supplement and clarify those aspects of the world's history which were otherwise obscure. As Burnet himself wrote:

This theory being Chiefly Philosophical, Reason is to be our first Guide; and where that falls short or any other just occasion offers itself, we may receive further light and confirmation from the Sacred writings.[28]

Burnet was reluctant, however, to invoke miracles or supernatural explanations too readily for the whole thrust of his work was to emphasise the extent to which God worked through the natural processes of nature by means of his ordinary rather than his extraordinary Providence; or, as Burnet himself put it, his aim was to bring out

the true Notion and State of *Natural Providence*, which seems to have been hitherto very much neglected, or little understand in the World.

It followed, then, that

if we would have a fair view and right apprehensions of Natural Providence, we must not cut the chains of it too short, by having recourse, without necessity, either to the First Cause, in explaining the Origins of things: or to Miracles, in explaining particular effects. This, I say, breaks the chains of Natural Providence ... Neither is anything gain'd by it to God Almighty; for 'tis but ... to take so much from his ordinary Providence, and place it to his extraordinary.[29]

It was no accident that Woodward produced not only a critique of Burnet's (and Spencer's) historical views but also a much more influential attack on his cosmogony. For, in Woodward's view, both Burnet's historical and cosmogonical systems served, however unintentionally, to weaken the foundations of true religion since both greatly reduced the extent of providential intervention in terrestrial affairs. Thus it was the aim of Woodward's *Essay towards a Natural History of the Earth* (1695) to establish that the form of the terraqueous world could only be explained by God's direct intervention as explicated in Scripture — an endeavour which predictably was enthusiastically received by John Edwards who saw it as confirming that 'what we behold in the world is a Proof of a Deity and Providence'.[30] Woodward placed particular emphasis on the significance of the biblical Deluge which he viewed as necessitating a virtual second Creation which could only be explained in miraculous terms. In explicit opposition to Burnet he affirmed that

the Deluge did not happen from an accidental Concourse of Natural Causes, as the above-cited [Burnet] is of Opinion; that very many Things were then certainly done, which never possibly could have been done without the Assistance of a *Supernatural Power*.[31]

Thus Woodward saw his account of the Deluge as demonstrating 'the Fidelity and Exactness of the Mosaic Narrative of the Creation'.[32]

Just as Woodward's emphasis on direct providential intervention was reflected both in his writings on cosmogony and history — the latter of which argued at length for the overriding historical importance of the Jews as God's Chosen People and the consequent cultural insignificance of the Egyptians — so, too, Burnet's historical work, like his cosmogony, argued for a more naturalistic interpretation of the role of Providence in history. Even in his *Sacred Theory* Burnet had implied agreement with the general tenor of Spencer's historical views by commenting that 'as to the *Jews*, 'tis well known that they have no ancient Learning, unless by way of Tradition, amongst them.'[33]

This point he developed further in his *Archaelogiae Philosophicae* (1692) in which he argued that the Mosaic law was influenced not only by the Egyptians but also by another Gentile people, the Zabians. In considering the 'Contention' between the Jews and the Egyptians 'about the Precedency and Antiquity in learned Discoveries' he sided with the Egyptians on the grounds that the Greeks freely acknowledged their intellectual debt to the Egyptians but made no mention of the Jews. Moreover, against those who viewed Moses as the source of all ancient learning he argued that

it appears from the sacred Scriptures, that the *Egyptian* Wisdom was more ancient than his [Moses'] and he was the Disciple rather than the Teacher of that learned Nation.

As for those who would go further back and claim Abraham as the source of Egyptian learning Burnet urged them to consider how short a time Abraham was in Egypt and how unlikely it was that 'a Stranger, of a different Language, [could] in less than two Years Space, instruct a Nation'. But Burnet did not altogether reject the biblical historical framework since he argued that 'the Original of the Barbaric Philosophy' could be traced back to 'the Deluge, and Noah the common Father of *Jews* and *Gentiles*'. For Noah as an 'Inhabitant of both Worlds' —

the ante-diluvian and post-diluvian — 'delivered the Lamp of Learning from one to the other' even though, with the passage of time, this learning 'very much declined, and . . . those seminal Doctrines were almost choaked by the prevailing Tares'. Indeed, the restoration of such ancient learning through 'the Principles of Nature and clear Reason' Burnet saw as paving the way for the coming of the millenium for

when the End of all Things approaches, Truth, being revived, may shine with double Lustre, as the Prelude of a future Renovation.[34]

It followed from such a view of world history that aspects of the true original Noachian religion and learning could be found in the different cultures of antiquity whether Jewish or Gentile — that there was not, in short, the great gulf between the civilization or even the religion of the Jews as God's Chosen People and that of their pagan neighbours which had been fundamental to the traditional understanding of the way in which God had shaped world history. This same tendency to downplay direct providential intervention in human affairs can also be seen in the way in which Burnet, both in the *Sacred Theory* and the *Archaeologiae Philosophicae*, attempted to demythologise aspects of the biblical record. In defending his departure from the Mosaic account of Creation in the former he argued that

if we should follow the Vulgar Style and literal sense of Scripture . . . we must renounce Philosophy and Natural Experience . . . [but] Scripture never undertook, nor was ever designed to teach us Philosophy, or the Arts and Sciences.[35]

But in the *Archaeologiae Philosophicae*, which he described as a 'Commentary, or Appendix to the *Theory of the Earth*,'[36] he took this principle further to imply that Scripture was not only a partial guide to natural philosophy but also to history. Thus he argued that the biblical account of the Fall in the Garden of Eden included 'some things Parabolical, and, which will not bear a construction altogether Literal'. He questions for example how Eve could have been made out of a rib and how a serpent could speak.[37] Hence a popular jingle of the time satirised Burnet and others for maintaining that

All the Books of Moses
Were nothing but supposes.[38]

Nonetheless, Burnet's protestation in his *Archaeologiae Philosophicae* that 'I call God to *Witness* that in this or any other Writing I never proposed more to my self, than the Promotion of Piety founded upon Truth'[39] was no doubt sincere. For Burnet saw himself as distinguishing the essential religious message of the Old Testament, which was still relevant to his own enlightened age, from the poetic setting in which it had been set to meet the needs of a primitive people.[40] Moses, he wrote, had 'followed the popular System; that most pleases the People ... And in so doing, he rightly consulted the Publick Safety.'[41] Not surprisingly, however, the *Archaeologiae Philosophicae* brought Burnet's clerical career to an abrupt halt. The eighteenth-century historian, John Oldmixon, reported that on the death of John Tillotson, the Archbishop of Canterbury, in 1694 Burnet's name was put forward as a possible successor

with some Prospect of success, till upon a Representation of Certain Bishops, that some of his Writings were too Sceptical, another Divine [Tenison] was pitch'd upon.[42]

However, Burnet was left undisturbed as master of Charterhouse [school] — an indication of the theological latitude that could exist in the Established Church, particularly after the revolution of 1688.

Burnet, like Spencer, had been at Cambridge (where he was a fellow of Christ's College before he left to become master of the Charterhouse in 1685) at the same time as Newton — indeed when Newton took his MA in 1668 it was Burnet who officiated as Senior Proctor.[43] Burnet and Newton had obviously become acquainted at Cambridge since, some time before publishing his *Sacred Theory*, Burnet consulted Newton who, in January 1680/1, sent a lengthy and generally sympathetic response to Burnet's overall line of argument. In particular Newton largely endorsed the form of scriptural interpretation which underlay the work: like Burnet, Newton believed that where necessary the miraculous elements of the Mosaic account of Creation should be reinterpreted since Moses had been writing for a primitive, pre-scientific people. As Newton himself put it:

As to Moses I do not think his description of creation either Philosophical or feigned, but that he described realities in a language artifically adapted to the sense of ye vulgar ... his business being not to correct the vulgar notions in matters philosophical, but to adapt a description of ye creation as handsomely as he could to ye sense & capacity of ye vulgar.[44]

Newton also concurred with Burnet's underlying belief in the importance of what he called 'Natural [or General] Providence' as against the traditional readiness to invoke God's extraordinary Providence — a sentiment pithily summarised in Burnet's italicised remark that '*The Course of Nature is Truly the will of God*'.[45] For Newton also argued that God's creative power could best be understood as working through the normal processes of nature even though he (like Burnet) acknowledged that there were rare occasions — such as at the original Creation — when God did intervene directly in the natural realm. As he wrote in his letter to Burnet:

Where natural causes are at hand God uses them as instruments in his works, but I do not think them alone sufficient for ye creation.[46]

Newton's historical writings also have much in common with those of Burnet and it is possible that the two men discussed matters historical as well as issues relating to natural philosophy. Newton's copy of Burnet's *Archaeologiae Philosophicae* indicates close study[47] although by the time that this work was published in 1692 Newton had already developed the general outlines of his historico-theological system.[48] Like Burnet, Newton regarded Noah, rather than Abraham or Moses, as the original source of true religion and learning; consequently, he, too, argued that vestiges of truth could be found among the ancient Gentile peoples as well as that of the Jews since all were descendants of Noah and his sons. Both also shared the belief that modern philosophy was contributing to the recovery of the ancient truths which had been distorted after Noah's death.

As Westfall has shown in his important article[49] on Newton's major (though unpublished) theologico-historical work, the *Theologiae Gentilis Origenes Philosophicae* Newton conceived of the true religion as that which most closely resembled that which prevailed at the time of Noah immediately after the Deluge, before the idolatry — which to Newton was the root of all evil not only in religion but also in politics and even philosophy — began to corrupt it. The subsequent religious history of humankind, as outlined in the Old and New Testaments, was, then, in Newton's views, a record of the attempts to purge religion of its idolatrous accretions. The nearest that Newton came to acknowledging publicly such a view of religious history was in a passage of his *Chronology of the Ancient Kingdoms Amended,* a work which he was preparing for publication when he died in 1727 and which was published by his executors in the following year.[50] In it he described the religion of Moses and the prophets as being based on

The precepts of the sons of Noah, which was the primitive religion of both Jews and

Christians, and ought to be the standing religion of all nations, it being for the honour of God, and good of mankind.[51]

In the privacy of his unpublished manuscripts Newton made it evident that he saw the task of restoring the true Noachian religion as continuing up to the Protestant Reformation and beyond since Newton took the view that even Protestant Christianity needed to be purged of what he considered its idolatrous trinitarianism.

Such a reading of human history, then, largely reduced the role of the Old Testament prophets, and even that of Christ, to that of religious reformers rather than the instruments of a wholly new divine revelation — for, to Newton, the true religion had been established long before either Christ or the prophets. In one of his jottings he came close to acknowledging explicitly such a considerable breach with orthodoxy for he noted as the theme of a proposed (but never written) chapter of a version of the *Theologiae Gentilis*:

What the true religion of the children of Noah was before it began to be corrupted by the worship of false Gods. And that the Christian religion was not more true and did not become less corrupted.[52]

Elsewhere, in a manuscript entitled 'A Short Scheme of the True Religion', Newton again hinted at his belief that Christianity did not constitute a wholly new revelation since residues of the true Noachian religion — or at least the ethical teaching which was to Newton the most important part of religion — could be found in cultures such as that of China which were far removed from Judaeo-Christian civilization. The principle that we should love our neighbours as ourselves, he wrote,

was the Ethics, or good manners, taught the first ages by Noah and his sons by their seven precepts, the heathens by Socrates, Confucius and other philosophers, the Israelites by Moses and the Prophets and the Christians more fully by Christ and his Apostles.[53]

Like Burnet, then, Newton departed from the traditional belief that there was in antiquity an unbridgeable gulf between the religion of the Jews and that of the Gentiles. However, Newton went considerably further than Burnet in downplaying the significance of Christ, for in Burnet's theologico-historical scheme 'the whole Hinge of Providence with respect to human Affairs, turns upon the Mystery of the Messiah'.[54]

Newton's conception of world history, then, reduced the role of Providence to intervening occasionally and, generally, through natural processes to re-establish the original true religion rather than to establish a wholly new religious order. Similarly, Newton's conception of the role of Providence in the natural order (like that of Spencer) was to establish a natural harmony at the beginning and to intervene only occasionally and, generally speaking, through natural agencies, to maintain the clock-like order established at Creation. As Newton remarked in the thirty-first (and last) query of his *Opticks* all material things were 'variously associated in the first creation by the counsel of an intelligent agent. For it became Him who created them to set them in order'. But, he continued, over time 'some inconsiderable irregularities' do arise

from the mutual actions of comets and planets upon one another, and which will be apt to increase, till this system wants a reformation.[55]

Such a 'reformation' could, of course, only be the result of providential intervention but in Newton's view, the Creator generally achieved these ends by working through secondary causes, particularly the comets.[56]

The same query (and the *Opticks* as a whole) concludes with a rather oblique reference to the theologico-historical *schema* that Newton had developed in his *Theologiae Gentilis*. As natural philosophy is perfected, he wrote, so too 'the bounds of moral philosophy will be also enlarged'[57] — this, in effect, meant a growth in true religion since for Newton true ethics, which included our duty to God, constituted virtually the sum and substance of true religion. Thus in his unpublished *Irenicum* he defined the true religion of Noah as 'The religion of loving God and our neighbour'.[58] To Newton, then, the true natural philosophy was inextricably linked with the true religion since, as he wrote in the same section of the *Opticks*:

For so far as we can know by natural philosophy what is the First Cause, what power He has over us, and what benefits we receive from Him, so far our duty towards Him, as well as that towards one another, will appear to us by the light of Nature.

Conversely, the loss of the true natural philosophy was therefore closely associated with the undermining of true religion and ethics, a loss that Newton saw as the outcome of what for him was the true Fall of Man: the growth of idolatry. Thus he continues the above passage by concluding the *Opticks* as follows:

And no doubt, if the worship of false gods had not blinded the heathen, their moral philosophy would have gone farther than to the four cardinal virtues; and instead of teaching the transmigration of souls, and to worship the Sun and Moon, and dead heroes, they would have taught us to worship our true Author and Benefactor, as their ancestors did under the government of Noah and his sons before they corrupted themselves.[59]

One of the major purposes of the *Theologiae Gentilis* was to trace the tragic history of the way in which the corruption of true religion by idolatry was accompanied by the corruption of true natural philosophy. For, as Newton's conclusion to the *Opticks* suggests, not only did he maintain that the religion of Noah and his sons was pure and undefiled so, too , was their understanding of the natural world. The central rite of the Noachian religion was the maintenance of a perpetual vestal flame at the centre of some form of shrine — what Newton called a 'prytaneum'. The prytaneum was a model of God's creation, the universe, and the fire a symbol of the sun and hence an affirmation of a heliocentric view of the universe. Newton discerned vestiges of this ancient rite in most human societies: thus he viewed such diverse structures as Stonehenge and the Jewish temple as examples of prytanea and he drew on accounts of contemporary India and the East Indies to provide instances of the survival of the cult of the vestal flame.[60] But with the growth of idolatry the true significance of this rite was lost and so, too, was the true natural philosophy. Gradually the sense of cosmic order was lost as

men were led by degrees to pay a veneration to these sensible objects and began at length to worship them as the visible seals of divinity . . . For tis agreed that Idolatry began in ye worship of ye heavenly bodies & elements.[61]

Among the Egyptians — the major instigators of such idolatry — the vestal fire came to be regarded as an image of a fire in the centre of the earth rather than the sun, a corruption which was accompanied by the change from a heliocentric to a geocentric cosmology.[62]

Thus the re-establishment of the true heliocentric natural philosophy and the accompanying stripping away of idolatrous animistic conceptions of the way in which the universe worked was, for Newton, both a scientific and a religious mission. His work in natural philosophy and his theologico-historical investigations all shared the same goal: the restoration of that true understanding of God's relations with humanity and the natural order which had been obscured by the growth of idolatry after the time of Noah and his sons. Like Thomas Burnet,[63] Newton's disciple, William Whiston (and, very probably, the considerably more reticent Newton himself) regarded the recovery of the true natural philosophy as something of major religious, as well as scientific significance — indeed as ushering in the millenium. Thus Whiston viewed the publication of the *Principia*

as an eminent Prelude and Preparation to those happy Times of the Restitution of all things which God has spoken of by the Mouth of all his Holy Prophets since the world began, Acts, iii, 21.[64]

As Schaffer has shown,[65] one specific instance of Newton's belief that false conceptions of theology and natural philosophy were intertwined was his treatment of the problem of the movement of the comets. The view that comets operated only in the sublunar sphere — which, in Newton's view, had originated with Eudoxus (c. 427—c. 347 BC), who had studied under Plato and travelled in Egypt — was a corollary of the growth of idolatry. For the assumption that there were solid heavenly spheres which were supposed to prevent the movement of comets in the superlunary sphere derived from the belief that each heavenly body was associated with a god or spirit who moved the celestial sphere. By restoring the true understanding of the comets, then, Newton was attacking not only false astronomy but also what was, for Newton, the gravest of sins: idolatry. By re-establishing an understanding of the movement of the comets on true principles of natural philosophy and theology Newton also helped to contribute to the same goal as had prompted the *Discourse Concerning Prodigies* of his Cambridge contemporary, John Spencer: the undermining of the popular divination and superstition associated with comets by showing that the movements of comets were predictable and that their capacity to influence human affairs was limited by the fact that they belonged to the distant realm of the superlunary sphere. Newton also wished to show that the comets, and the heavenly bodies more generally, were merely inanimate servants of an all-wise and all-powerful Providence rather than, as idolatrous ancient astronomers had argued, the outward manifestations of a pagan pantheon.

Though it is highly probable that Newton knew and approved of Spencer's work on prodigies Newton does not appear to have made any specific reference to it. However, Newton's debt to Spencer's historical work — particularly his emphasis on the importance of Egyptian civilization for understanding the development of Judaism — is manifest. Newton plainly discussed issues related to the interpretation of the Old Testament with Spencer since, in a letter to Locke of 13 December 1691, he (Newton) remarked that Spencer had perused Le Clerc's controversial new Latin edition 'and likes ye design very well'.[66] After Spencer's death in 1693 Newton referred to him, in a letter to Otto Mencke, a professor at Leipzig, of 30 May as 'our friend' and also talked in some detail about the plans for publication of a second edition of Spencer's *De Legibus*.[67] Newton's library also included a well-thumbed copy of his work.[68] Most significantly, it was Spencer whom Newton cited in his *Theologiae Gentilis* to support the view that the religion of Moses was largely derived from that of the Egyptians 'for', as Newton wrote

Dr. Spencer has shown yt Moses retained all ye religion of ye Egyptians concerning ye worship of ye true God, & rejected only what belonged to ye worship of their false Gods . . . & that ye Mosaical religion concerning ye true God contains little else besides what was then in use amongst the Egyptians.[69]

Newton then used Spencer's painstaking scholarship in support of his basic hypothesis: that the true religion was that of Noah and his sons and that Judaism (and even Christianity) represented partial restorations of this pristine religion rather than wholly new religious dispensations. The Egyptians, he argued, had preserved an adultered form of the Noachian religion which they derived from Ham, one of the sons of Noah who had settled in the area.[70] Having invoked Spencer's evidence Newton continued the above passage by arguing that it followed that

it's certain that ye old religion of the Egyptians was ye true [Noachian] religion tho corrupted before the age of Moses by the mixture of fals Gods with that of ye true one.

Newton then concluded by vigorously reasserting the basic theme of the whole work — that it followed that the religion 'wch Moses taught ye Jews was no other than ye religion of Noah purged from the corruptions of ye nations'.[71]

Egypt, then, assumes a central importance in Newton's conception of world history for it had preserved the Noachian religion — albeit in an adulterated form — and transmitted it to the Jews who then went some way towards restoring it to its original form. Like Spencer's or Burnet's it is a conception of history which greatly restricts the extent of direct providential intervention in human affairs. God, in Newton's account, largely relies on his ordinary providence to achieve his purposes in history just as Newton also portrays Providence as chiefly working through secondary causes in nature. For both Newton and Spencer the religion of Moses was not, as a literal interpretation of the Old Testament would depict it, solely the result of direct divine revelation but rather was in large measure the product of the cultural assimilation by the Jews of certain aspects of Egyptian ritual practices. True, Judaism purged itself of many idolatrous Egyptian customs but this was a slow and by no means complete transformation — a further example of the way in which Providence was content to work through the natural processes of human history.

Elsewhere Newton, like Spencer, also suggested that aspects of Christianity developed out of Judaism by a similar process of historical evolution. Thus Newton emphasised the gradual way in which such Jewish practices as the appointment of a president of the synagogue were transformed into the Christian office of bishops.[72] The relationship of Christianity to Judaism, then, was to Newton akin to the relation between the religion of Moses and that of the Egyptians. As David Gregory noted after studying Newton's 'tract on the origin of nations, the Theologiae Gentilis' at Cambridge in 1694 Newton was of the view that 'Moses began a reformation but retained the indifferent elements of the Egyptians . . . Christ reformed the religion of Moses'.[73]

For Newton, then, the Egyptians played almost as central a role in the maintenance of the true Noachian religion as the Jews. This same tendency to emphasise the historical importance of the Egyptians in the transmission of religion and culture rather than, as had traditionally been the case, the Jews, can also be seen in other aspects of Newton's work. Where many sixteenth and seventeenth century scholars (including Theophilus Gale) argued that all languages derived from Hebrew, the original Adamic tongue,[74] Newton argued that the other languages of the Middle East derived from Egyptian rather than Hebrew.[75] Again Newton departs from the common apologetic device of tracing Greek learning back to the Jews either directly or via the Egyptians when he asserted in a draft introduction of Book III of the *Principia* that it was as a result of the astronomical observations of the Egyptians that the true, heliocentric natural philosophy

was spread abroad among other nations; for from them it was, and the nations about them, that the *Greeks*, . . . derived their first, as well as soundest, notions of philosophy.[76]

Where Newton's Cambridge contemporaries, the Platonists, Ralph Cudworth and Henry More, attempted to trace the atomic philosophy back to Moses Newton followed them only as far back as the Egyptians and the Phoenicians and, despite his close study of Cudworth's work, Newton was not persuaded by the Cambridge Platonist's identification of Moschus the Phoenician — the alleged founder of atomism — with Moses.[77] He did, however, draw heavily on the material which Cudworth assembled to demonstrate that the Egyptians 'excel[led] almost all other nations in reputations for wisdom, antiquity, and almost every other advantage'.[78]

But such admissions of Egypt's historical importance were made rather grudgingly by Newton for, in his historical *schema*, Egypt was not only central in the transmission of some elements of the true religion and the true natural philosophy but also central in the propagation of both religious and philosophical error. For Newton agreed abundantly with Cudworth's claim that

there is no question to be made, but that the Greeks and Europeans generally derived their polytheism and idolatry from the Egyptians.[79]

As David Gregory noted after reading the *Theologiae Gentilis* in 1694 it was Newton's view that

it was the Egyptians who most of all debased religion with superstition and from them it spread to the other peoples.[80]

In his other historical manuscripts Newton attributed the growth of such idolatry to the overweening ambitions of the Egyptian kings who sought to overawe their subjects by deifying their predecessors. It was this that led to what virtually takes the place of the Fall of Adam in Newton's theologico-historical system: the corruption of the true religion and, with it, the true natural philosophy of Noah. 'Here then', he writes of the actions of the Egyptian kings,

we have the true original of ye corruption of the religion of Noah and the true cause of its spreading so early and so generally. For this policy of ye kings of Egypt soon took with ye kings of other nations.[81]

This corruption was compounded by the Egyptian priests who pandered to the ambitions of the kings and consolidated their own position by taking further this process of royal deification since they made reigns of those Gods very long & very ancient and separated them from the reign of the deified man.[82]

This idolatry also infected their study of astronomy for as idolatry became more and more established in Egypt as a result of royal and priestly ambitions so, too, priests began to attribute the movement of the heavens to the influences of the gods. As Newton wrote in his *Theologiae Gentilis*:

And by means of these fictions the souls of ye dead grew into veneration with ye stars & . . . were taken for ye Gods, which governed this world.[83]

In another draft of the same work Newton caustically remarked

And so Astrology and gentile Theology were introduced by astute priests to promote the study of the Stars and to increase the power of the priesthood.[84]

Eventually, the result was the decline of the true, heliocentric philosophy and its replacement by the geocentric system which was subsequently codified by the Egyptian, Ptolemy.[85]

However, the significance which Newton attributes to the Egyptians for ill as well as good only serves to emphasise further their centrality in his conception of world history. The fall of the Egyptians into idolatry and their departure from the true natural philosophy provides Newton with virtually a partly secularised alternative to the biblical myth of the expulsion of Adam and Eve from Eden. For in Newton's account the original Fall has little significance — true religion, morality and philosophy could be found after Adam, and even after the Flood, in the age of Noah and his sons. The real fall from grace comes with the corruption of the Noachian religion as a result of the growth of idolatry — something for which Newton held the Egyptians to be chiefly responsible. In Newton's theologico-historical system, then, the Egyptians assume something of the central significance traditionally accorded to the Jews, except that in Newton's account the positive and negative features of Egypt's impact on the ancient world are explained largely in naturalistic terms. Thus the vital role that Newton attributes to the Egyptians in the transmission of the true Noachian religion to the Jews (albeit in an adulterated form) to some extent takes the place of a direct divine revelation to Moses while the original sin of idolatry is largely explained in terms of kingly and priestly ambitions rather than by recourse to the role of Satan.

However, the Egyptians loom so large in Newton's account almost against his will, for it was the aim of much of his historical writing to deflate the claims of the Egyptians and, conversely, to magnify the historical importance of the Jews. Indeed, Manuel sees most of Newton's chronological writings as a reaffirmation of the traditional historical *schema* in which the Jews were regarded as the true source of the civilization of the ancient world.[86] Where possible Newton did attempt to demonstrate the historical primacy of the Jews by undermining the Egyptian chronology as the work of ambitious priests working in league with kings wishing to instil obedience in their subjects by exaggerating the antiquity of their dynasty.[87] In his *Chronology of Ancient Kingdoms Amended* Newton devoted particular attention to demonstrating that the reign of Solomon was an historical watershed marking the beginnings of the great empires — a development which he argued occurred first in Israel.[88] Since this and a short chronological abstract based on it were the only historical works of Newton to be published he naturally came to be regarded as an advocate of the traditional claims for Jewish as against Egyptian priority. This the more so since in the *Chronology* Newton set out to argue that many of the attributes of civilization developed in Israel not long before the reigns of Solomon or his father, David, with the clear implication that other nations derived such skills from the Jews. Thus in discussing the age of mankind he wrote that:

And the original of letters, agriculture, navigation, music, arts and sciences, metals, smiths, and carpenters, towns and houses was not older in Europe than the days of Eli, Samuel and David; and before those days the earth was so thinly peopled, and so overgrown with woods, that mankind could not be much older than is represented in Scripture.[89]

But Newton's ambivalence in the last sentence of this passage is revealing. Where possible his historical writings are intended to bolster the biblical chronology but, in history, as in natural philosophy, Newton did not regard Scripture as providing an account which the true believer was obliged to follow without question or interpretation. And, as Westfall and Manuel point out,[90] the approach of Newton and other would-be reconcilers of biblical and classical chronologies, with their elaborate calculations designed to intermesh the biblical histories with data derived from the records of Gentiles and (in the case of Newton) astronomical phenomena, tended to undermine the privileged position of Scripture. Thus, in Newton's hands, the Bible frequently became but one ancient source among others to be used in a similar manner to the way one used the records of the Egyptians or the Greeks. Fabulous events in the Bible, as in the Gentile histories, were to be regarded — as Newton regarded the episode of the sun standing still at the battle of Jericho — as 'poetic expressions' which needed to be translated into secular historical terms. True, Newton's whole aim was to establish the ultimate authenticity of the biblical history but in doing so he drained it of much of its theological content by reducing it to the same level as other historical records. Thus in his *Chronology of the Ancient Kingdoms Amended* there is no hint of the biblical drama of salvation as Newton incorporates the scriptural history into a lifeless succession of kings and battles. Though Newton's purely chronological scripturally-based writings may have sought to assert the primacy of the Jews where, as in his *Theologiae Gentilis*, he moved beyond narrative to provide some sort of interpretative framework to explain the religious history of humankind the Eygptians — almost against Newton's wishes — are accorded a degree of historical significance which almost rivals that of the Jews.

Newton's ambivalent attitude to the Egyptians — like his somewhat cautious use of Scripture as an historical record — indicates the extent to which he, like Spencer, Burnet and other scholars of the age, reflects what Hazard[91] calls the 'crisis of the European conscience' of the late seventeenth century: the transition from a civilization based on an unquestioning acceptance of Christianity to one in which government, and social institutions more generally, were increasingly justified in more secular terms. This, in turn, reflects a growing wariness about linking such institutions too closely to a specifically religious conception of the world in order to avoid the confessional strife unleashed by the Reformation. Newton's almost instinctive hostility to the Egyptians derived from a long-established form of Christian apologetic — the argument that Gentile learning (particularly that of the Egyptians and the Greeks) ultimately derived from Jewish sources. As Bernal points out,[92] this long-engrained tendency to downplay the importance of Egyptian culture was further strengthened for Newton and many of his contemporaries by the fact that anti-Christian writers ranging from Giordano Bruno in the sixteenth century to the English Deists of the early eighteenth century had looked to ancient Egypt to provide their beliefs with the seal of antiquity.

On the other hand, however, Newton's natural tendency to extend to the history of the Judaeo-Christian tradition the emphasis on the role of secondary agents which had produced such explanatory success in his natural philosophy brought with it, however reluctantly, an ineluctable tendency to accord the Egyptians great significance in the history of religion and civilization. Moreover, as a close student of the ancient chronicles Newton found it difficult to downplay the role of the Egyptians to the extent demanded by the traditional interpretation which accorded the Jews almost total historical primacy. In any case, as Spencer, Burnet and others had pointed out, such a view was not clearly stated in Scripture and, indeed, could be challenged on the basis of Scripture itself — for did not the Bible speak of Moses drinking deeply at the founts of Egyptian wisdom? Newton's ambivalent attitude towards the Egyptians was even reflected in his discussion of points of historical detail: sometimes he suggests that writing derived from the Jews but elsewhere he attributes it to the Egyptians;[93] though in his *Chronology of the Ancient Kingdoms Amended* he makes much of the significance of the fact that Solomon's reign preceded that of the great dynasties of Egypt, in the *Theologiae Gentilis* he appears to question this by describing Egypt as 'ye oldest of kingdoms'.[94]

Newton's tendency — however reluctantly — to accord the Egyptians greater significance in world history indicated an attitude to the role of Providence which was also reflected in his natural philosophy. Thus his greater attention to the role of the Egyptians, rather than the Jews, in the transmission of the true religion and of the arts of civilization betokened a greater emphasis on the extent to which Providence worked in history through secondary causes rather than by means of direct intervention. Similarly, in his natural philosophy he and, *a fortiori*, his clerical popularisers such as Bentley, Clarke and Whiston, argued that his natural philosophy demonstrated the way in which Providence used such secondary agents as gravitation or the movement of the comets to achieve that harmony and order which, to Newton and his disciples, was the hallmark of the work of the Creator.

However, just as the Deists discarded the role of a continuing and active Providence in Newton's system of natural philosophy leaving God to act merely as the First Cause — the distant and remote Watchmaker God — so, too, some of the Deists argued for an understanding of world history devoid of providential direction. The Deists also sought to undermine the significance of Scripture by relegating the role of the Jews in the ancient world to one of almost total insignificance and, conversely, by magnifying the role of the Egyptians. The most influential statement of this overtly anti-Judaeo-Christian historical interpretation was to be Voltaire's *Philosophy of History* written as an explicit attack on the theocentric conception of world history embodied in Bishop Bossuet's influential *Discours sur L'Histoire Universelle* (1681)[95] — a work which took as its guiding theme for the history of the ancient world the biblical account of the history of the Jews as 'a palpable account of his [God's] external providence'.[96] By contrast, Voltaire implicitly rejects both the antiquity and the historical reliability of the Old Testament by writing of the Chinese chronicles that 'It is incontestable that [they are] the most ancient annals of the world' and praising them for providing 'an uninterrupted succession, circumstantial, complete, judicious, without any mixture of the marvellous'. With characteristic impishness he dismisses Genesis' account of the peopling of the world after the Flood with the aside that

I shall leave to men more learned than myself, the trouble of proving that the three children of Noah, who were the only inhabitants of the globe, divided the whole of it amongst them.

In a chapter entitled 'Whether the Jews taught other Nations, or whether they were taught by them' Voltaire, predictably, sets out to show the extent to which the Jews derived their culture from the Gentiles and, particularly, the Egyptians.[97]

But just as Voltaire used Newton's natural philosophy — which Newton and most of his English disciples considered as a natural aid to Christian apologetics — as a Deistic weapon against the Church so, too, Voltaire's overtly Deistic system of world history ultimately largely derived from the work of English Christian scholars such as Spencer. Another important late seventeenth century scholar who arrived at similar conclusions to those of Spencer about the relationship between ancient Egyptian and Jewish culture was Sir John Marsham, author of *Chronicus Canon Aegyptiacus, Ebraicus, & Graecus* (1672). Like so much historical scholarship of the period this work aimed to achieve a peaceful reconciliation between the biblical, classical and Egyptian chronologies thus undercutting what Bishop Stillingfleet called one of

the most popular pretences of the Atheists of our Age . . . the irreconcilableness of the account of Times in Scripture, with that of the learned and ancient Heathen Nations.[98]

In the preface to this work Marsham, at pains to emphasise his orthodoxy, argued that his view of the relations between the Jews and the Egyptians was in accord with that of many of the Church Fathers and with the *Dissertatio de Urim et Thummim* (1669) — which he praises for its exhaustive scholarship[99] — Spencer's early work (which was later incorporated into the *De Legibus*) in which he demonstrated the extent of Egyptian influence on Jewish divinatory practices. Marsham's work did much to provide scholars with a coherent survey of Egyptian history and was used extensively: Spencer drew on it in his *De Legibus* and Newton often cites his work — it was Marsham, for example, who was Newton's authority for the (false) identification of the purportedly early pharoah, Sesostris, with the figure of Sesac who is mentioned in the Bible as dating from after the death of Solomon.[100]

Marsham's work — along with other Christian apologists like Spencer and Burnet — was, however, put to very different uses by a number of the English Deists as part of their assault on the Old Testament. Along with Burnet's *Archaeologia Philosophica* — the influence of which he enthusiastically acknowledges — it was very likely Marsham and Spencer on whom Toland was drawing when, in his *Letters to Serena* (1704), he attacked the view that 'the Jews and a world of Christians pretend that the Egyptians had all their learning from Abraham'. By contrast, Toland argued that 'the Jews were of all Eastern People the most illiterate' and that it was the Egyptians who were 'the Fountains of Learning to all the East'.[101] In his *Christianity as Old as Creation* (1730) Toland's fellow Deist, Matthew Tindal, cited 'Marsham, and Others' as his authority for asserting that the Jews derived the ritual of circumcision from the Egyptians. For Tindal this was one illustration of the thesis embodied in his title: that neither Moses nor Christ instituted a new religious order since the true religion should have been evident from the beginnings of humankind to anyone of good will and sound reason — in short, that 'True Religion, in all Places and Times, must ever be the same; Eternal, Universal and Unalterable'.[102] Such was also the thesis of Newton's *Theologiae Gentilis* (though there is nothing to suggest any mutual influence) with the difference that, in Newton's work, Moses and Christ are still accorded a significant (albeit much reduced) role in the restoration of the true, original religion which, for Newton, was that of Noah and his sons. Tindal's reference to 'Marsham, and Others' probably embraced Spencer since, although the author of *De Legibus Hebraeorum* does not definitely state that the Jews derived the practice of circumcision from the Egyptians, he assembled a great deal of evidence which suggested that this was highly probable.[103]

Tindal's work prompted a sermon entitled 'The Wisdom of the Ancients Borrowed from Divine Revelation, or, Christianity Vindicated Against Infidelity' (1731) from that tireless defender of orthodoxy, Daniel Waterland. As the title suggests this was a restatement — though couched in very cautious terms — of the ancient argument that the Gentiles, whether Egyptian or Greek, had derived their learning from the Jews. Though Waterland is gentle in his treatment of two such venerable scholars as Marsham and Spencer he makes it clear that he regards their work as having provided ammunition for Deistic attacks on the Old Testament such as that of Tindal. While the ancient argument for the priority of Jewish learning may have been overstated, he writes,

Two very considerable writers, Sir John Marsham and Dr. Spencer, appear to have slighted it too much. They have not only called in question the prevailing opinion of the ancient Apologists, but they have run directly counter to it; pretending that the Pagans did not borrow from the Jews, but that the Jews rather copied after the Egyptians or other Pagans, in such instances as both agree in: a strange way of turning the tables, confounding history, and inverting the real order of things.[104]

Waterland's work prompted, in turn, a series of pamphlets from his Cambridge colleague, Conyers Middleton, whose religious beliefs appear to have hovered uncertainly between a form of enlightened Christianity and a mild Deism — Gibbon wrote of him that he 'rose to the highest pitch of scepticism, in any wise consistent with religion'.[105] In the first of these pamphlets, *A Letter to Dr. Waterland* ... (1731), Middleton vigorously asserted that the defence of Christianity was ill-served by Waterland's form of apologetics for anyone familiar with the ancient world knew that '*Aegypt was a great and powerful nation*' when the Jews were '*an obscure contemptible people*, famed for no kind of literature; scarce known to the polite world till the Roman Empire dispersed them'. Moreover, the Greek historians clearly favoured the view that it was the Jews who borrowed such ritual practices as circumcision from the Egyptians rather than, as Waterland maintained, the other way around

and twas the authority of these that induced the *learned Marsham*, and the no less *learned Spencer* too, to favour the opinion of your adversary [Tindal].[106]

In a subsequent pamphlet directed at one of Waterland's advocates Middleton reasserted his view that the fact that the Jews derived many of their rituals from the Egyptians had been so clearly demonstrated by the learned Spencer that

'no Man, unless supremely credulous', as he says, 'can believe it have been the Egyptians'. And indeed both he and Marsham derive in a manner the whole ritual Law from this very source of Egypt.[107]

Spencer's and Marsham's works also provided much of the scholarly foundation for Warburton's ill-digested compilation of learning, *The Divine Legation of Moses Demonstrated* (1741). This was written, as Rossi[108] points out, in an attempt to turn the attacks of the Deists (and especially John Toland) on Christianity back on themselves by using their claims about the relationship between the Jews and the Egyptians in the defence of orthodoxy. In the *Divine Legation* Warburton advances the paradoxical argument that the Jews plainly received a Divine Revelation which instilled in them a strong belief in an active Providence who rewarded good and punished evil since — unlike the Gentiles and, in particular, the Egyptians — they were able to maintain social and political order without the sanction of a belief in an after-life. This was particularly remarkable since in other respects, writes Warburton (echoing Spencer and Marsham),

the Jews were extremely fond of Egyptian manners and did frequently fall into Egyptian superstition, and that many of the Laws given to them by the Ministry of Moses were instituted, partly in compliance to their prejudices and partly in opposition to those superstitions.[109]

Elsewhere Warburton makes explicit his debt to Spencer by praising 'the learned SPENCER' as one who

hath fully exhausted this subject in his excellent work . . . and thereby done great service to divine revelation: for the RITUAL LAW, when thus explained, is seen to be an institution of the most beautiful and sublime contrivance.

Ultimately, he argued, Spencer's thesis could be reduced to the proposition that '*Moses gave the ritual law to the Jews because of the hardness of their hearts*; the very hypothesis of Jesus Christ'. Like Spencer, then, Warburton saw the influence of the Egyptians on the Israelites as having been providentially ordained though it was an example of Providence's use of secondary agents rather than of direct divine intervention. Similarly, he saw the preservation of Egypt 'from total destruction', despite its role in promoting idolatry, as part of

Providence's design to ensure the diffusion 'of civil life and polished manners, which were to derive their source from thence'.[110]

Far from arguing in the traditional manner that the Egyptians derived their learning from the Jews it was an important part of Warburton's highly ingenious form of historical apologetics that Egyptian civilization was older than that of the Jews and that the Jews drew heavily upon it. For, in Warburton's opinion, this made it all the more remarkable that the Jews did not adopt the Egyptian belief in an afterlife and, consequently, strongly suggested that they were the beneficiaries of a special revelation which ensured a belief in an active Providence and thus enabled them to maintain good behaviour and civil order without such an essential sanction as fear of retribution in a life to come. Warburton therefore devoted much space to confuting Newton as one of the chief exponents (at least in his published works) of the view that Jewish civilization was older than that of the Egyptians. Such a position, Warburton acknowledged, put himself, as a Christian apologist, in some rather odd company since 'the present turn, indeed, of free-thinking is to extol the high antiquity of Egypt, as an advantage to their cause'. Nonetheless, Warburton was confident that as his contemporaries came to appreciate the force of his strikingly original argument and the orthodox emphasised the debt of Israel to Egypt so, too 'we shall see the contrary notion, of the low antiquity of Egypt become the fashionable doctrine' among non-believers.[111]

The aim of Warburton's work, like that of Spencer's, was, then, to illustrate the way in which God's general Providence could use the natural processes of history — such as the influence of the Egyptians on the Jews — to achieve his ends. However, the basic apologetic intent of the work is also evident in the way in which Warburton emphasises the extent to which such historical events were not, as the Deists claimed, essentially random but rather were firmly under the direction of Providence, even though it might be acting through secondary causes. Similarly, in his depiction of the relationship between God and the natural world Warburton was at pains to emphasise that, although God generally worked through what he called 'the stated laws of Physics', nonetheless these were also part of a divinely-planned, providential order. Thus in his sermon, 'Natural and Civil Events the Instruments of God's Moral Government' — preached in 1755 on the Fast Day called in response to the Lisbon earthquake — Warburton argued that it was a 'disrespectful notion of divine Wisdom' to maintain that God could only manifest his system of rewards and punishments through miracles rather than through the natural order since

God, by the admirable direction of his general providence, so adjusts the circumstances of the natural and moral system, as to make the events in the former to serve for the regulation of the latter.[112]

Although Warburton's understanding of God's relation to history and nature emphasised the extent to which God worked through his general or ordinary Providence he was also anxious to leave room for God's exercise of his extraordinary Providence — in short, for miracles. For in the *Divine Legation* the fact that the Jews maintained civil order without a belief in an after-life is, to Warburton, a clear indication of the fact that they were exposed to overt manifestations of God's extraordinary Providence which made them obey the moral law without any other sanction. In another of his works revealingly entitled, *Julian: or A Discourse concerning the Earthquake and Firey Eruptions, which defeated that Emperor's Attempt to rebuild the Temple at Jerusalem. In which the Reality of a Divine Interposition is shewn* ... (1750), Warburton set out to document a miracle of the fourth century AD 'worked by the immediate hand of God, and not through the agency of his servants'[113] to counter the claims of Middleton that any purported miracles after the apostolic period were likely to be spurious. Consistent with such an understanding of history, in which the miraculous still played a significant role, Warburton was reluctant to treat many of the historical passages of the Bible as being allegorical in the way in which passages dealing with natural philosophy had been reinterpreted for, as he wrote in *The Divine Legation* in relation to Newton's historical work,

though the end of the sacred history was certainly not to instruct us in astronomy yet it was, without question, written to inform us of the various fortunes of the people of God; with whom, the history of Egypt was closely connected.[114]

Nonetheless, though Warburton was wary of allowing the Deists an opening to eliminate the role of direct providential intervention in history altogether, the overall drift of his historical writing was to give greater emphasis to God's ordinary rather than his extraordinary providence. Appropriately, Warburton's youthful work (his second) — *A Critical and Philosophical Enquiry into the Causes of Prodigies and Miracles as Related by Historians* (1727) — was concerned to cleanse

profane history of many supposed instances of supernatural intervention while providing rules for distinguishing true miracles from the
false.[115]

But, ironically, just as Warburton had attempted to turn the tables on
his Deistic opponents by using their claims about the greater antiquity
of Egyptian culture as against that of the Jews in defence of orthodoxy
so, too, Voltaire attempted to use Warburton's apologetic labours as
part of his Deistic assault on Christianity. For it was largely from
Warburton and other English popularisers of the scholarly labours of
Marsham and Spencer that Voltaire drew his material on the relations
between the Egyptians and the Jews which, in his *Philosophy of History*,
formed part of his Deistic assault on the traditional Judaeo-Christian
conception of history which had received its last great statement in
Bossuet's *Discours sur L'Histoire Universelle*. As Torrey and
Brumfitt[116] have shown, Voltaire had closely studied Warburton's
Divine Legation and Middleton's pamphlets against Waterland before
writing the *Philosophy of History* in 1765. He was also greatly influenced by Tindal's critical approach to the Bible in *Christianity as Old as
Creation*. By such an indirect route, then, did the scholarship of
Spencer and Marsham, which had been intended to draw together into
a harmonious whole the biblical and classical historical sources, come
to serve the purposes of the eighteenth century's most influential
opponent of revealed religion. By a similar irony the natural philosophy
of Newton which its author and his early English disciples saw as a
confirmation of the existence of a Christian Providence was also to be
used by Voltaire as another weapon in the crusade to '*écrasez l'infâme*'.

But Voltaire's work represented the response of a French intellectual
class which was profoundly alienated from its society and keen to press
into service any potential means of attack on an ossified Church and
State. In eighteenth century England, however, as the anti-clericalism
which lay at the root of the Deists' work abated once the Whigs took
firm control of the Church so, too, did the tendency to use history (or
natural philosophy) in order to undercut the power of priestcraft also
begin to wane. The belief that the normal processes of nature were
controlled by a directing Providence became so closely associated with
the dominant Newtonian natural philosophy that science and Christian
apologetics came to be seen as natural allies.[117] As such a view gained
acceptance so, too, did the principle that the scriptural account of

natural processes frequently needed to be treated as being couched in metaphorical language appropriate to the needs of a pre-scientific people — for such a view had been advocated by Newton himself and was reflected in his natural philosophy. The increasing sway of Newtonian natural philosophy thus helped to strengthen two principles which had considerable importance for a more secular understanding of history: an emphasis on God's ordinary, as against his extraordinary Providence, and a willingness to interpret Scripture metaphorically. As such principles gained ground such ancient problems as the relationship between the Jews and the Egyptians came less and less to be seen as central issues in the defence of revealed religion. For once it was accepted that Providence worked in history, as in natural philosophy, primarily through secondary agents — such as the influence of the Egyptians on the Jews — and that the creation of the Mosaic law, like the Mosaic cosmology, should not be interpreted literally as the immediate result of divine intervention, than such issues lost much of their urgency. Lord Monboddo, for example, prompted little controversy (at least, of a theological kind) when he argued in the late eighteenth century in his *Of the Origin and Progress of Language* (1773—92) that Egypt was the native country of all

Arts, Science, and Philosophy, and . . . from thence they have been derived to all the Nations[118]

— a claim that would once have been regarded as undermining the central historical importance of the Jews.

Furthermore, as it came to be appreciated the extent to which the religion of the ancient Israelites was influenced by their Gentile neighbours and, in particular, by the Egyptians, so too the view was strengthened that true religion, like other areas of human experience, had developed historically and was, in large measure, the daughter of time. William Robertson, who saw no conflict in combining his role as Moderator-General of the Church of Scotland with his pioneering work in establishing history as a secular discipline, discoursed on this theme in his sermon entitled 'The Situation of the World at the Time of Christ's Appearance, and its Connexion with the Success of his Religion, Considered' delivered in 1775. In it he emphasised the extent to which Providence 'conducteth all his operations by general laws' arguing that 'The Almighty seldom effects by supernatural means anything which could have been accomplished by such as are natural'.[119] For

Robertson an important illustration of this general theme was the extent to which God had relied on natural historical processes in the development and promulgation of Christianity. 'The light of revelation', he writes, 'was not poured in upon mankind all at once, and with its full splendour'. Rather the development of true religion was conditioned by such natural phenomena as the rise and fall of empires for

In proportion as the situation of the world made it necessary the Almighty was pleased further to open and unfold his scheme. And men came by degrees to understand this progressive plan of Providence.[120]

In Robertson's account the outstanding example of the way in which Christianity had benefited from such natural causes was the fact that it had largely been disseminated through the agency of the Roman Empire. Conversely natural causes could also play a role in retarding true religion: Robertson partly attributed what he saw as the persistence of pagan residues in ancient Judaism to the effects of climate along with the influence of the Israelites' Gentile neighbours.[121] As such an historically-based understanding of the development of Christianity gained ground so, too, the literal account of world history outlined in the Bible came to be seen as being itself a reflection of particular historical circumstances which required the skills of an historian to explain and to illuminate.

Consequently, although historical issues still remained of considerable importance for an historically-based religion such as Christianity it increasingly came to be acknowledged that history — including history of the ancient world which traversed similar ground to that outlined in the Bible — like natural philosophy, was a field which ought to be distinguished from theology. Of course it was assumed by believers that ultimately the fruits of secular history would be in accord with Scripture, rightly understood, and that secular history, like natural history, revealed the natural mechanisms by which Providence achieved its ends. Nonetheless, the Bible itself was less and less assumed to be the ultimate foundation on which histories of humankind should be based. Just as it came to be accepted that the Bible had not been intended to provide a system of natural philosophy so, too, it became widely, though by no means universally, acknowledged that the biblical narrative was not meant to supply the sort of historical framework which the age of the Enlightenment came to demand — one which provided an explanation of the origins of human society and institutions in terms of natural causes and which, as far as possible, used the methods so successfully employed in the natural sciences. Increasingly, then, biblical history became the pattern which was invoked to give meaning to an individual's religious experience rather than to provide a framework for explaining the development of the ancient world.[122] Moreover, as knowledge expanded, the traditional impulse to unite all fields of learning under the aegis of theology began to weaken, thus strengthening the autonomy of history as a discipline with different methods and goals to that of biblical theology. Thus that movement of ideas to which we give the name the Scientific Revolution involved more than a change in humanity's attitude to the natural world for it also helped to tranform our attitude to our past and in so doing helped to bring to birth the discipline of history in its modern secularised form.

NOTES

Note Added in Proof.
J. E. Force and R. H. Popkin (eds), *Essays on the Context, Nature and Influence of Sir Isaac Newton's Theology*, Dordrecht (1990), which includes a number of valuable articles relevant to this essay, appeared too late to be used in its preparation. For some comments see my review of this work in *Physis* (forthcoming).

[1] Funkenstein, A. *Theology and the Scientific Imagination*, Princeton (1986).
[2] For a useful review of the literature on this subject see Preston, J., 'Was There an Historical Revolution', *Journal of the History of Ideas* 38, 353—64 (1977).
[3] Smith Fussner, F., *The Historical Revolution: English Historical Writing and Thought 1580—1640*, London (1962), especially pp. 25, 297, 306—8 and Sypher, G. W., 'Similarities between the Scientific and the Historical Revolutions at the End of The Renaissance', *Journal of the History of Ideas* 26, 353—68 (1965).
[4] Manuel, F., *Isaac Newton Historian*, Cambridge, Mass. (1963), p. 29 and McGuire, J. E. and Rattansi, P. M., 'Newton and the "Pipes of Pan"' *Notes and Records of the Royal Society of London* 21 (1966), pp. 128—9. An early eighteenth-century sermon entitled 'The Wisdom of the Ancients Borrowed from Divine Revelation or, Christianity Vindicated against Infidelity' by the Anglican divine, Daniel Waterland, provides a useful survey of this form of historical interpretation complete with detailed references on this point in the works of Church Fathers such as Justin Martyr, Clement of Alexandria, Tertullian, Origen and Augustine. Van Mildert, W. (ed.), *The Works of the Reverend Daniel Waterland, D. D.* 2nd ed. 6 Vols, Oxford (1843), V, pp. 4—13.
[5] McGuire, J. E., 'Neoplatonism and Active Principles: Newton and the Corpus Hermeticum', in R. S. Westman and J. E. McGuire, *Hermeticism and the Scientific Revolution*, Berkeley (1977), p. 128.
[6] Manuel, *Newton Historian*, p. 101.
[7] Gale, T., *The Court of the Gentiles: or a Discourse Touching the Original of Human*

Literature, both Philologie and Philosophie, from the Scriptures and Jewish Church 2nd ed., Oxford (1672—82), Part I, p. [i], heading to Ch. X; Part II, pp. 39, 44.
8 Wallis, J., Three Sermons Concerning the Sacred Trinity, London (1691), p. 99.
9 Spencer, J., De Legibus Hebraeorum Ritualibus et Earum Rationibus, Libri Quatuor 2nd ed., Cambridge (1727), II, pp. 640—1. Translated in Woodward, J., 'Of the Wisdom of the Antient Egyptians; a Discourse concerning their Arts, their Sciences, and their Learning: their Laws, their Government, and their Religion. With occasional Reflections upon the State of Learning, among the Jews; and some other Nations' Archaeologia 4, p. 268 (1777).
10 Spencer, De Legibus, II, p. 649. Translated Woodward, 'Wisdom of the Antient Egyptians', p. 269.
11 Spencer, De Legibus, II, p. 1089.
12 Literal translation from ibid. II, p. 741.
13 Spencer's sermon attracted considerable attention to judge from the comments of that connoisseur of sermons, Samuel Pepys, who, on 1 June 1664, recorded reading 'Mr. Spenser's Book of Prodigys' and finding it 'most ingeniously writ, both for matter & style'. Pepys later notes that on 25 May 1666 he was out walking with a friend and 'discoursing & admiring of the learning of Dr. Spenser'. Latham, R. and Matthews, W. (eds), The Diary of Samuel Pepys 10 Vols, London (1970— 83), V, p. 165 and VII, p. 133. Spencer's sermon was republished in a second and considerably enlarged edition in 1665. In the same year he also published A Discourse Concerning Vulgar Prophecies: Wherein the Vanity of Receiving Them as the Certain Indications of any Future Events is Discoursed; And Some Characters of Distinction Between True and Pretending Prophets are Laid Down which, as the name suggests, was again directed at popular enthusiasm and its threat to political and religious order.
14 Spencer, J., A Discourse Concerning Prodigies: Wherein the Vanity of Presages by Them is Reprehended, and their True and Proper Ends Asserted and Vindicated, Cambridge (1663), pp. [vi], [iii].
15 Ibid., [i].
16 Ibid., pp. 14—6.
17 Ibid., p. 43.
18 Ibid. 2nd. ed., London (1665), p. 136.
19 Aegyptica . . . sive de Aegyptiacorum Sacrorum cum Hebraicis Collatione Libri Tres . . ., Amsterdam (1683). Cited in translation in Warburton, W. The Divine Legation of Moses Demonstrated 2 Vols, London (1837), II, p. 167.
20 Edwards, J. Compleat History, pp. 249, 252.
21 Edwards, J. Cometomantia, London (1684), pp. 94, 98, 28.
22 In 1682, Bayle published (in French) his Letter to M. L. A. D. C., Doctor of the Sorbonne. Wherein it is Proved in the Light of Various Arguments Derived from Philosophy and Theology that Comets Are in No Sense Portents of Disaster . . . . A sequel followed in 1683 with further supplements in 1694 and 1705. On the significance of Boyle's works see Hazard, P. The European Mind 1680—1715, (Cleveland, Ohio (1963), pp. 155—161.
23 Edwards, J. Cometomantia, p. 131. Among the arguments which Edwards attacks is one developed by Spencer (though Edwards, understandably, does not explicitly cite his distinguished, senior Cambridge colleague on the point) — namely, Spencer's conten-

tion that comets could not be intended as a divine warning to a particular nation since comets were visible to many different countries simultaneously. *ibid.*, p. 158 and Spencer, *Prodigies* (1663), p. 17.

24 Woodward, 'Of the Wisdom of the Antient Egyptians', pp. 280, 225.

25 *Ibid.*, pp. 218, 228—9, 238, 281.

26 *Ibid.*, p. 271.

27 This work was published in two parts. The Latin first part appeared in 1681 (and was translated into English in 1684) and the Latin second part was published in 1689 and translated in 1690. Jacob, M. C. and Lockwood, W. A., 'Political Millenarianism and Burnet's *Sacred Theory*', *Science Studies* 2, p. 265 (1972).

28 Burnet, T. *The Theory of the Earth* 2nd ed., London (1691), p. 6.

29 *Ibid.*, pp. 289, 314—5.

30 Edwards, J. *A Demonstration of the Existence and Providence of God* . . . (1696), p. 257. Cited in Levine, J. M. *Dr. Woodward's Shield: History, Science and Satire in Augustan England*, Berkeley (1977), p. 265.

31 Woodward, J. *An Essay Toward a Natural History of the Earth*, London (1695), p. 165.

32 Cited in Porter, R. S. *The Making of Geology. Earth Science in Britain 1660—1815*, Cambridge (1977), p. 76.

33 Burnet, *Theory of the Earth*, p. 281.

34 Burnet, T. [*Archaeologiae Philosophicae sive*] *Doctrina Antiqua de Rerum Originibus: Or, An Inquiry into the Doctrine of the Philosophers of all Nations, Concerning the Original of the World* Translated by Mr. Mead and Mr. Foxton, London (1736), pp. 45, 56, 241, 244, 246. For an important discussion of Burnet's writings see Rossi, P. *The Dark Abyss of Time. The History of the Earth and the History of Nations from Hooke to Vico*, Chicago (1984), pp. 33—41, 66—74, 89—94.

35 Burnet, T. *An Answer to the Late Exceptions Made by Mr. Erasmus Warren Against the Theory of the Earth*, London (1690), p. 84.

36 Burnet, [*Archaeologiae Philosophicae*], p. 2.

37 *The Seventh and Eighth Chapters of Dr. Burnet's Archiologiae [sic] Philosophicae, together with his Appendix to the Same . . . Rendered into English, by Mr. H. B.* [Henry Brown] in Blount, C. *The Miscellaneous Works*, London (1695), pp. 29, 32—3, 39.

38 Cited in Redwood, J. *Reason, Ridicule and Religion: The Age of Enlightenment in England, 1660—1750*, London (1976), p. 119.

39 *The Seventh and Eighth Chapters of Dr. Burnet's Archiologiae Philosophicae*, p. 50.

40 Redwood, *Reasons, Ridicule and Religion*, p. 121.

41 *The Seventh and Eighth Chapters of Dr. Burnet's Archiologiae Philosophicae*, p. 54.

42 Cited in Kubrin, D. *Providence and the Mechanical Philosophy: The Creation and Dissolution of the World in Newtonian Thought*, PhD thesis, Cornell University (1968), pp. 145—6. The extent of clerical hostility to Burnet's work is indicated by the remark in 1693 of Humphrey Prideaux, canon of Norwich and author of an historical account of the period between the Old and the New Testaments, that the coffee-house atheists made much use of the *Archaeologiae Philosophicae* 'to confute ye account ye Scriptures give us of ye creation of ye world'. Thompson, E. M. (ed.), *Letters of Humphrey Prideaux, Sometime Dean of Norwich to John Ellis* . . . . Camden Society Publications, n.s. XV, London (1875), pp. 162—3.

[43] Edleston, J. (ed.), *Correspondence of Sir Isaac Newton and Professor Cotes* ... , London (1850), p. xliv.

[44] Turnbull, H. W. *et al.* (eds), *The Correspondence of Sir Isaac Newton* 7 Vols, Cambridge (1959—77), II, p. 331.

[45] Burnet, *Theory of the Earth*, p. 315.

[46] Turnbull, *Correspondence of Newton*, II, p. 334.

[47] Harrison, J. *The Library of Isaac Newton*, Cambridge (1978), p. 112.

[48] Westfall dates the beginnings of Newton's most important theologico-historical work, the *Theologiae Gentilis* from the mid 1680s although he returned to it in the early 1690s (at much the same time as Burnet's work appeared). Newton thereafter continued to revise the work until at least 1716. Westfall, R. S. 'Isaac Newton's *Theologiae Gentilis Origines Philosophicae* in W. W. Wagar, (ed.), *The Secular Mind. Transformations of Faith in Modern Europe*, New York (1982), pp. 15—34.

[49] *Ibid.*

[50] Westfall, R. S. *Never at Rest. A Biography of Isaac Newton*, Cambridge, (1983), p. 812.

[51] Newton, I. *Opera Quae Exstant Omnia* ed. S. Horsley, 5 Vols, London (1785), V, pp. 140—1.

[52] Cited in Westfall, 'Newton's *Theologiae Gentilis*', p. 30.

[53] McLachlan, H. (ed.), *Sir Isaac Newton: Theological Manuscripts*, Liverpool (1950), p. 52.

[54] Burnet, T. *The Faith and Duties of Christians* translated into English by Mr. Dennis, London [1728], p. 57.

[55] Newton, I., *The Opticks*, Great Books, Chicago (1952), p. 542.

[56] Kubrin, D. 'Newton and the Cyclical Cosmos: Providence and the Mechanical Philosophy', *Journal of the History of Ideas* 28, 325—46 (1967).

[57] Newton, *Opticks*, p. 543.

[58] McLachlan, *Newton: Theological Manuscripts*, p. 28.

[59] Newton, *Opticks*, pp. 543—4.

[60] Jewish National and University Library, Yahuda MS 41, fols 1—3.

[61] *Ibid.* fol. 8.

[62] Westfall, 'Newton's *Theologiae Gentilis*', p. 26.

[63] See ref. 34 above.

[64] Whiston, W. *Memoirs of the Life and Writings of Mr. William Whiston* ... 2 Vols., London (1749), I, p. 38. *Acts* 3:20—1 reads: 'And he shall send Jesus Christ ... whom the heavens must receive until the times of restitution of all things, which God hath spoken by the mouth of all his holy prophets since the world began'.

[65] Schaffer, S. 'Newton's Comets and the Transformation of Astrology', in P. Curry (ed.), *Astrology, Science, and Society: Historical Essays*, Woodbridge (1987), pp. 219—43.

[66] Turnbull, *Correspondence of Newton*, III, p. 185.

[67] *Ibid.*, p. 292. The second edition did not, however, eventuate until 1727 when, interestingly enough, it was published with the assistance of a bequest from the latitudinarian Archbishop Tenison of Canterbury, a former fellow of Spencer's college — an indication that Spencer's work was by no means universally regarded with suspicion by his fellow clergy.

[68] Harrison, J. *Library of Isaac Newton*, p. 242.

[69] Yehuda MS 41, fol. 5.

[70] *Ibid.* fol. 10. Newton's Cambridge contemporary, Ralph Cudworth, argued that not only had Ham settled in Egypt but that the name of the Egyptian supreme god, Ammon (or Hammon), was derived from Ham. Cudworth, R. *True Intellectual System* 3 Vols, London (1845), I, p. 572, (first edition, 1678).

[71] Yehuda MS 41, fol. 5.

[72] Manuel, *Newton Historian*, p. 148.

[73] Turnbull, *Correspondence of Newton*, III, p. 338.

[74] On this point see Allen, D. C. 'Some Theories of the Growth and Origin of Language in Milton's Age', *Philogical Quarterly* 28, 5—16 (1949), and Kottman, 'Fray Luis de Léon and the Universality of Hebrew: An Aspect of Sixteenth and Seventeenth Century Language Theory', *Journal of the History of Philosophy* 13, 297—310 (1975).

[75] Westfall, 'Newton's *Theologiae Gentilis*', p. 23.

[76] Newton, I. *The Mathematical Principles of Natural Philosophy and his System of the World* translated by Andrew Motte. 2 Vols, Berkeley (1974), II, p. 549.

[77] Thus in his unpublished 'classical scholia' to the *Principia* Newton traces the atomic philosophy only as far back as Moschus the Phoenician from whom it passed to the Egyptians and thence to the Greeks. Casini, P. 'Newton: The Classical Scholia', *History of Science* 22, p. 36 (1984). In his extensive notes 'Out of Cudworth' (now in the William Andrews Clark Memorial Library, University of California, Los Angeles) Newton copied much of Cudworth's material on Moschus but ignores his identification of Moschus with Moses. On this identification see Sailor, D. B., 'Moses and Atomism', *Journal of the History of Ideas* 25, 3—16 (1964).

[78] Cudworth, *True Intellectual System*, III, p. 185. In one respect Newton went even further than Cudworth in praising the Egyptians for in the Clark MS he wrote of the Egyptians' natural philosophy (which he interpreted as embodying a mystical form of atomism) that 'Dr. Cudworth therefore is much mistaken when he represents this Philosophy as Atheistical' (lines 31—2). A recent study of this manuscript concludes that:

Newton was very much convinced of the priority and thus the importance of Egyptian learning in the ancient world, and this is reflected in almost an entire page of the references and quotes which he took from Cudworth.

Sailor, D. B. ' Newton's Debt to Cudworth', *Journal of the History of Ideas* 49, p. 549 (1988).

[79] Cudworth, *True Intellectual System*, I, p. 519.

[80] Turnbull, *Correspondence of Newton*, III, p. 338.

[81] Cited in Manuel, *Newton Historian*, p. 115. Newton makes a similar point in the *Theologiae Gentilis* where he argues that idolatry began in Egypt with 'annual solemnities in honour of their first king & queen Osiris & Isis'. From thence such practices spread to Greece and other countries 'by the colonies of the Egyptians which were very many & the commerce wch ye nations had with one another'. Yehuda MS 41, fol. 10.

[82] Cited in Manuel, *Newton Historian*, p. 91.

[83] Yehuda MS 41, fol. 9v.

[84] Cited in Westfall, ' Newton's *Theologiae Gentilis*', p. 26.

[85] *Ibid.*

[86] Manuel, *Newton Historian*, pp. 39, 93.

[87] *Ibid.*, p. 91.

[88] *Ibid.*, p. 89 and Wilcox, D. J. *The Measure of Times Past. Pre-Newtonian Chronologies and the Rhetoric of Relative Time*, Chicago (1987), p. 209.

[89] *Chronology of the Ancient Kingdoms Amended* in I. Newton, *Opera Omnia* (ed.) S. Horsley, p. 14.

[90] Westfall 'Newton's *Theologiae Gentilis*', p. 23 and Manuel, *Newton Historian*, pp. 58–9, 140.

[91] Hazard, *The European Mind 1680–1715*.

[92] Bernal, M. *Black Athena. The Afroasiatic Roots of Classical Civilization*. Vol. 1 *The Fabrication of Ancient Greece 1785–1985* (London, 1987), p. 191. As Bernal points out, his overall thesis — that the Greeks derived much of their learning from the Egyptians — was a commonplace of seventeenth-century scholarship.

[93] Manuel, *Newton Historian*, pp. 98, 119.

[94] Yehuda MS 41, fol. 11.

[95] Brumfitt, J. H. (ed.), *La Philosophie de L'Histoire*, Vol. 59 *The Complete Works of Voltaire* (ed.) T. Besterman *et al.*, Geneva (1969), pp. 16–7, 32–5.

[96] Bossuet, J. *An Universal History from the Creation of the World to the Empire of Charlemange* translated James Elphinson, London (1778), p. 154. Such a providentalist view of history also coloured Bossuet's brief survey of modern history in which he focusses on the history of the Christian Church as 'a perpetual miracle, and a shining testimony of the immutability of the counsels of God' (p. 370).

[97] Voltaire, *The Philosophy of History*, London (1822), pp. 155, 69, 147–9. On the Jews' debt to Egypt see also Chapter XXII.

[98] Stillingfleet, E. *Origines Sacrae: Or a Rational Account of the Grounds of Natural and Revealed Religion* Seventh edition, Cambridge (1702), Preface. (first edition, 1662).

[99] Marsham, J. *Canon Chronicus Aegyptiacus, Ebraicus, Graecus* ... , Franeker (1696), [ii].

[100] Spencer, *De Legibus*, II, p. 655 and Manuel, *Newton Historian*, p. 101; for references to Marsham in the *Theologiae Gentilis* see Yehuda MS 41, fols. 5, 23. On Marsham and his influence see Rossi, *Dark Abyss of Time*, pp. 125–6.

[101] Toland, J. *Letters to Serena* ..., London (1704), pp. 26, 39, 40.

[102] Tindal, M. *Christianity as Old as Creation*, London (1730), pp. 90, 282.

[103] Spencer, *De Legibus Hebraeorum* ... I, pp. 54–61.

[104] Waterland, 'The Wisdom of the Ancients ...', *Works* (ed.) Van Mildert, V, p. 14.

[105] Cragg, G. R. *Reason and Authority in the Eighteenth Century*, London (1964), pp. 32–3. On Middleton's religious beliefs see Gascoigne, J. *Cambridge in the Age of the Enlightenment. Science, Religion and Politics from the Restoration to the French Revolution*, Cambridge (1989), pp. 138–41.

[106] Middleton, C. *A Letter to Dr. Waterland* In Middleton, C. *The Miscellaneous Works* 4 Vols, London (1752), II, pp. 154, 153.

[107] Middleton, C. *A Defence of the Letter to Dr. Waterland* In *ibid*, II, p. 216. Among those whom Middleton thought ought to take more account of such findings was

Newton — thus he took him to task for his contention that the Egyptians 'had not even the use of Letters till about Solomon's Reign' (pp. 231—2).

[108] Rossi, *Dark Abyss of Time*, p. 237. The first part of the *Divine Legation* appeared in 1737 and the second in 1741. The third part was published in fragmentary form in 1788 nine years after Warburton's death.

[109] Iversen, E. *The Myth of Egypt and its Hieroglyphs*, Copenhagen (1961), p. 103.

[110] Warburton, W. *The Divine Legation of Moses Demonstrated* 2 Vols, London (1837), II, pp. 150, 183, 138—9.

[111] *Ibid.*, pp. 119, 91.

[112] Warburton, W. *The Works* ed. Bishop Hurd, 7 Vols, London (1788), V, pp. 294, 288, 290.

[113] *Ibid.*, IV, p. 362,

[114] Warburton, *Divine Legation*, II, p. 91.

[115] Evans, A. W. *Warburton and the Warburtonians*, Oxford (1932), p. 22.

[116] Torrey, N. L. *Voltaire and the English Deists*, New Haven (1930), pp. 9, 128, 170—4 and Brumfitt, *La Philosophie de L'Histoire*, pp. 20, 61, 315—6.

[117] For a review of the literature on this subject see Gascoigne, J. 'From Bentley to the Victorians: The Rise and Fall of British Newtonian Natural Theology', *Science in Context* **2**, 219—56 (1988).

[118] Slotkin, J. S. *Readings in Early Anthropology*, London (1965), p. 229.

[119] Robertson, W. *The Works*, (ed.) D. Stewart, London (1837), pp. lii—iii.

[120] *Ibid.*, p. lii.

[121] *Ibid.*, p. liv.

[122] Frei, H. W. *The Eclipse of Biblical Narrative. A Study in Eighteenth and Nineteenth Century Hermeneutics*, New Haven (1974), p. 152.

From Bentley to the Victorians: The Rise and Fall of British Newtonian Natural Theology

The Argument

The article explores the reasons for the rise to prominence of Newtonian natural theology in the period following the publication of the *Principia* in 1687, its continued importance throughout the eighteenth and first half of the nineteenth centuries, and possible explanations for its rapid decline in the second half of the nineteenth century. It argues that the career of Newtonian natural theology cannot be explained solely in terms of internal intellectual developments such as the theology of Newton's clerical admirers or the impact of the work of Hume or of Charles Darwin. While such intellectual movements are undoubtedly of considerable importance in accounting for the rise and fall of Newtonian natural theology, they do not of themselves explain why British society was more receptive to particular bodies of thought in some periods rather than in others. Hence this article – in common with a number of recent studies – attempts to draw some connections between the growth of Newtonian natural theology and the character of Augustan society and politics; it also attempts to link the decline of this tradition with such nineteenth-century developments as the growing separation between church and state and the secularization of the universities and of scientific and intellectual life more generally.

One of the most distinctive features of British intellectual life in the eighteenth century, and in much of the nineteenth, was the extent to which science was seen to be allied to the cause of religion. The tradition of scientifically based natural theology was not altogether absent in Continental countries but, by and large, outside Britain science was more likely to be used as a weapon by the opponents of the old regime in church and state than in Britain. In Britain, by contrast, the tradition of natural theology provided reassurance that the ecclesiastical order and the political regime with which it was closely associated were consistent with the highest reaches of human reason and were proof against the assaults of the infidel. As Cannon puts it:

How different the English Enlightenment was from the French. Sheltered under Newton's great name, science and religion had developed a firm alliance in

England, symbolized by that very British person, the scientific parson of the Anglican Church. (Cannon 1978, 2)

Such respectable credentials also made the tradition of natural theology a ready vehicle for the popular dissemination of novel scientific developments (Gillispie 1959, 227) and, well into the nineteenth century, it provided a common intellectual context which gave unity to the growing number of scientific disciplines by providing them with a shared theological sheet anchor (Young 1980).

The elastic and pliable nature of natural theology meant that it could take different forms according to the scientific and theological presuppositions of its various practitioners. However, the longevity and popularity of the genre in Britain owed most, as Cannon's remarks suggest, to the prestige which it had acquired by the successful recruitment of Newton's scientific achievement as an ally in the cause of religious apologetic. This association between Newton and the cause of religion also had the effect of helping to immortalize the reputation of Newton among his countrymen. As George W. Hemming put it in his review of David Brewster's monumental *Life of Newton*:

Yet another influence of incalculable strength [in the growth of Newton's reputation] was derived from the obvious association of the discoveries of Newton with the teachings of religion, and with the theological speculations of the philosopher himself. . . . Through the teaching of the pulpit the humblest classes of English society were constantly reminded that their country could boast of a natural philosopher with whom none of the infidel teachers of Paris could compete, and who did not disdain to apply his powers to the reverent study of the mysteries which they affected to despise.([Hemming] 1861, 406)

Not that it was Newton alone who helped father the tradition of natural theology – Sprat, Wilkins, Boyle, John Ray and others all played an important part in linking the cause of science with religious apologetic. However, it was in the wake of Newton's *Principia* that the tradition of natural theology became firmly established and it continued thereafter to bask in the reflected glory of the great Sir Isaac. Conversely, as the Newtonian form of religious apologetic began to lose public acceptance so, too, did the whole genre of natural theology lose much of its intellectual vitality. By charting the rise and fall of Newtonian natural theology, then, this essay will attempt to shed light on two more general questions: firstly, why it was that in Britain, in contrast with most Continental countries, religion achieved a rapprochement with those new intellectual currents to which we give the name, the Enlightenment, and, secondly, why in the course of the nineteenth century British science became less and less linked with the cause of religion.

In tracing the fortunes of an intellectual tradition such as that of Newtonian natural theology we can distinguish between those factors which affected its coherence as a system of ideas and those which help to explain why it was accepted or rejected by the larger society – to put it crudely between *internal* and *external* considerations. Both levels of explanation are important: unless a system of ideas is popularly perceived to be internally consistent, it is unlikely to command widespread support; on the other hand, internal intellectual consistency alone does not explain why some systems of ideas attract a widespread popular following and others do not. In the case of Newtonian natural theology, it has to be explained both how the forbidding mathematical and scientific exposition contained in the *Principia* came to be integrated into a system of popular religious apologetic and how, too, this type of theology came to command so widespread a following even among those incapable of reading the *Principia*. Similarly, the decline of Newtonian natural theology cannot be explained solely in internal intellectual terms, important though these were. Thus, although the challenge to the argument from design, which Darwin's work posed, undoubtedly played a major role in the decline of natural theology, there remains the obstinate fact that Hume, in his *Dialogues Concerning Natural Religion,* had presented as far back as 1776 a damaging full-scale assault on the intellectual foundations of natural theology (and, in particular, on its Newtonian form) with little effect on popular belief. The decline of Newtonian natural theology cannot, therefore, be explained solely in terms of a challenge to its intellectual presuppositions. It requires, too, some explanation of why the external social conditions, which had once favored its rise and dissemination, were ultimately to change to the point where such intellectual critiques came to be widely perceived as having undermined a previously well-established tradition.

Newton was reported by Derham "to have made the *Principia* abstruse to avoid being baited by the little smatterers in mathematics"(Keynes MS. 130, King's College, Cambridge). Indeed, the form and mathematical rigor of Newton's *Principia* necessarily restricted it to a highly select audience and thereby, as Newton hoped, reduced the possibility of the public controversy which he so detested. Yet Newton and his deliberately abstruse *Principia* had, by the time of his death in 1727, become the object of a public cult which was epitomized in the valedictory elegy of James Thompson in which Newton was praised as one:

> Who, while on this dim spot, where mortals toil
> Clouded in dust, from motion's simple laws,
> Could trace the secret hand of Providence,
> Wide-working through this universal frame
>
> (Quoted in Buchdahl 1961, 49–50)

This apotheosis of Newton, and of the form of natural theology with which he became so closely associated, underlies the difficulty of explaining the rise to prominence of Newtonian natural theology solely in internal intellectual terms. Newton's work was accessible to few, yet it became an object of widespread public acclaim. In the first edition of the *Principia* Newton made only one passing reference to God (Manuel 1974, 31), yet soon after its publication his work became closely associated with the cause of Christian apologists. True, once the *Principia* was taken up by the theologians, Newton gave this apologetical enterprise his guarded blessing, but in its original form his work provided no obvious stimulus for such an undertaking.

The means by which aspects of Newton's *Principia* became linked with Anglican apologists of the late seventeenth and early eighteenth centuries are well-known, thanks particularly to Metzger (1938) and, more recently, Jacob (1976). Though Bentley, the inaugural Boyle lecturer for 1692, was barely competent to follow the *Principia,* he nonetheless saw in this work a useful source of illustrations of the argument from design. In his famous correspondence with Newton, Bentley obtained the Master's enthusiastic support for the basic goal of using Newtonian natural philosophy as an ally in the cause of religion and his rather more qualified support for taking up particular aspects of the *Principia* as a means of bolstering the design argument. Thus Newton assented to the view that providential intervention could alone explain the fact that the heavenly bodies were so positioned that their movement counterbalanced the gravitational force which would otherwise cause all celestial matter to coalesce in one great mass – an argument that Newton was later to develop further following Bentley's initial prompting (Hoskin 1977). Newton also responded to Bentley's inquiries by maintaining that the regular movements of the planets "were impressed by an intelligent Agent" (Thayer 1974, 47), since celestial bodies did not necessarily conform to such ordered orbits, as the erratic behavior of comets testified. Newton also gave some highly qualified support to Bentley's contention that gravity could be regarded as an instance of the workings of Providence in the universe. However, Newton had to restrain Bentley's attempts to press as many aspects of the *Principia* as possible into theological service vetoing, for example, his suggestion that there was something "extraordinary in the Inclination of the Earth's Axis for proving a Deity" (ibid., 49).[1]

Why such enthusiasm for a work which Bentley was barely qualified to read? Here Bentley's political and religious loyalties are relevant together with those of other, later, Boyle lecturers such as John Harris, Samuel and John Clarke, William Whiston, and William Derham who carried on Bentley's work of linking Newtonian natural theology and Anglican apologetics. Jacob has shown the importance of these early Boyle lecturers in using Newtonian natural philosophy as a means of countering threats to the established church in the troubled conditions that followed the 1688 revolution. It should be added, too, that the concern of Bentley and other like-minded Boyle lecturers was not only to strengthen the position of the established church as a whole but also to consolidate the position of themselves and their fellow Low Churchmen in a church which remained deeply divided in its response to the events of 1688. Though Jacob tends to treat the Boyle lecturers as an ideologically consistent school, such divisions of opinion

---

[1] Bentley's theological borrowings from Newton also had an implied statistical character: that it was much more probable that a divine intelligence had imposed order on the universe than that such apparent design was the result of chance. This protostatistical reasoning was later to be further developed by Abraham de Moivre (1667–1754) and John Arbuthnot (1667–1735) (Sheynin, 1970–71).

were even reflected in the differing outlook of the Boyle lecturers themselves – as Bowles points out, by no means all of the lecturers took the same enthusiastic attitude to the alliance of Newtonianism and Anglicanism which Bentley had endeavored to foster. John Hancock, for example, in his Boyle lectures for 1706 described Newton's law of attraction as "a late Notion and Assertion in Philosophy, that every thing attracts every thing; which is in effect to say, that nothing attracts any thing" (Bowles 1976, 314).

What, then, did Bentley and his Low Church allies see in Newtonian natural philosophy which could help provide a bulwark against their enemies both within and outside the established church? In the first place the emphasis on natural theology, which was naturally associated with Bentley's selective exposition of Newton's work, meant that he and his allies were helping to focus public attention on those aspects of theology about which there was widespread agreement rather than on those contentious areas of revealed theology which were a source of division within the church and in the nation in general. More generally Bentley and his allies, as firm upholders of the post-1688 constitution and, subsequently, of the Hanoverian succession against the claims of the Stuarts with their appeal to traditional notions of divine right, were naturally anxious to promote those forms of Christian theology which emphasized the accord between Christianity and reason in opposition to their High Church opponents who favored a more transcendental view of both the nature of Christianity and the nature of kingship. The defenders of the Revolutionary Settlement were to place increasing emphasis on its accord with reason, using such devices as the Lockean social contract to argue for the rational basis of the eighteenth-century constitution. Such an intellectual milieu favored the dissemination within the established church of a form of theology which also placed a great deal of emphasis on the harmonious relations between Christianity and reason using Newtonian natural theology as one of its most potent illustrations.

Secondly, the Newtonian concept of gravity – or at least that providential explanation of it which Bentley and his followers popularized – was seen as a useful weapon against metaphysical systems which challenged Christianity. It provided a middle path between the pantheistic monism of philosophers such as Spinoza and the dualism of Descartes (Gay 1963, 93) – the latter being a system which, in English eyes, was likely to degenerate into another form of monism, the materialism of Hobbes. Gravity, proclaimed Bentley (1724, 127) in the Boyle lectures, "the great Basis of all Mechanism, is not itself mechanical; but the immediate Fiat & Finger of God" – an argument which, in his view, "will undermine & ruin all the Towers & Batteries that the Atheists have raised against Heaven." Bentley's concern at the influence of Hobbes among the leisured classes is apparent in a letter (written at the same time that he was giving his Boyle lectures) in which he remarked that "Not one English Infidel in a hundred is any other than a Hobbist; which I know to be rank Atheism in the private study & select conversation of those men, whatever it appear to be abroad." (Wordsworth 1842, 1:39). Though Bentley regarded Spinoza as less of a threat within England than Hobbes, Samuel Clarke, when he came to give his

Boyle lectures in 1706, was sufficiently concerned about the influence of the Dutch metaphysician to castigate "the Vanity, Folly, and Weakness of Spinoza: who ... concludes from thence, that *the whole World, and every thing contained therein, is one Uniform Substance, Eternal, Uncreated, and Necessary ...*" (quoted in Colie 1963, 207).

The concern of the Boyle lecturers, as Jacob has emphasized, was that opponents of revealed religion, such as Hobbes and Spinoza, would weaken not only the church's intellectual dominance but would also undermine its authority as a custodian of morality and social order. Jacob's pioneering work has done much to draw attention to the sociopolitical context in which Newtonian natural philosophy and natural theology came to prominence. However, in reacting against the traditional, highly internal, and intellectualist accounts of the reasons for the diffusion of systems of ideas such as Newtonianism, Jacob gives rather more weight to external social considerations than her evidence will readily bear. In arguing that Newton's Low-Church followers "synthesize[d] the operations of a market society and the workings of nature in such a way as to render the market society natural" (1976, 51), Jacob attributes to natural philosophy a cultural and even political centrality that it lacked in the early eighteenth century. Scientific concerns were still too much at the periphery of intellectual life to provide the norm and authority by which fundamental institutions such as the state or the character of the economy might be justified (Porter 1978, 249). The ideological significance of Newtonianism was exercised more indirectly and less potently as an aid to a particular theological and ecclesiastical party – that of the Low Churchmen – which in turn played a significant part in the political-cum-ecclesiastical debates which followed the Glorious Revolution. Even those Boyle lecturers who did draw on Newton's work treated it as but one weapon – albeit an increasingly important one – from among an elaborate armory of theological defenses. It was only *after* some of the Boyle lecturers and other popularizers of Newton's work had helped to invest it with particular authority – largely derived (at least at first) from its perceived theological importance – that it could be invoked by those, like Adam Smith, who wished to draw a parallel between the workings of a laissez-faire economy and those of the universe more generally.

External social considerations, then, are useful in helping to explain the rise to prominence of Newtonian natural theology, but they have their limits. The language of Newtonianism (or, for that matter, political economy) was still too unfamiliar to contemporaries in the decades following the *Principia* to make the claim that "the Newtonian vision of the natural world provided irrefutable justification for the public order and controlled self-interest sanctioned and maintained by church and state" (Jacob 1976, 269). For Newtonian natural philosophy to become culturally important it was first necessary to establish that it could be amalgamated convincingly with already established and authoritative sources of belief, notably theology – something which, of course, Newton's early clerical disciples succeeded in doing. In short, a novel system of ideas such as Newtonianism or Newtonian natural theology had first to demonstrate its internal, intellectual coherence before it could become part of the ideological equipment of Augustan England.

It might be argued, however, that not only the reception of such ideas but also their very formulation was shaped by the social and political system of the time. Certainly, the parallel between the workings of the macrocosm of the universe and the microcosm of the political order had a long history and was still deeply rooted in the habits of mind of Newton and his contemporaries. Conservative social theorists had, for example, long used the highly hierarchical Ptolemaic system as a metaphor for the need for social deference in the world of man. It is to be expected, then, that perceptions of the way in which the natural world worked should be to some extent determined by the character of the sociopolitical system. But such a parallel was, of course, often very imperfect since the internal logic of a system of ideas and its need for consistency could lead to an emphasis on concepts which had no very clear social analogue – the Ptolemaic system, for example, demanded the use of equants and epicycles even though these marred the clarity and the religious and social significance of the system.

Though the cosmology of Newton and his disciples was inevitably intertwined with their theological and political views, such social considerations only very partially explain the development of the Newtonian world view. For it is doubtful whether English society changed sufficiently in the course of the seventeenth century to generate the marked break with traditional cosmological theory which Newtonianism constituted. Despite the Civil War, the aristocratic and clerical elites were firmly reestablished at the Restoration and it was on their favor that Newton, a member of a highly clerical university and, subsequently, a client of the aristocratic Montague family, depended for advancement. One might question, then, Freudenthal's contention that Newton was so greatly influenced by bourgeois concepts of freedom and individualism that these helped shape his "fundamental postulate, that a system consists of equal elements whose essential properties are attributable to each single element, independently of the system" (Freudenthal 1986, 189–90) – a postulate on which he erected his concept of absolute space.

Given the traditionally close links between theology and political theory it is to be expected that conceptions of God and conceptions of kingship were often closely related. Schaffer (1980b) and Shapin (1981), then, present a generally convincing case for the argument that the view of the role of the Deity taken by Newton's chief philosophical spokesman, Samuel Clarke, was largely shaped by his desire, as a Court Whig, to affirm both the majesty of the House of Hanover and the need for the power of the monarch to be in accordance with the rule of law. Both Clarke and his opponent Leibniz, do, after all, use the language of politics to describe the role of God. Thus, when objecting to Leibniz's argument that God should not need to intervene in the universe Clarke responds:

If a king had a kingdom, wherein all things would continually go on without his government or interposition it would be to him, merely a nominal kingdom; nor would he in reality deserve at all the title of king or governor. (Quoted in Shapin 1981, 201)

Nonetheless, though Clarke's exposition of the role of God in the Newtonian system may well have been influenced by his political views, one may still wish to qualify Shapin's claim that "considerations of social use did not *follow* autonomous processes of evaluation but were in fact *central* to natural philosophers' judgments" (ibid., 215). Though Newton was a determined opponent of James II and in 1689 wrote a letter to his (Newton's) fellow dons justifying James's deposition on the grounds that kings must comply with the laws of the land (Turnbull et al. 1959–77, 3, 12), the conception of God he outlines in the General Scholium to the second edition of the *Principia* could readily be equated with a defence of the powers of an absolute monarchy – especially since Newton employs political vocabulary such as "governs," "dominion," and "Ruler" : "This Being governs all things, not as the soul of the world, but as Lord over all; and on account of his dominion he is want to be called 'Lord God' . . . or 'Universal Ruler'" (Thayer 1974, 42).

True, in Newton's cosmology God generally works through a system of laws like a good constitutional monarch but in Newton's (and Clarke's) voluntarist theology these laws depend for their existence on God's will and could be suspended at his pleasure – as in the case of miracles. Though Clarke and subsequent British Newtonian popularizers largely succeeded in making the Newtonian conception of God mesh with the political realities of Hanoverian England, it does not follow that Newton's original system complied so readily with such social imperatives. Thus criteria of social use can often better explain the popularization and dissemination of ideas than their original formulation.

The success of the efforts of Bentley, Clarke, and others of Newton's early clerical disciples to enlist Newton's work as a means of defending the position of the established church and, indirectly, the political order with which the church was inextricably linked, appears to have made Newton himself more inclined to make explicit the theological assumptions with which his scientific work had always been closely associated. The most obvious example of Newton's growing willingness to foster publicly the alliance between his scientific work and Christian apologetic is, of course, the General Scholium to the second edition of the *Principia* (which may have been written at the suggestion of Bentley [Hurlbutt 1965, 58]). Its theological section concludes with an affirmation of the general enterprise of Newton's clerical disciples, if not of their specific arguments: "And thus much concerning God, to discourse of whom from the appearances of things does certainly belong to natural philosophy" (Thayer 1974, 44–45), a text that William Whewell was later to take as the motto for his Bridgewater treatise in 1833. In the *Opticks* (1704), too, Newton at a number of points endorsed (with habitual caution) the kind of natural theology propagated by Bentley. Drawing on the arguments for the need for a Providential design to ensure the regular movement of the planets and the prevention of a cosmic collapse into a single mass which he had outlined in the correspondence with Bentley, Newton asked in his twenty-eighth query:

To what end are comets, and whence is it that planets move all one and the same way in orbs concentric while comets move all manner of ways in orbs very eccentric, and what hinders the fixed stars from falling upon one another? (Thayer 1974, 155)

In the thirty-first query Newton's adherence to a providential explanation of the movement of the planets becomes even more explicit when he writes: "Such a wonderful uniformity in the planetary system must be allowed the effect of [divine] choice" (ibid., 177). In the subsequent Latin edition of the *Opticks* (1706), as Kubrin has pointed out, Newton suggested a new argument for the divine supervision of the cosmos by arguing that, left to itself, the solar system would tend to dissolution – a fate that was avoided thanks to the providentially controlled mechanism of the comets, the periodicity of which had only recently been established (Kubrin 1973, 152).

At the time of Newton's death in 1727, then, the tradition of Newtonian natural theology was well established. It was a form of natural theology which attempted to maintain a balance between two images of the Deity – a general Providence who created the world *ex nihilo* and established and kept in being the laws by which it continued to operate, and a special Providence who continued to intervene in the workings of the universe in the manner suggested by Newton. One example of the tension caused by these two different theological emphases was the problem of miracles – if miracles occurred frequently, they would upset the order and stability which underlay the whole cosmic system; on the other hand, if the possibility of miracles was not allowed for, then it was difficult to maintain that God continued to direct and govern the universe. As Leibniz pointed out, the Newtonian system was caught between the Scylla of continual divine intervention and the Charybdis of a form of naturalism which minimized God's activity:

If God is oblig'd to mend the course of nature from time to time, it must be done either supernaturally or naturally. If it be done supernaturally, we must have recourse to miracles, in order to explain natural things: which is reducing an hypothesis ad absurdum: for, every thing may easily be accounted for by miracles. But if it be done naturally, then God will not be *intelligentia supramundana;* he will be comprehended under the nature of things; that is, he will be the soul of the world. (Alexander 1956, 20)

As McGuire (1968) and Heimann (1978) have stressed, Newton and his early disciples sought a way out of this dilemma by portraying the regular operation of the laws of nature as dependent on the continuing exercise of God's will. But Newton's work, as its subsequent reception was to show, was liable to be regarded as providing evidence for the existence of a watchmaker God who created the universe and then left it to operate according to its own laws. Newton and his clerical followers appear to have been aware of this danger since they gave prominence to those aspects of the *Principia* which could provide an appropriate analogy for the activities of an active Providence who did more than create the universe and establish the laws by which it continued to operate. Thus Samuel Clarke, Newton's unofficial philosophical spokesman, was acutely conscious of the dangers of placing so much emphasis on the regular workings of the laws of nature that the role of God might be undermined. As he wrote in his correspondence with Leibniz:

> The notion of the world's being a great machine, going on without the interposition of God, as a clock continues to go on without the assistance of a clockmaker; is the notion of materialism and fate, and tends, (under the pretence of making God a *supra-mundane intelligence*), to exclude providence, and God's government in reality out of the world. (Alexander 1956, 14)

The phenomenon of gravity which, in the Newtonian system, could not be explained in mechanical terms, offered a rich source of such providential interpretations of the continued workings of nature (as Metzger's classic work has shown) even though Newton himself was ambivalent on the matter. There were also other aspects of the Newtonian system which could be regarded as indications of an active Providence, such as the "continued miracle" which, wrote Newton, "is needed to prevent the sun and fixed stars from rushing together through gravity" (quoted in Force 1985, 124), or the need periodically to correct irregularities in the movement of the planets which arise from the attraction between the planets and the comets and which, wrote Newton in the *Opticks,* "will be apt to increase, till this System wants a Reformation" (quoted in Kubrin 1973, 147).

However, these aspects of Newtonian natural theology which emphasized God's special providence showed less resilience and longevity than the image of the watchmaker God. There are a number of possible explanations for the decline in the association between Newtonian natural theology and an active Providence. Firstly, the Newtonian conception of gravity lost much of its inexplicability as natural philosophers abandoned the expectation that it should have some mechanical explanation and simply accepted it as a given. As gravity lost its mystery so, too, it became a less and less appropriate basis for a voluntarist natural theology. Secondly, the anomalies which, in the view of Newton and his clerical followers, necessitated the continuing intervention of the Deity were, largely thanks to Lagrange and Laplace, explained away as explicable in terms of Newton's own fundamental laws of motion. As Laplace put it in the preface to his great work, *A Treatise of Celestial Mechanics* (1799–1825):

> Newton published, towards the end of the seventeenth century, the discovery of universal gravitation. Since that period, Philosophers have reduced all the known phenomena of the system of the world to this great law of nature.... (Quoted in Odom 1966, 546)

Indeed, Laplace regarded an appeal to theological rather than physical considerations as foreign to the true Newtonian method. When discussing the formation of
the universe he pointedly remarked: "I cannot forgo noting here how Newton
strayed on this point from the method that he otherwise used so effectively" (Hahn
1986, 272).

Important though they are, however, such internal developments within Newtonian
celestial mechanics do not explain why the voluntarist elements within Newtonian
natural theology appear to have been in decline well before the work of Laplace or
even before the nature of gravity had ceased to be a matter of controversy. Even
around the time of Newton's death in 1727 there are indications that the public
imagination was far more preoccupied with the way in which Newton's work provided evidence for a Creator-Mechanic than for an active Providence. Pope's famous
epitaph to Newton

Nature and Nature's Law lay hid in Night
God said, let Newton be and all was Light
(Pope 1963, 808)

emphasizes Newton's achievement in dispelling mystery rather than the ways in
which Newton's work revealed unexplained aspects of Nature which suggested the
need for the intervention of God's special providence. Note, too, that Pope speaks of
Nature's laws rather than of God's laws. When Lord Cobham came to construct his
temple of British worthies at Stowe at about the same time, he, like Pope, emphasized Newton's achievement in revealing the workings of the watchmaker God,
Newton being proclaimed as one "whom the God of Nature made to comprehend his
works; and from simple principles, to discover the laws never known before, and to
explain the appearance, never understood, of this stupendous universe" (quoted in
Haskell 1970, 306). So strongly did Newton become associated with this image of a
Creator-Mechanic that William Blake coined a special term – "Nonbodaddy"
[Nobody's daddy?] – to describe what he regarded as the lifeless Deity which was
popularly associated with Newton's work. He also attempted to construct an alternative poetic vision of the workings of the universe to replace that which, in the public
mind, was linked with Newton (Ault 1974, 7–8).[2]

The public preoccupation with the Newtonian watchmaker God may be partly
explained by the fact that an understanding of the irregularities that offered scope for
the workings of an active Providence required a reasonable degree of scientific
competence; by contrast, the concept of a universe which conformed to a machine-
like regularity was more directly comprehensible. However, even mid-eighteenth-
century scientific expositions of Newton's work (as Odom points out) could down-

---

[2] Though Ault also makes the point that Newton himself – as opposed to the public perceptions of his
work – was antagonistic in many respects to Cartesian mechanism. However, Blake regarded even those
antimechanistic elements of Newton's thought as hostile to the life of the imagination. Significantly,
Blake's character, Urisen, figures variously as an embodiment of the Newtonian world view and of Satan
(Ault 1974, 98).

play the importance of such irregularities or their theological significance. Benjamin Martin in his *A Panegyrick on the Newtonian Philosophy* ... (1754) viewed the universe as being largely the product of a set of uniform laws even though he argued that true (i.e., Newtonian) natural philosophy could assist our understanding of Revelation:

> For by acquainting us with the Manner in which primary and secondary Causes act, the first absolutely and independently, the last mechanically and consequentially, we are brought to see that the first may interpose to produce any of the phenomena of Nature, without interrupting the Course of her Operations in the ordinary way. (Quoted in Odom 1966, 544)

The increasingly restricted role accorded to direct providential intervention in the workings of the universe is also reflected in the change in the conception of the laws of nature which Heimann sees as one of the characteristics of the second half of the eighteenth century. By the latter part of the century Newton's conception of the laws of nature as divinely imposed active principles increasingly gave way to the view of those, like Hutton, who held that the laws of nature were self-regulating (Heimann 1973, 2, 24).

In summary, even by the mid–eighteenth century, the voluntarist aspects of Newton's work had been largely overshadowed by the increasing emphasis on the way in which the Creator worked through the laws of nature – a development which cannot be directly explained in terms of scientific changes within this period. One may speculate, then, that the social conditions of Britain as the century progressed offered a more and more fertile environment for ideas of a God of order rather than an interventionist Deity. The more self-confident Britons of the mid–eighteenth century could look back on the upheavals of 1688 and the succession crisis of 1714, which accompanied the establishment of the House of Hanover, as increasingly distant memories. They may, therefore, have been more inclined than their compatriots of the troubled decades immediately after the publication of the *Principia* in 1687 and the Revolution of 1688 to emphasize those features of the Newtonian heritage which gave prominence to concepts of order and regularity and to downplay suggestions that either the laws of the land or the laws of nature might be suspended. It may be suggested, too, that the increasing constitutional restrictions on the power of the Hanoverian kings made mid-eighteenth-century Britons less sympathetic to voluntarist conceptions of a God capable, like an absolute monarch, of ruling independently of established laws. Certainly, one of the most characteristic features of Hanoverian Britain was its emphasis on the majesty and mystique of the law – so much so that, as Hay (1975) and others have pointed out, eighteenth-century Britain had a remarkably weak system of state coercion since it relied on the public rituals of the law to keep order. In such a social setting it is not altogether surprising that Newtonian natural theology became increasingly associated with the observance of law and order rather than with an interventionist Deity.

Though Newtonian natural theology tended to narrow its focus in the course of the eighteenth century, it, nonetheless, became deeply entrenched in British intellectual life as one of the major defenses of the established church and the social order associated with it. So closely was Newton's work identified with the argument from design that it was this form of the design argument that was, as Hurlbutt has pointed out, the chief target of Hume's *Dialogues Concerning Natural Religion* (1776) (Hurlbutt 1965, xii). McPherson also makes the point that Hume was so preoccupied with the Newtonian version of the design argument that he employed a rather strained analogy between cosmic phenomena and the generation of plant life in order to argue that one might as well compare the structure of the universe with the generation of a plant as with a machine such as a watch (McPherson 1972, 58–59). As mentioned previously, Hume's arguments, for all their philosophical acuteness, had little immediate impact on the continuing popularity of forms of natural theology based on the design argument. It is true that later exponents of the design argument, like Paley or Whewell, did give greater emphasis to the view that the design argument confirmed the faith of the believer rather than providing a means of converting the skeptic, but in general Hume's arguments in the *Dialogues* were largely unheeded. By contrast his argument in the "Essay upon Miracles" (1751), that miracles could not be believed because it would take testimony greater than a miracle to persuade someone that the universe had departed from its accustomed lawlike behavior, prompted a flurry of replies – an indication that Hume had touched a sensitive nerve when he had drawn attention in such forceful terms to the tensions that had always existed in Newtonian natural theology between an interventionist God capable of miracles and a God who was the source of the universe's order and predictability. When Hume attacked the design argument, by contrast, his criticisms were too remote from the way of thinking of most of the eighteenth-century educated classes to produce much argument. Perhaps, too, his largely destructive comments on natural theology had little force until some alternative system – such as that of Darwin – could be produced to account for the apparent design in nature.

But while Newtonian natural theology still remained an integral part of the late eighteenth-century intellectual landscape, its association with the image of God the Creator and Lawgiver rather than an active Providence made it less suited to the needs of the troubled age that followed the American Revolution and, a fortiori, the French Revolution – an age that was associated with such religious revivals as the growing influence of the Evangelical Clapham Sect or the rise of the High Church Hackney Phalanx (predecessor of the Oxford Movement). In such a climate there was a greater demand for forms of religious apologetic which gave greater emphasis to the immediate and observable hand of Providence protecting the individual and the larger society from the perils that assailed them. Natural theology came more and more to be associated with a recourse to the more accessible sciences such as natural history rather than to the remote astronomical phenomena that were the foundation of Newtonian natural theology. Such a biologically based form of natural philosophy was, of course, no novelty. It had existed since at least Galen's time and, thanks to the work of Ray and Derham, it had continued to command a widespread following throughout the eighteenth century. However, the great prestige of Newton's work had tended to give prominence to astronomically rather than biologically based systems of natural theology (Gillespie 1987, 49) – or, to use the titles of Derham's Boyle lectures, to astro-theology rather than physico-theology. This vogue of astro-theology during much of the eighteenth century represented a reversal of the tendency which was apparent in English natural theology before the *Principia* commanded widespread public attention – for in pre-*Principia* England, the apologetic potential of astronomy tended to be downplayed (Gillespie 1987, 2). Boyle, for example, wrote "that the situations of the celestial bodies do not afford, by far, so clear and cogent arguments of the wisdom and design of the author of the world, as do the bodies of animals and plants" (quoted in Brooke et al. 1974, 20).

By the end of the eighteenth century, however, physico- rather than astro-theology was again predominant. The most successful work of natural theology at the turn of the eighteenth century – Paley's *Natural Theology* (1802) – drew most of its evidence from natural history and anatomy even though Paley, as a Cambridge senior wrangler and former college tutor, was well qualified to provide a restatement of the traditional themes of Newtonian natural theology. But, on the contrary, Paley argued that astronomy was "*not* the best medium through which to prove the agency of an intelligent Creator," though he did concede that once the existence of God was established, astronomy "shows beyond all other sciences, the magnificence of his operations." As a successful teacher and popularizer, Paley regarded astronomy as lacking the immediacy and general appeal of natural history and anatomy: plants and animals could be minutely dissected by any intelligent layperson while all that such a person could see of the heavens was "nothing, but bright points, luminous circles, or the phases of spheres." Astronomy, then, did not so readily lend itself to that analysis of "relation, aptitude and correspondence of *parts*" which was the best defense of the argument from design. Paley was even rather lukewarm about the apologetical potential of one of the earliest and most powerful forms of Newtonian natural theology: the phenomenon of gravity. "The motions of the heavenly bodies," he wrote, "are carried on without any sensible intermediate apparatus; whereby we are cut off from one principal ground of argumentation, analogy" (Paley 1845, 517). Paley's critique of astronomy as a basis of natural theology received the enthusiastic endorsement of Francis Jeffrey in the *Edinburgh Review* who argued that Paley's comments on this point "may serve to point out the superiority which his systematic argument possesses over the pious learnings of his predecessors" ([Jeffrey] 1802–1803, 299).[3] Nor was Paley's work the only example of the increasing reliance of natural theology on the biological rather than the mathematical sciences. Thus Jones (1965, 232) has noted the way in which

---

[3] The authorship of this, and other anonymous nineteenth-century articles, has been identified by the use of Houghton (1966–79). In this section on Paley and in some of the concluding remarks of this paper I am drawing on material in Gascoigne 1989, 280–81, 300–308.

late eighteenth-century poets "changed from celestial systems to English flowers and birds" when praising the works of God.

This growing popularity of natural history rather than astronomy as an aid for the argument from design also reflects the growing interest in the Baconian sciences in the late eighteenth century. "Natural history," wrote Sir Leslie Stephen of the late eighteenth century, "in the earlier part of the century had been regarded with good-humoured contempt as a pursuit of bugs, beetles, and mummies.... Now it was beginning to be recognized that such pursuits might be a credible investment of human energy ..." (Stephen 1962, 1:322).

The growth of sciences such as botany, zoology, chemistry, and geology can, in turn, be related to the growth of provincial culture and with it the number of scientific societies which acted as a focus for such forms of scientific inquiry. Metropolitan scientific life was also becoming more diverse. After 1781 the Royal Society's monopoly was broken and within sixty years there were sixteen new London scientific bodies in the provinces and over two dozen specialist ones (Thackray 1974, 674). The growth of such societies challenged the traditional scientific predominance of both the Royal Society and the two ancient English universities, with which Newton and his early popularizers had been associated, and helped promote a greater diversity of scientific activities which was reflected in the increasing diversity of natural theology. Because of such changes and perhaps, too, because natural theology based on natural history had been overshadowed for much of the eighteenth century by the Newtonian variety, physico-theology showed, at the end of the century, a greater freshness and ability to capture the public imagination than Newtonian astro-theology which, as Metzger writes (1938, 167), by the late eighteenth century, "*avait perdu de sa fraîcheur spontanée, pour se réduire le plus souvent à une rhétorique quasi officielle, banale et inefficace.*"[4] Metzger's comments perhaps exaggerate the ineffectiveness of Newtonian natural theology in this period but they do underline the growing ossification of this intellectual tradition.

The waning fortunes of Newtonian natural theology in this period also owed much to the impact of the French Revolution – for this had cast a cloud over the traditional harmony between Newtonianism and Christianity by bringing home to the educated classes in Britain the fact that across the Channel Newton's works had been appropriated by the infidel. The 1801 supplement of the third edition of the *Encyclopedia Britannica* contained articles by John Robison, professor of natural philosophy at Edinburgh (1770–1805), and author of *Proofs of a Conspiracy against all the Religions and Governments of Europe* ... (1797), in which, after a critique of French materialists like La Mettrie and British skeptics like Hume, he deplored the "absurd and shocking consequences of the mechanical philosophy now in vogue" and urged all to "abandon its bloodstained road, and return to the delightful paths of nature, to survey the works of God ... which offer themselves on every hand in designs of the

---

[4] "had lost some of its spontaneous freshness, and was most often reduced to a near-official rhetoric, banal and ineffectual."

most extensive influence and the most beautiful contrivance" (quoted in Hughes 1951, 369). The target of such criticism was not the great Sir Isaac himself but rather the way in which (in the view of Robison) others had misused his work. Thus when discussing the manner in which the solar system maintained its equilibrium by oscillating about a mean state, Robison wrote that this

> strikes the mind of a Newton, and indeed any heart possessed of sensibility to moral or intellectual excellence, as a mark of wisdom prompted by benevolence. But De LaPlace and others, infected with the *Theophobia Gallica* engendered by our licentious desires, are eager to point it out as a mark of fatalism.(Quoted in Morrell 1971, 49–50)

He later sadly added that "Newton, one of the most pious of mankind, was set at the head of the atheistic sect" (ibid., 51). Such an association between Newton and infidelity was due, suggested Robison in his earlier *Proofs of a Conspiracy . . .* , to the way in which Newton's work had been misused by materialists like Priestley who wished to argue for the self-sufficiency of matter and the laws of motion. By contrast, Robison appealed to Newton's own General Scholium, with its emphasis on God's continuing superintendence over the universe, in order to revive those voluntarist elements of Newtonian natural theology (ibid., 48) which had been overshadowed by the emphasis on the watchmaker God in the more complacent atmosphere of the mid–eighteenth century. But some of Robison's contemporaries went further and rejected Newton altogether turning instead to the biblically based Hutchinsonian natural philosophy which enjoyed something of a revival in the 1790s (Thackray 1970, 246).

By the early nineteenth century, then, there was an ambivalent attitude to Newton's role as a defender of Christianity. There was a small minority who regarded Newton as no longer the friend of belief but rather as a source of infidelity. Some, like Shelley, who wrote in *Queen Mab* that "the consistent Newtonian is necessarily an atheist" (quoted in Force 1985, 155), rejoiced at this but others, like Coleridge, regarded Newton's influence with some dismay. Thus Coleridge commented that:

> It has been asserted that Sir Isaac Newton's philosophy leads in its consequences to Atheism; perhaps not without reason, for if matter by any powers or properties *given* to it, can produce the order of the visible world, & even generate thought; why may it not have possessed such properties by inherent right? & where is the necessity of a God? (Brinkley 1955, 402)

Predictably, Coleridge also found the tradition of Newtonian natural theology, which emphasized God the divine watchmaker rather than an ever-present Providence, a source of infidelity: "Sir Isaac Newton's Deity seems to be alternatively operose & indolent to have delegated so much power as to make it inconceivable what he can have reserved. He is dethroned by Vice-regent second causes" (ibid., 403). Elsewhere he wrote of the Hebrew poets that "In God they move and live and *have* their being; not had, as the cold system of Newtonian Theology represents, but *have*" (ibid., 407).

Despite such dissenting voices Newton was too closely identified with Christian apologetic to be discarded so readily. More typical was the attempt on the part of a number of prominent early nineteenth-century theologians and scientists to revitalize those elements of Newtonian natural theology which could be regarded as evidence for the existence of an active Providence. In his influential Bridgewater treatise – *Astronomy and General Physics Considered with Reference to Natural Theology* (1833) – Whewell acknowledged that Newtonian principles had been used by those who maintained that the workings of the universe could be explained solely in naturalistic terms. "And the follower of Newton," he wrote,

> may run into the error with which he is sometimes charged, of thrusting some mechanic cause into the place of God, as if he do not raise his views, as his master did, to some higher cause, to some source of all forces, laws, and principles. (Whewell [1833] 1862, 286)

To counter such a tendency Whewell emphasized the continued dependence of the universe and its laws on a sustaining Deity. Much of his work is given over to illustrations of the way in which the laws of nature must have been framed by a purposeful Creator. Like Bentley he argued that "the regularity ... of the solar system excludes the notion of accident in the arrangement of the orbits of the planets" (ibid., 134). Whewell also put forward a more novel argument when he suggested that the "agreement in the form of the laws that prevail in the organic and inorganic world" – apparent in the manner in which the cycle of vegetable and animal life conformed to that of the plant – was evidence of providential design (ibid., 25). But such systems of laws, insists Whewell, suggest the existence of a Deity who is not only "the author" but also "the governor of the universe" (ibid., 307).

This emphasis on the role of an active Providence can also be seen in the passage where Whewell sets out to refute Laplace's argument that the workings of the universe can be explained without reference to final causes – an enterprise Whewell compares with a savage marveling at a steam engine who "should cease to consider it a work of art, as soon as the self-regulating part of the mechanism had been explained to him" (ibid., 301). For Whewell regarded the workings of the laws of nature as a more secure resting place for natural theology than the argument from design based on particular phenomena. The occurrence of such phenomena, he concluded, quoting from Laplace's comments on Newton, might eventually be explained in scientific terms and "are, therefore, in the eyes of the philosopher nothing more than the expression of the ignorance in which we are of the real causes." However, Whewell continued, such objections had little force if one emphasized the teleological nature of the concept of natural laws since thereby "the notion of design and end is transferred by the researches of science, not from the domain of our knowledge to that of our ignorance, but merely from the region of facts to that of laws" (ibid., 300). For Whewell, then, the operation of the laws of nature was predicated on the existence of a Deity who not only created the laws of nature but also kept them in being – or, as Whewell himself put it, such laws entailed "the reality of Final Causes and consequent belief in the personality of the Deity" (ibid., 299).

Not that Whewell was averse also to using specific instances of the laws of nature as illustrations of the argument from design – particularly if such illustrations derived from the work of Laplace himself and so could be turned back against the infidel principles of the French. Like Robison, Whewell was greatly attracted to Laplace's finding that though the solar system underwent periodic fluctuations, these were so balanced that they kept the system in basic equilibrium. Whewell regarded this finding as so potent an illustration of the continuing supervision of the universe by an active Providence that he added a further refinement of his own. For Whewell suggested that the fact that the smaller planets conformed to the overall pattern of periodic oscillation by means of fluctuations which were so great that if these were required of the larger planets the whole system would be liable to fly apart was "a mark of provident care in the Creator" (ibid., 141). Though other early Victorian natural theologians ignored Whewell's ingenious embellishment, they made much of Laplace's original finding. Thus Lord Brougham maintained that this

> discovery which makes the glory of Lagrange and Laplace . . . may most justly be classed as a truth both of the Mixed Mathematics and of Natural Theology – for the theologian only adds a single link to the chain of the physical astronomer's demonstration, in order to reach the great Artificier from the phenomena of his System. (Brougham 1835, 42)

The same argument was also advanced by Thomas Turton (1836, 49), a living embodiment of the traditional alliance between Newtonianism and Anglicanism since he served both as the Lucasian professor of mathematics at Cambridge (1822–27) and as bishop of Ely (1845–64). It also recurs, with greater scientific and theological caution, in Thomas Chalmers' *On Natural Theology* (1835–41, 1, 208) and, as late as 1855, was referred to by Brougham and Chalmers' fellow Scot, John Tulloch (1855, 95).

Another instance of the attempt to integrate new astronomical investigations into a form of Newtonian natural theology which would give greater emphasis to the continuing and active role of Providence was advanced by the Oxford theologian, James Gabell. In his work of 1847, Gabell argued that the discovery that some stars disappeared while other new stars made their appearance was – together with the geological evidence that old species of plants and animals had given way to new ones – an indication

> that the Creator of the Universe has not (as some have supposed) delegated the government of it to a system of general laws, thenceforth dismissing all care about it himself: for He has manifestly, from time to time, interposed with His Almighty power for the purpose of working certain changes in it: whence it is a just inference, that He is *constantly* watching and presiding over it. . . . (Gabell 1847, 342–43, 298–99)

A more predictable response on the part of those who still looked to Newton's work to provide a source of evidence for an active Providence as well as a remote Creator was simply to restate some of those arguments for continued divine intervention in cosmic affairs which had been advanced by Newton and his early clerical followers but which had been partly forgotten in the course of the eighteenth century. One such argument was to point to the erratic behavior of the comets as an instance of the need for some providential direction of the universe. Thus the Scottish divine, Thomas Dick, wrote, in a work that took a portrait of Newton as its frontispiece, that if the comets struck the earth "the consequences would be awful beyond description. But we may rest assured that that almighty Being who at first launched them into existence directs all their motions, however complicated ..." (Dick 1846, 1:301). In the same work Dick (like Bentley and his followers) cited the phenomenon of gravity – along with other examples of natural forces such as magnetism – as instances of his general thesis that "the investigations of natural philosophy *unfold to us the incessant agency of God*" (ibid., 2:124). Another argument which had played an important part in the early establishment of Newtonian natural theology was raised by Thomas Chalmers in his *Discourses on the Christian Revelation viewed in Connection with the Modern Astronomy ...*, in which he argued that the hand of Providence was evident in the fact that the force of gravitation was prevented by counterbalancing forces from "consolidat[ing], into one stupendous mass, all the distinct globes of which the universe is composed." Chalmers continued his discussion of this point to instance another example of Providential design revealed by the recent discovery that the nebulae were "arranged into distinct clusters" (1835–41, 7:34–35).

A more novel departure in the forms of natural theology which were linked with Newton's work was the argument, advanced by Brougham, that the fact that man's intellect could explain the workings of the universe was an indication that the human mind could not be explained solely in terms of matter and therefore pointed to "the great First Cause, which alone can call both matter and mind into existence, has alone the power of modulating intellectual nature" (Brougham 1835, 72). To Brougham the work of Newton (whose *Principia* he regarded as "the greatest work of man" [ibid., 149]) or of other mathematicians like Clairaut, or of Euler, Lagrange, or Laplace were ultimate examples of the separation of mind from matter since their work was based on "the force of the mind itself, when it acts wholly without external aid, borrowing nothing whatever from matter, and relying on its own powers alone" (ibid., 70).

Like Brougham, Thomas Chalmers in his Bridgewater treatise – *On the Power, Wisdom and Goodness of God as Manifested in the Adaptation of External Nature to the Moral and Intellectual Constitution of Man* (1833) – argued for the religious significance of the nature of man's intellect, placing considerable emphasis on the importance of Newton's work and on what Chalmers called the "physico-mathematical" sciences. However, Chalmers, as the title of his work suggests, put forward a slightly different case than Brougham did, maintaining that the fact that the results of the abstract intellectual process and the realities of external nature should so strikingly harmonize . . . can only be explained by the intervention of a Being having supremacy over all and who had adjusted the laws of matter and the properties of mind to each other. (Chalmers 1835–41, 2:176–77)

Chalmers' reference to the "intervention" of God and his "adjustment" of nature indicates the emphasis that early nineteenth-century natural theology accorded to the role of an active Providence. Like Newton and his early followers, Chalmers regarded the laws of nature as lacking any autonomy since they continually depended for their existence on the exercise of God's will (Smith 1979, 62).

There was the difficulty that – together with developments in the astronomical sciences (such as the discovery of the oscillation of the solar system about a mean) which, it might be argued, favored natural theology – there were also new astronomical theories that were difficult to reconcile with natural theology as traditionally understood. In the first place there was the elimination of many of the irregularities which had traditionally been seen as providing evidence for the continuing role of an active Providence. In his *Observations on the Hypotheses which have been assumed to Account for the Cause of Gravitation* (1806), Samuel Vince (Plumian professor of astronomy at Cambridge, 1796–1821) had attempted to turn this development to the advantage of natural theology by arguing that the perfection of the heaven's motion showed the hand of the Creator since "imperfection is always found in the operation of mechanical causes" (p. 26); but it is an indication that his views attracted little enthusiasm that this theme does not appear to have been further developed by other nineteenth-century theologians. On the contrary, Chalmers, for example, continued to maintain that if Laplace were correct and "all the beauties, and benefits of the astronomical system be referred to the single law of gravitation, it would greatly reduce the strength of the argument for a designing cause" (Chalmers 1835–41, 1:207). Such a view helps to explain Chalmers' rather reserved attitude toward the utility of astronomy as a basis for natural theology and his reference to the "doubtfulness of evidence there may be in the mechanism of the heavens" (ibid., 1:212).

Another difficulty was the growing interest in what Whewell (a great coiner of neologisms) termed the nebular hypothesis of Laplace (Yeo 1984, 18). Stated simply, this theory proposed that the development of the cosmic system could be explained in naturalistic terms as the result of the application of fundamental laws of motion working on a primordial mass. Though neither Laplace nor, still less, John Herschel – on whose astronomical data Laplace drew heavily – viewed the nebular hypothesis as providing a cosmic evolutionary mechanism, by about the 1820s it had come to acquire such a directionalist significance (Brooke 1979b, 202–3; Schaffer 1980a). Not suprisingly, then, it came to be associated with theories of biological evolution. Thus in 1844 Robert Chambers in his highly controversial *Vestiges of Creation* brought the theories of cosmic and global evolution together as part of his general thesis that the natural and the human world could be explained in terms of a basic law of development (Ogilvie 1975). Chambers did allow for some divine agency in this process but the role of God was reduced to something like the watchmaker God who simply creates the system and then leaves it to operate by its own laws. As Brewster remarked in his scathing review of the *Vestiges*:

> The Divinity which they recognise is little more than the electric spark which disappears for ever, when it has lighted the train of causes and effects by which the planetary systems are to be framed, and all the living beings fashioned and perpetuated. ([Brewster] 1845, 472)

Though Brewster had once seen in the nebular hypothesis the "brightness of truth," Chambers' use of the theory to suggest a continuity in the development of animals and humans appears to have prompted him to regard the hypothesis in a more hostile spirit (Brooke 1979, 204) – hence his assertion that it reduced God to the position of an absentee lawgiver. By 1854 Brewster's aversion to the hypothesis had become even more pronounced and in his *More Worlds Than One* . . . he referred to the nebular hypothesis as "at once presumptuous and fanciful, subversive of every principle of the inductive philosophy, degrading to science, incompatible with religious truth and dishonouring to the great Author of the material universe" (Brewster 1867, 124). He also invoked the great name of Newton as part of this assault on the hypothesis. Citing the passage in Newton's letter to Bentley where Newton argued that matter could not randomly form itself into an ordered universe, Brewster argued that the Master "considered the nebular hypothesis, though in his time not known by that name, as not only absurd, but verging on atheism" (ibid., 249).

Whewell in his Bridgewater treatise of 1833 took a rather less hostile view of the hypothesis but argued that even if it were valid, nonetheless "a prior purpose and intelligence" must have controlled the processes which produced the universe as presently constituted. Thus, continued Whewell, the theory did not affect "the view of the universe as the work of a wise and good Creator" (1862, 158, 163). A similar set of arguments was advanced by Tulloch in his *Theism* . . . , in which he asked, apropos of what he called the "Laplacian cosmogony":

> Whence the existence of the nebulous mass itself? Whence the peculiar character which enabled it to separate and contract in the fitting way, and in no other? Whence the determinant velocity of the primitive movement, destined to such results, and no other? (Tulloch 1855, 102)

Tulloch concluded his discussion of this point by invoking "the conclusion in which the great intellect of Newton rested" that "we can only rest in an original self-subsistent Mind, in which the whole cosmical order lives, and from which it ever proceeds."

The energetic response of British natural theologians to the nebular hypothesis is but one example of the more general vitality of Newtonian physico-theology in the early Victorian period after its ossification in the late eighteenth century. The bases of this renewal largely derive from the 1830s, the decade when Chalmers' and Whewell's Bridgewater treatises both appeared (in 1833) and when Brougham's *Discourse of Natural Theology* and Chalmers' *On Natural Theology* were published (in 1835). In its own small way Newtonian natural theology shared in the same process of reform and revival which convulsed church and state in the decade of the Great Reform Bill of 1832.

Indeed, a number of the reviews of Brougham's work do link his reexamination of natural theology with the momentous changes that were occurring in both church and state. *Tait's Edinburgh Magazine* greeted the appearance of Brougham's book with the words:

> We cannot doubt but its appearance will have a strong tendency to advance the cause of Church Reform, which is one vast stride towards general amelioration. In proportion as Natural Theology becomes studied as an inductive science, the reign of dogmatic tyranny, ceremonious superstition, and the power that grasps the soul that it may devour the body will cease. . . . (Anon. 1835, 807)

The *Westminster Review,* which shared *Tait's Edinburgh Magazine*'s hostility to the traditional privileges of the English established church, also referred in its review (by R. H. Horne) to the fact that "the whole country is in a state of excitement on the question of Church property and influence" and commended Brougham's book since, thanks to it,

> the public mind which has so long been almost exclusively absorbed in politics and the means of effecting practical reformations, is thus suddenly opened, and at a critical moment, to considerations which have hitherto been confided to a comparatively insulated class of abstruse thinkers. . . . ([Horne] 1835, 333)

In an age when reform of the state and the reform of the established church were inseparably linked, a renewed emphasis on a nondogmatic form of natural philosophy which could appeal to most varieties of Protestantism helped weaken the claims of those, like the Tractarians, who maintained that the Church of England should be preserved in its traditional form since its ecclesiastical forms were divinely sanctioned. Significantly, Brougham's work on natural theology was published at much the same time as he was also personally and actively involved in campaigning for political reform and for a system of national education which meant some weakening of Anglican influence. Natural theology, then, could be used as a means of altering as well as of shoring up the established order.

But though the Newtonian brand of natural theology enjoyed something of a revival in the early Victorian period, it had to compete with a growing number of other varieties of the same theological genre. This reflected both the growing number of scientific disciplines and the increasing religious heterogeneity of Britain – the latter being a product of such religious movements as the Evangelical Revival and the Oxford Movement – and the growing self-confidence of English Dissent after the repeal of the Test and Corporation Acts in 1828, acts which had reduced Nonconformists to the status of second-class citizens since 1662. As a result of these influences natural theology became increasingly fragmented and, as Brooke puts it, "the natural theology which Darwin is supposed to have destroyed was already a house divided against itself" (Brooke 1977, 222). Moreover, the general public continued to show a preference for forms of natural theology which drew on the empirical sciences, particularly that of geology, which accorded with the increasing historical preoccupations of an age experiencing rapid change. It is significant that only one of the eight Bridgewater treatises on natural theology – Whewell's *Astronomy and General Physics Considered with Reference to Natural Theology* (1833) – was devoted directly to astronomy and that even a keen student of astronomy like Thomas Chalmers concluded that one "could draw so much weightier an argument for a God, from the construction of an eye than from the construction of a planetarium" (Chalmers 1835–41, 1:212), supporting his view with arguments suspiciously similar to those of Paley.

Nonetheless, astronomy and, by extension, the forms of natural theology associated with it, still continued to enjoy a particular prestige. Though other sciences may have enjoyed a greater popular following, astronomy retained its reputation as *the* science par excellence – it was, for Whewell, the "only perfect science" and even the chemist, botanist, and geologist Charles Daubeny could write of the nonastronomical sciences: "the summit of their ambition, and the ultimate aim of the efforts of their votaries, is to obtain their recognition as the worthy sisters of the noblest of these sciences – Physical Astronomy" (Cannon 1964, 488). This preeminence of astronomy meant that Newtonian natural theology still maintained something of its traditional importance as a means of defending religious belief despite the growing competition from other varieties of natural theology. In his *Discourse on the Studies of the University* (1834, 14), the Cambridge geologist Adam Sedgwick was still sufficiently in awe of Newtonian natural theology to assert that "a study of the Newtonian philosophy . . . teaches us to see the finger of God in all things animate and inanimate." Some even attempted to reassert the former preeminence of Newtonian natural theology. In his review of Whewell's Bridgewater treatise the physician John Abercrombie conceded that the astronomical sciences lacked the immediacy of what he called (following Whewell) "terrestrial adaptations" but, he continued,

in another respect, however, the testimony given by the mechanism of the heavens, is more irresistible than that which issues from any one limited province of the creation. For, thanks to the inventive genius of Newton, and the prodigious sagacity and industry of his followers – our knowledge of the system is far more complete and exact than that which has hitherto been attainable in any other department of natural philosophy: As exhibiting a collection of simple laws, adapted to the accomplishment of a vast variety of purposes, the science of astronomy stands before us in a state of unrivalled perfection. ([Abercrombie] 1833, 95–96)

Thus, just as astronomy remained the paradigmatic science, so, too, Newtonian natural theology still could claim to be the form of natural theology on which other, rival, forms ought to be based. Though an early devotee of Paley's form of natural theology with its emphasis on the biological sciences, the young Darwin, in a notebook jotting written in 1837, invoked the analogy of God's use of gravitational attraction in controlling the workings of the heavens as a means of justifying his own view that natural selection could be combined with a providential view of the workings of nature. Rather than invoking God as the designer of each planet and its movements and as the Creator of every different species of animal, it was, he wrote, a "much more simple and sublime [view of God's] power" to argue: "let attraction act according to certain law, such are inevitable consequences, – let animal[s] be created, then by fixed laws of generation, such will be their successors" (quoted in Mandelbaum 1971, 86) – a passage which underlines both the continuing prevalence and authority of the Newtonian form of the design argument and Darwin's own considerable debt to the tradition of natural theology (Durant 1985).[5]

Furthermore, as the argument from the design of particular species or particular objects became more and more subject to scientific and philosophical objections, there was an attempt by avant-garde theologians like Baden Powell to claim for an astronomically based natural theology a renewed importance because it was the best means of displaying the underlying order of the universe and therefore of the need for a Supreme Intelligence. In his 1843 review of Rigaud's essay on the *Principia* and his edition of the correspondence of seventeenth-century English scientists, Powell urged theologians to turn away from the sort of natural theology associated with Paley and to cultivate instead

a just apprehension of the real nature of final causes, not merely in the limited sense of *means* and *end,* but in the extended meaning of the evidences of design and mind, in the *order,* arrangement, and harmony, of the laws of the material universe. ([Powell] 1843, 437)

When he returned to this subject in 1857, Powell commented with satisfaction that natural theologians were now more inclined to emphasize the idea of a fundamental order in nature rather than particular examples of design, something which, in his view, naturally led to a greater emphasis on the importance of astronomy (Yeo 1977, 368; Ruse 1975, 509, 513).

The most simplistic, but nonetheless most popularly effective manifestation of the Newtonian natural theological tradition was the appeal to the person of Newton as

---

[5] For a recent discussion of the influence of Darwin on Newtonian natural theology and, in particular, the concept of the divine arrangement of universal laws, see Cornell 1986.

the great exemplar of the union of scientific genius and Christian belief – an ad hominem argument that enjoyed a considerable vogue in the nineteenth century despite the caveats of those like B. H. Malkin who wrote in his review of Brewster's popular life of Newton: "In truth, however, it is but an ill compliment to religion to consider the testimony of any individual, even Newton himself, as of importance to its interests. It is not on such evidence that its reception or authority will ever depend" ([Malkin] 1832, 9). More typical, however, was the view of David Brewster in his review of Brougham's *Discourse of Natural Theology* that one of the best defenses of Christianity was to point to "that cloud of witnesses which is resplendent with the names of Milton and Locke, – of Bacon, Newton, and Boyle" ([Brewster] 1836–37, 264). As this comment suggests, appeals to the authority of Newton were often combined with other figures from the pantheon of the alliance of science and religion. In particular Bacon and Locke were often joined with Newton in an Anglican trinity – so much so that the anticlerical radical Richard Carlile in 1821 bemoaned the fact that these three were "claimed as the patrons of superstition" (Yeo 1985, 285).

Appropriately, it was Brougham who was chosen to give the oration at the erection of Newton's statue at Grantham in 1858, following a public subscription that had the support of the Royal Society and the patronage of Victoria and Albert. It was an occasion on which Newton's services to religion were emphasized, Brougham praising him as "a man whose talents had never been exercised but for the extension of truth, for the instruction of mankind, and with a view to illustrate the wisdom and power of the Creator" (Brougham 1858, 382). Prominent at the proceedings were the bishop of Lincoln and delegates from the clergy of the diocese whose presence was a reminder of the importance of Newton's early clerical disciples in the dissemination of his work.

However, the most important nineteenth-century votary of Newton's scientific and religious reputation was Brougham's fellow Scot, the scientist-churchman Sir David Brewster. His 1831 popular biography of Newton was enthusiastically received, passing through seven subsequent English editions, sixteen American, two German, and one French (Wallis and Wallis 1977, 228–30), and it did much to strengthen the association between Newton and Christian belief. In it Brewster appealed to the example of Newton as one who "by thus uniting philosophy with religion, ... dissolved the league which genius had formed with scepticism, and added to the cloud of witnesses the brightest name of ancient or of modern times" (Brewster 1831, 295).

When Brewster widened his researches in preparation for his magisterial two-volume life of Newton, he was grieved to find that Newton's religious beliefs were not orthodox (Gordon 1870, 261–62), though a well-developed Presbyterian conscience compelled him to admit this. Nonetheless, Brewster regarded Newton's Christian beliefs – heterodox though they might be – as a cause for "the apostle of infidelity [to] cower beneath the implied rebuke" and for "the timid and the wavering [to] stand firm in the faith" (Brewster 1855, 2:314).

However, the publication of Brewster's full life of Newton in 1855 did lead to some public questioning of Newton's religious credentials. One reviewer of the work enthusiastically commended it for illustrating that "the great name of Newton stands out in firm opposition to the ranks of proud infidel philosophers," but then sadly acknowledged that

> in regard to some of the leading doctrines of the gospel, Newton's views were at least dubious. . . . How small, in such a case the value of even his achievements in the field of knowledge! What if he had traversed its entire extent and gathered its choicest fruits, if this divine knowledge were not his! (Anon. 1856, 156)

But such Evangelical reservations about Newton's doctrinal soundness did little to weaken the apologetic importance of Newton as the supreme example of the way in which scientific eminence could be combined with Christian belief.

In summary, then, in the first half of the nineteenth century natural theology based on the work (or reputation) of Newton remained an important and venerable part of natural theology – still one of the major forms of intellectual discourse – even though the Newtonian form of natural theology no longer had the dominance it had once enjoyed. But despite Brewster's complacent assertion that "at no period in the annals of religion and science have these spring tides of civilization advanced with a more irresistible energy, and a less mutual disturbance than in the present day" (1845, 470), the links between science and religion were to weaken greatly in the second half of the century. As they did so, the tradition of scientifically based natural theology was to be more and more swept to the margins of intellectual life. By 1894 the Oxford theologian William Wallace could write dismissively of natural theology in the course of his Gifford lectures (the successors of the Boyle lectures): "At the present time, Natural Theology is apt to seem a belated stranger, if not even an impertinent intruder, in the circle of the sciences . . ." (Wallace 1898, 5). Wallace went on to attribute much of the historical importance of this form of natural theology to "the authority of Sir Isaac Newton [which] contributed, in England at least, to subordinate science for certain presuppositions from theology." But in his opinion, Newton's work was of doubtful use for theology since a reliance on what theologians were later to call a "God of the gaps" – a Deity who is invoked when no scientific explanation is immediately available – "only means that the man of science is unable to construct a scheme of evolution without *lacunae* from the assumed primordial state of matter down to the ordered system of the present epoch" (ibid., 7). Natural theology, argued Wallace, reflected the presuppositions of the utilitarian eighteenth century and of the less complex conceptions "of matter and of nature" that prevailed in that period. By the time of Paley and, still more, the Bridgewater treatises, "it had lost in naturalness what it professed to gain in logical plausibility" for it had become increasingly apparent that "the great mechanician is only a mode, and an insufficient mode of conceiving God's supremacy" (ibid., 9).

Even by the middle of the century there were indications that the tradition of natural theology was no longer commanding the authority it had once enjoyed. An essayist, writing in *The British Quarterly Review* of 1848, commented that the lack of any proper treatment of the "great problem of physical evil ... renders our Bridgewater Treatises so little valuable as works of natural theology" (Anon. 1848, 224). More damning still was a reviewer of John Tulloch's *Theism* ... , who wrote of the Bridgewater treatises:

> With regard to these, we believe we are but speaking the general impression in saying that, so far as their specific purpose was concerned, they were a failure. They did not win a single convert from the negative to the affirmative. They did not by a hair's-breadth approach solution of the awful difficulties environing the theme, for a single mind and heart on which these difficulties were weighing. ....
> We question whether, by their meagerness of conclusion, they did not suggest in many a mind doubts and perplexities that might have been otherwise escaped. (Anon. 1855)

Moreover, the reviewer makes it clear that such problems were intrinsic to the very nature of natural theology rather than being due to the individual authors since "it is doubtful if greater ability on the whole could have been brought to bear upon their several subjects " (ibid., 322–23).

Such comments again underline the point that the tradition of natural theology was under considerable strain even before the appearance of *The Origin of Species* in 1859. True, the impact of the *Origin* did much to weaken further the standing of natural theology and particularly those varieties of it which, in the manner of Paley, were based on an appeal to particular examples of design drawn from the animal or plant kingdoms. More and more educated Victorians came to accept Darwin's own view that "the old argument of design in nature, as given by Paley, which formerly seemed to me so conclusive, fails, now that the law of natural selection has been established" (Darwin 1958, 87).

However, there were other forms of natural theology: notably those, like much of Newtonian astro-theology, which argued that the harmony and regularity of the universe as a whole – and, in particular, its conformity to a set of basic laws – pointed to the hand of a Designer. As Darwin had indicated in his 1837 notebook jotting, such an argument could be extended from the realm of astronomy to that of biology and the principle of evolution could, like that of gravitation, be viewed as a means of providential control. In the famous ending of the *Origin* Darwin again outlined such an approach to natural theology by referring to "several powers, having been originally breathed ['by the Creator' added in second edition (Ospovat 1980, 173)] into a few forms or into one" and by implying a comparison between the mechanisms of natural selection and gravitation in the passage where he writes of the evolution of "endless new forms" "whilst this planet has gone cycling on according to the fixed law of gravity" (Darwin [1859] 1968, 459–60). Moreover, along with a motto drawn from Bacon urging the study of both "the book of God's word" and "the book of God's work," the *Origin* was prefaced by a quotation drawn from Whewell's Bridgewater treatise which referred to the operation of "Divine power" through "the establishment of general laws" (ibid., 50). Though Darwin had abandoned belief in revealed religion before 1859, such an appeal to the concept of a universe controlled by general laws implanted by the Creator appears to have been genuinely meant (Ospovat 1981, 72), even though Darwin later became increasingly agnostic about the issue. As he wrote to Hooker in 1870: "My theology is a simple muddle; I cannot look at the universe as the result of blind chance, yet I see no evidence of beneficent design, or indeed of design of any kind, in the details" (quoted in Brown 1986, 25).

Despite his own subsequent agnosticism, Darwin had indicated that it was possible to reconcile the theory of evolution with a system of natural theology based on general laws in much the same manner as Newtonian natural theology had married together providential design and a law-abiding universe. Nor were Darwin's doubts about the validity of a Paleyan-style design argument altogether new – as we have seen, in 1843 Baden Powell had urged natural theologians to turn from a preoccupation with the design of particulars to an emphasis on "the *order,* arrangement, and harmony, of the laws of the material universe" ([Powell] 1843, 437). Moreover, Powell had, in 1855, endeavored to extend such a line of argument to the biological realm by reference to Owen's view that anatomical structures demonstrated archetypal or general characteristics even though these were adapted by secondary causes in particular instances (Ospovat 1981, 141–42; Desmond 1982, 46–47, 64). A similar preoccupation with general laws and evidence of order in nature can be found in the work of admirers of Darwin in Britain and the United States who wished to reconcile his scientific findings with theistic belief – though, just as Newton's early clerical followers sought to avoid portraying God as the absentee ruler of a clockwork universe by investing gravitation with providential significance, so, too, Darwin's Christian followers generally portrayed evolution as a divinely directed rather than as a random force (Moore 1979, 217–98; Bowler 1983, 44–57).

Like Darwin, some of these natural theologians drew attention to the parallel between the long-established appeal to the divinely ordained character of general laws in astronomy, and the more recent one based on biological phenomena. The astronomer and scientific popularizer R. A. Proctor even went so far as to argue that Darwin's work was as suited to natural theology as that of Newton, writing in 1882 that,

when the theory of universal gravitation became thoroughly established, it was found to be in perfect accordance with the idea of a universal lawgiver. . . . Yet when the Newton of our own time advanced a theory which bears to biology (so far as is possible in matters so unlike) the same relation that the law of gravity bears to astronomy . . . an unreasoning fear possessed many lest this natural sequel of the universe should alter men's conceptions of the government of the universe. (Proctor 1882, 999)

Another article of the same year entitled "Newton and Darwin" (which was probably also by Proctor) argued that, just as Newton's work had revealed "the grandeur of the universe," so, too, had Darwin made apparent "the vastness of past and future time" and, by so doing, had made: "*that* religion, in which all men may (in which all reasoning men *must*) agree, has been rendered infinitely grander – infinitely more impressive by our new knowledge. It has also been rendered infinitely more reasonable" (Anon. 1882, 546).

But such attempts to draw Darwin's work within the pale of the natural theological tradition had only limited success and the enterprise largely died out by the end of the century (Bowler 1983, 54). This can, of course, be largely attributed to the intellectual difficulties of combining Darwin's account of evolution, with its emphasis on nature working by wasteful and frequently cruel methods, with a natural theology emphasizing a purposeful and benevolent Creator. Yet Newtonian natural theology had also faced considerable (though perhaps lesser) intellectual difficulties in combining aspects of the mechanical philosophy with a belief in a Deity that was both all-powerful and all-free. The contrast between the long history of Newtonian natural theology and the largely abortive attempts to construct a Darwinian natural theology suggests, then, that not only intellectual difficulties stood in the way of such an amalgamation but also a wider social and cultural climate which was less conducive to an alliance between science and religion than that which had prevailed in the eighteenth and early nineteenth centuries. The historian G. M. Trevelyan (1876–1962) records that as a schoolboy he abandoned his belief in Christianity because he had heard that a man called Darwin had disproved it (Chadwick 1970, 2:1) – an anecdote that underlines the readiness of some within the British educated classes, who barely knew anything of the *Origin,* to assume, nonetheless, that the alliance between science and religion was at an end. While the internal development of an intellectual tradition is certainly important, public perceptions and habits of mind also play a major part in both the preservation and abandonment of systems of ideas – such public responses being largely shaped, in turn, by the larger, social and political movements of the age.

During the eighteenth century the alliance between secular learning (including science) and religion had much greater consequences for society as a whole than it was to have in the nineteenth century – particularly after the reforms of the nineteenth century dissolved many of the traditional links between church and state. In a society such as eighteenth-century Britain it was important for the stability of society that the rational processes on which the state (however inappropriately) claimed to be based should have their reflection in the theology and practice of the established church. A church which emphasized the mysterious and other-worldly to the detriment of the capabilities of the human intellect would have been an inappropriate ally of a state which prided itself on having abandoned such sacerdotal pretensions of the Stuart monarchy as touching for the King's Illness and which based the theoretical foundation of the king's power not on divine right but on such rational arguments as the social contract. The eighteenth-century English established church and its Scottish counterpart were, then, both characterized by an emphasis on the accord between Christianity and reason – an intellectual environment which was naturally conducive to the growth of natural theology.

After the 1830s, however, the importance of the church as a support for the state and the wider social fabric steadily declined even though individual religious belief remained an important feature of Victorian society (or at least, of its literate classes). In 1828 it became possible for Protestant Dissenters to become full citizens and, in 1829, the still more controversial Catholic Emancipation Bill was passed – acts which effectively dissolved the traditional theory that, as "the judicious Hooker" had put it in the sixteenth century, "one and the selfsame people are the Church and the Commonwealth" (Hooker 1863, Vol. II: Bk. 7, iii, 6). Such changes were followed by the reform of the political system and of the church and other sectors of society, thus further weakening the traditional associations that had linked the established church with the fabric of everyday life. As Brooke (1985, 69) suggests, too, such changes also undermined that parallel between a relatively static sociopolitical order and a static God-given natural order which underlay much traditional natural theology. After the 1830s, then, the state and society more generally became progressively more secular in character, and as it did so one of the major traditional impulses to harmonize sacred and secular learning was weakened. One specific instance of this general process was the gradual declericalization of Oxford and Cambridge, the traditional function of which as clerical seminaries had hitherto provided a strong institutional imperative to blend science and theology (Yeo 1985, 287) – a tradition which had produced many of Newton's early clerical followers together with such illustrious early nineteenth-century scientist-divines as Whewell, Sedgwick, and Henslow at Cambridge, and Buckland and Baden Powell at Oxford.

The progressive secularization of the state meant, too, that the established church was less subject to political control than in the past – something which helps to explain the revival of the Church of England's traditional assembly, Convocation, in 1852, after it had been dissolved by the monarch in 1717. The revival of Convocation is a specific instance of the way in which, after the 1830s, the established church could pursue its own goals without being as circumscribed by the social and political functions it traditionally had performed – even if the cost of this greater freedom of movement was a loss of many of its constitutional privileges. This can be seen, too, in the way that the Church of England placed an increasing emphasis on the need for its clergy to receive a professional education suited to their pastoral and theological functions rather than, as in the past, minimizing such specifically clerical training lest its ministers acquire habits of mind and behavior which would set them too much apart from their lay counterparts (Heeney 1976). This growing divorce between lay and clerical education was naturally to weaken further the incentive to harmonize natural and revealed knowledge.

The change in the relations between church and state provided the background for the formation of the Oxford Movement, a religious revival which owed its immediate origins to what its proponents claimed was the cavalier way the reformed state was dealing with ecclesiastical issues – grievances which led to the proposal, by some members of the Oxford Movement that the church should formally disassociate itself from the state. Appropriately, the Oxford Movement was not only associated with a questioning of the traditional alliance of church and state but also with the traditional alliance of science and religion. Newman's friend John William Bowden criticized the deliberately broad and uncontroversial form of natural theology associated with the British Association for the Advancement of Science as that of a "rationalising latitudinarian" (Orange 1975, 287), while Newman himself also wrote of "the undue exaltation of the reason" associated with such scientific associations (Morrell and Thackray 1981, 231). Newman was also to dismiss the whole genre of natural theology – or what he called "physical theology" – describing it as "no science at all, for it is ordinarily nothing more than a series of pious or polemical remarks upon the physical world viewed religiously" (quoted in Cannon 1978, 12). For Newman, science and revealed theology occupied separate spheres which should be kept apart. "Theology and Physics," he later wrote, "cannot touch each other, have no intercommunion, have no ground of difference or agreement, of jealousy or of sympathy" (ibid., 11).

The other earlier major religious revival that affected the Victorians – that of the Evangelicals – was bitterly opposed to many of the aims of the Oxford Movement but, nonetheless, shared with it the view that too much emphasis had been traditionally accorded to natural rather than revealed theology. Sedgwick was plainly seeking to counter the influence of the growing number of Cambridge Evangelicals when, in his *Discourse on the Studies of the University,* he set out to refute those who "asserted within these very walls there is no religion of nature, and that we have no knowledge of the attributes of God or even of his existence, independently of revelation" (Sedgwick 1834, 18). Sedgwick's was but one of a number of pleas by those still committed to the tradition of natural theology to their fellow believers not to divorce the cause of religion from that of science. Brougham adopted the ad hominem approach, arguing that "Boyle and Newton were as sincerely attached to Christianity as any man in any age, and they are likewise the most zealous advocates of Natural Religion" (1858, 203). Thomas Dick urged that "it must, therefore, have a pernicious effect on the minds of the mass of the Christian world, when preachers in their sermons endeavour to undervalue scientific knowledge, by attempting to contrast it with the doctrines of Revelation" (Dick 1846, 2:180–81), while Brewster warned of the dangers to both religion and science "when the votaries of either place them in a state of mutual antagonism" (Brewster 1867, 138).

But even some of the more notable nineteenth-century advocates of natural theology were more inclined than their eighteenth-century counterparts to contrast natural theology unfavorably with Revelation as a source of Christian belief – an indication that they, too, were affected by the general climate of religious revival and greater emphasis on ecclesiastical autonomy. In his Bridgewater treatise, Whewell prefaced his remarks with the observation:

> Yet, I feel most deeply . . . that this, and all that the speculator concerning Natural Theology can do, is utterly insufficient for the great ends of Religion; namely, for the purpose of reforming men's lives, of purifying and elevating their characters, of preparing them for a more exalted state of being. (Whewell [1833] 1862, vi)

These sentiments were also echoed by Chalmers in his Bridgewater treatise: "Natural Theology may see as much as shall draw forth the anxious interrogations, 'What shall I do to be saved?' The answer to this comes from a higher theology" (quoted in Dillenberger 1961, 209). Elsewhere he described natural theology as serving merely "as a harbinger to the higher lessons of the gospel" (Smith 1979, 61–62).

Moreover, scientists as well as theologians were increasingly inclined to emphasize the autonomy of their discipline. As science grew in complexity and social standing in the course of the nineteenth century, its practitioners were less content to justify their activities by recourse to some other goal, such as the advancement of natural theology (Turner 1978). In any case, the growing demands of both science and theology made it increasingly difficult to achieve preeminence in both fields. As scientists perceived themselves more and more in professional terms, another traditional function of natural theology became less relevant. Both the early Royal Society and the early British Association for the Advancement of Science had looked to natural theology to provide common ground for members drawn from different denominations (Morrell and Thackray 1981, 229). But as those involved in the advancement of science defined themselves more and more in terms of their particular speciality rather than by reference to their individual religious beliefs, then so, too, did the need to find a common form of theology become less pressing. Moreover, as political and social divisions based on religious differences began to weaken with the gradual secularization of the state and of society, so, too, the irenic and ecumenical importance of natural theology began to fade (Brooke 1979a, 42).

Thus, although the decline in what Brewster called "the holy alliance" between science and religion – an alliance that found its embodiment in natural theology – can be partly explained by such clashes between science and religion as that symbolized by the famous confrontation between Thomas Huxley and Bishop Wilberforce at Oxford, its decline probably owed more to the result of more fundamental social and intellectual changes which weakened the incentives that had previously existed to harmonize sacred and secular learning. But the eclipse of natural theology in the later Victorian era should not blind us to this tradition's earlier importance in fostering the social recognition and dissemination of British science, nor should it obscure the long-lived and fruitful association between natural theology and the work of Isaac Newton.

Acknowledgments

An earlier version of this paper was delivered at a workshop commemorating the 300th anniversary of the publication of Newton's *Principia* held at Tel Aviv University and at the Van Leer Jerusalem Institute, 27–30 April 1987. I am grateful to the participants of the workshop for their comments. My thanks also go to my colleague, Dr. James Franklin, and to three anonymous referees of this paper.

References

[Abercrombie, J.], 1883. "Review of W. Whewell's Bridgewater Treatise, *Astronomy and General Physics* . . . (1833)," *British Critic* n.s. 14:72–113.

Alexander, H. G., 1956. *The Leibniz-Clarke Correspondence*. Manchester: Manchester University Press.

Anon., 1835. "Review of H. Brougham's *A Discourse of Natural Theology*," *Tait's Edinburgh Magazine* n.s. 2:806–19.

Anon., 1848. "Chemistry and Natural Theology," *British Quarterly Review* 7:204–38.

Anon., 1855. "Review of John Tulloch, *Theism* . . . (1855)," *Westminster Review* 64:319–53.

Anon., 1856. "Review of D. Brewster, *Memoirs of . . . Sir Isaac Newton* (1855)," *Leisure Hour* 5:156.

Anon., 1882. "Newton and Darwin," *Knowledge* 1:545–46.

Ault, D. D., 1974. *Visionary Physics: Blake's Response to Newton*. Chicago: The University of Chicago Press.

Bentley, R., 1724. *Eight Sermons at the Hon. Robert Boyle's Lectures*. Cambridge.

Bowler, 1983. *The Eclipse of Darwinism: Anti-Darwinian Evolution Theories in the Decades around 1900*. Baltimore: Johns Hopkins University Press.

Bowles, G., 1976. "The Place of Newtonian Explanations in English Popular Thought 1687–1727," D. Phil. thesis, University of Oxford.

Brewster, D., 1831. *The Life of Sir Isaac Newton*. London.

——, 1836–37. "Review of Brougham's *Discourse of Natural Theology* (1835) and his edition of Paley's *Natural Theology*," *Edinburgh Review* 64:263–302.

——, 1845. "Review of *Vestiges of the Natural History of Creation*," *North British Review* 3:470–515.

——, 1855. *Memoirs of the Life, Writings, and Discoveries of Sir Isaac Newton*, 2 vols. Edinburgh.

——, 1867. *More Worlds Than One. The Creed of the Philosopher and the Hope of the Christian*. London.

Brinkley, R. F., ed., 1955. *Coleridge on the Seventeenth Century*. Durham: Duke University Press.

Brooke, J. H., 1977. "Natural Theology and the Plurality of Worlds: Observations on the Brewster-Whewell Debate," *Annals of Science* 34:221–86.

——, 1979a. "The Natural Theology of the Geologists: Some Theological Strata," in *Images of the Earth. Essays in the History of the Environmental Sciences*, ed. L. J. Jordanova and R. S. Porter, British Society for the History of Science Monographs, no. 1, 39–66.

——, 1979b. "Review of R. L. Numbers, *Creation by Natural Law: Laplace's Nebular Hypothesis in American Thought*," *British Journal for the History of Science* 12:200–11.

——, 1985. "The Relations Between Darwin's Science and his Religion," in Durant 1985, 40–75.

Brooke, J. H., Hooykaas, R., and Lawless, C., 1974. *New Interactions between Theology and Natural Science*. Milton Keynes: Open University Press.

Brougham, H., 1835. *A Discourse of Natural Theology*. London.

——, 1858. Speech at Erection of Newton's Statue at Grantham, *Littell's Living Age* o.s. 59:374–82.

Brown, F. B., 1986. "The Evolution of Darwin's Theism," *Journal of the History of Ideas* 19:1–45.

Buchdahl, G., 1961. *The Image of Newton and Locke in the Age of Reason*. London: Sheed and Ward.

Cannon, S. F., 1978. *Science in Culture: the Early Victorian Period*. New York: Science History Publications.

Cannon, W. F., 1964. "The Normative Role of Science in Early Victorian Thought," *Journal of the History of Ideas* 25:487–502.

Chadwick, O., 1970. *The Victorian Church*, 2 vols. London: Black.

Chalmers, T., 1835–41. *The Works*, 25 vols. Glasgow.

Colie, R. L., 1963. "Spinoza in England, 1665–1730," *Proceedings of the American Philosophical Society* 107:183–219.

Cornell, J. F., 1986. "Newton of the Grassblade? Darwin and the Problem of Organic Teleology," *Isis* 77:405–21.

Darwin, C., 1958. *The Autobiography of Charles Darwin*, ed. N. Barlow. London: Collins.

——, [1859] 1968. *The Origin of Species*, ed. J. W. Burrow. Harmondsworth: Pelican.

Desmond, A., 1982. *Archetypes and Ancestors. Palaeontology in Victorian London 1850–1875*. London: Blond and Briggs.

Dick, T., 1846. *The Christian Philosopher; or, the Connection of Science and Philosophy with Religion*, 2 vols. Glasgow.

Dillenberger, J., 1961. *Protestant Thought and Natural Science*. London: Collins.

Durant, J. 1985. "Darwinism and Divinity: A Century of Debate," in Durant 1985, 9–39.

——, ed., 1985. *Darwinism and Divinity. Essays on Evolution and Religious Belief.* Oxford: Blackwell.

Force, J. E., 1985. *William Whiston. Honest Newtonian.* Cambridge: Cambridge University Press.

Freudenthal, G., 1986. *Atom and Individual in the Age of Newton. On the Genesis of the Mechanistic World View.* Dordrecht: Reidel.

Gabell, J. H. L., 1847. *The Accordance of Religion with Nature.* London.

Gascoigne, J., 1989. *Cambridge in the Age of the Enlightenment: Science, Religion and Politics from the Restoration to the French Revolution.* Cambridge: Cambridge University Press.

Gay, J. H., 1963. "Matter and Freedom in the Thought of Samuel Clarke," *Journal of the History of Ideas* 24:85–105.

Gillespie, N. C., 1987. "Natural History, Natural Theology, and Social Order: John Ray and the 'Newtonian Ideology,'" *Journal of the History of Biology* 20:1–50.

Gillispie, C. C., 1959. *Genesis and Geology. The Impact of Scientific Discoveries upon Religious Beliefs in the Decades Before Darwin.* New York: Harper and Row.

Gordon, M. M., 1870. *The Home Life of Sir David Brewster.* Edinburgh.

Hahn, R., 1986. "Laplace and the Mechanistic Universe," in *God and Nature. Historical Essays on the Encounter between Christianity and Science,* ed. D. C. Lindberg and R. L. Numbers, 256–76. Berkeley: California University Press.

Haskell, F., 1970. "The Apotheosis of Newton in Art," in *The Annus Mirabilis of Sir Isaac Newton,* ed. R. Palter, 302–21. Cambridge, Mass.: MIT Press.

Hay, D., 1975. "Property, Authority and the Criminal Law," in *Albion's Fatal Tree,* ed. D. Hay, P. Linebaugh, and E. P. Thompson, 17–64. London: Allen Lane.

Heeney, B., 1976. *A Different Kind of Gentleman. Parish Clergy as Professional Men in Early and mid-Victorian England.* Hampden, Conn.: Archon.

Heimann, P. M., 1973. "'Nature is a Perpetual Worker': Newton's Aether and Eighteenth-Century Natural Philosophy," *Ambix* 20:2–24.

—— , 1978. "Voluntarism and Immanence: Conceptions of Nature in Eighteenth-Century Thought," *Journal of the History of Ideas* 39:271–83.

[Hemming, G. W.], 1861. "Newton as a Scientific Discoverer," *Quarterly Review* 110:401–35.

Hooker, R., 1863. *Of the Laws of Ecclesiastical Polity,* in *The Works of Richard Hooker,* ed. J. Keble, 3 vols. Oxford.

[Horne, R. H.], 1835. "Review of H. Brougham's *Discourse of Natural Theology* (1835) and Related Works," *Westminster Review* 23:333–62.

Hoskin, M. A., 1977. "Newton, Providence and the Universe," *Journal for the History of Astronomy* 8:77–101.

Houghton, W. E., ed., 1966–79. *Wellesley Index to Victorian Periodicals, 1824–1900,* 3 vols. Toronto: Toronto University Press.

Hughes, A., 1951. "Science in English Encyclopaedias 1704–1875," *Annals of Science* 7:340–70.

—— , 1952. "Science in English Encyclopaedias 1704–1875," *Annals of Science* 8:323–67.

Hurlbutt, R. H., 1965. *Hume, Newton and the Design Argument*. Lincoln: University of Nebraska Press.

Jacob, M. C., 1976. *The Newtonians and the English Revolution 1689–1720*. Hassocks: Harvester.

[Jeffrey, F.], 1802–1803. "Review of W. Paley's *Natural Theology*," *Edinburgh Review* 1:287–305.

Jones, W. P., 1965. *The Rhetoric of Science*. London: Routledge and Kegan Paul.

Kubrin, D., 1973. "Newton and the Cyclical Cosmos: Providence and the Mechanical Philosophy," in *Science and Religious Belief*, ed. C. A. Russell, 147–69. London: University of London Press.

McGuire, J. E., 1968. "Force, Active Principles, and Newton's Invisible Realm," *Ambix* 15:154–208.

McPherson, T., 1972. *The Argument from Design*. London: Macmillan.

[Malkin, B. H.], 1832. "Review of D. Brewster, *The Life of Sir Isaac Newton* (1831)," *Edinburgh Review* 56:1–37.

Mandelbaum, M., 1971. *History, Man and Reason*. Baltimore: Johns Hopkins University Press.

Manuel, F., 1974. *The Religion of Sir Isaac Newton*. Oxford: Oxford University Press.

Metzger, H., 1938. *Attraction universelle et religion naturelle chez quelques commentateurs anglais de Newton*. Paris: Hermann.

Moore, J., 1979. *Post-Darwinian Controversies. A Study of the Protestant Struggle to Come to Terms with Darwin in Great Britain and America, 1870–1900*. Cambridge: Cambridge University Press.

Morrell, J. B., 1971. "Professors Robison and Playfair, and the *Theophobia Gallica*: Natural Philosophy, Religion and Politics in Edinburgh, 1789–1815," *Notes and Records of the Royal Society* 26:43–63.

Morrell, J. B., and A. Thackray, 1981. *Gentlemen of Science. The Early Years of the British Association for the Advancement of Science*. Oxford: Oxford University Press.

Odom, H. H., 1966. "The Estrangement of Celestial Mechanics and Religion," *Journal of the History of Ideas* 27:533–58.

Ogilvie, M. B., 1975. "Robert Chambers and the Nebular Hypothesis," *British Journal for the History of Science* 8:214–32.

Orange, A. D., 1975. "The Idols of the Theatre: the British Association and its Early Critics," *Annals of Science* 32:277–94.

Ospovat, D., 1980. "God and Natural Selection: The Darwinian Idea of Design," *Journal of the History of Biology* 13:169–94.

——, 1981. *The Development of Darwin's Theory. Natural History, Natural Theology and Natural Selection, 1838–1859*. Cambridge: Cambridge University Press.

Paley, W., 1845. *The Works*. Edinburgh.

Pope, A., 1963. *The Poems of Alexander Pope*, ed. J. Butt. London: Methuen.

Porter, R. S., 1978. "Review of M. C. Jacob, *The Newtonians and the English Revolution 1689–1720 (1976)*," *Social History* 3:246–49.

[Powell, Baden], 1843. "Review of S. P. Rigaud's *Historical Essay on the . . . Principia* (1838) and his *Correspondence of Scientific Men of the Seventeenth Century . . .* (1841)," *Edinburgh Review* 78:402–37.

Proctor, R. A., 1882. "Newton and Darwin," *Contemporary Review* 41:994–1002.

Ruse, M., 1975. "The Relationship between Science and Religion in Britain, 1830–1870," *Church History* 44:505–22.

Schaffer, S., 1980a. "Herschel in Bedlam: Natural History and Stellar Astronomy," *British Journal for the History of Science* 13:211–39.

——, 1980b. "Newtonian Cosmology and the Steady State," Ph.D. thesis. University of Cambridge.

Sedgwick, A., 1834. *Discourse on the Studies of the University*. London.

Shapin, S., 1981. "Of Gods and Kings: Natural Philosophy and Politics in the Leibniz-Clarke Disputes," *Isis* 72:187–215.

Sheynin, O. B., 1970–71. "Newton and the Classical Theory of Probability," *Archive for the History of the Exact Sciences* 7:217–43.

Smith, C., 1979. "From Design to Dissolution: Thomas Chalmers' Debt to John Robison," *British Journal for the History of Science* 12:59–70.

Stephen, L., 1962. *History of English Thought in the Eighteenth Century*, 2 vols. London: Harbinger.

Thackray, A., 1970. *Atoms and Powers. An Essay on Newtonian Matter-Theory and the Development of Chemistry*. Cambridge: Cambridge University Press.

——, 1974. "Natural Knowledge in Cultural Context: the Manchester Model," *American Historical Review* 79:672–709.

Thayer, H. S., ed., 1974. *Newton's Philosophy of Nature: Selections from his Writings*. New York: Free Press.

Tulloch, J., 1855. *Theism: The Witness of Reason and Nature to an All-Wise and Beneficient Creator*. Edinburgh.

Turnbull, H. W., J. F. Scott, A. R. Hall, and L. Tilling, eds., 1959–77. *The Correspondence of Sir Isaac Newton*, 7 vols. Cambridge: Cambridge University Press.

Turner, F. M., 1978. "The Victorian Conflict Between Science and Religion: a Professional Dimension," *Isis* 69:356–76.

Turton, T., 1836. *Natural Theology with Reference to Lord Brougham's Discourse. . . .* Cambridge.

Vince, S., 1806. *Observations on the Hypotheses Which Have Been Assumed to Account for the Cause of Gravitation*. Cambridge.

Wallace, W., 1898. *Lectures and Essays on Natural Theology and Ethics*. Oxford.

Wallis, P., and R. Wallis, 1977. *Newton and Newtoniana, 1672–1975*. Folkstone: Dawson.

Whewell, W., [1833] 1862. *Astronomy and General Physics Considered with Reference to Natural Theology*. London.

Wordsworth, C., 1842. *The Correspondence of Richard Bentley,* 2 vols. London.

Yeo, R., 1977. "Natural Theology and the Philosophy of Knowledge in Britain, 1819–1869," Ph.D. thesis. University of Sydney.

———, 1984. "Science and Intellectual Authority in Mid-Nineteenth-Century Britain: Robert Chambers and *Vestiges of the Natural History of Creation,*" *Victorian Studies* 28:5–31.

———, 1985. "An Idol of the Market-place: Baconianism in Nineteenth-Century Britain," *History of Science* 23:251–98.

Young, R. M., 1980. "Natural Theology, Victorian Periodicals and the Fragmentation of a Common Context," in *Darwin to Einstein. Historical Studies on Science and Belief,* ed. C. Chant and J. Fauvel, 69–106. London: Longman.

III

Sensible Newtonians: Nicholas Saunderson and John Colson

A telling anecdote is recorded by Martin Folkes, Cambridge graduate and president of the Royal Society (1741–53): 'After Sir Isaac printed his *Principia*', Folkes recalled what was said about Newton 'as he passed by the students at Cambridge'. Apparently one witty undergraduate remarked, 'there goes the man who has writt a book that neither he nor any one else understands'. The task of rendering the abstruse *Principia* into the common coin of Enlightenment Britain largely fell to the Lucasian professors who followed Newton: William Whiston, Nicolas Saunderson and John Colson. Though the period when Saunderson and Colson held office as Lucasian professors (1711–39 and 1739–60) was not a heroic one in the annals of Cambridge University, it was, none the less, an era when the mathematical tradition for which the university became so well known was consolidated. It was, moreover, a period when the Newtonian version of the workings of the natural world still had rivals who had to be vanquished, an intellectual contest in which the Lucasian professors played a major role. Just as Cambridge's divinity professors saw it as their function to combat heretical assaults on the Church of England so, too, the university's mathematical professors saw their duty as defending the pure gospel of Newtonianism. The analogy extends to the fact that, just as Anglican orthodoxy required the rejection of deviant forms of Christianity, so the Cambridge professors played an important role in establishing what constituted the canonical version of the Newtonian system of the world.[1]

When the blind Nicholas Saunderson was elected to the Lucasian chair in 1711, the university was still digesting the great Newtonian achievement. Newton's formidable mathematical masterpiece took some time to be embraced by the great minds of Europe so it is not surprising that it was a slow and arduous process to put it into a form which undergraduates or even, indeed, college tutors could understand. Making the great peaks of the *Principia Mathematica* accessible to gownsmen was a weighty burden that was put on the shoulders of the early-eighteenth-century Cambridge Lucasian professors since Newton himself, always anxious to avoid controversy, reportedly 'made the *Principia* abstruse to avoid being baited by the little smatterers in mathematics'.[2]

The first substantial Cambridge Newtonian textbooks of natural philosophy were produced by Newton's successor as Lucasian professor, William Whiston, who held the office from 1702 to 1710. But reorientating a curriculum which had, since the high Middle Ages, been dominated by Aristotelian philosophy was, inevitably, a slow process. True, the traditional scholastic curriculum was in rapid decline following the achievements of the Scientific Revolution of the seventeenth century. Moreover, Cartesian philosophy had made rapid progress in late-seventeenth-century Cambridge partly because, as an all-embracing philosophical schema deduced from first principles, it readily filled the void created by the waning of scholasticism. The result, however, was that Newton's early disciples in Cambridge had to fight a battle on two fronts: demolishing the ailing Aristotelianism and combating the influence of Cartesianism which was still vigorous on the continent.

The task was further complicated by the character of Cambridge teaching. The foundation of the Lucasian professorship in 1663 and the lustre of its first two incumbents, Barrow and Newton, indicate that the professorial mode of instruction still played a major role in late-seventeenth-century Cambridge. Professorial teaching was

NICHOLAS SAUNDERSON *LLD*
Lucafian Profeffor of Mathematicks in
the University of Cambridge
Died 19 Ap. 1739 Aged 56

I.Vanderbanck pinx.1718. From the Original painted for Martin Folkes Esq:. G:Vander Gucht. Sculp.

Figure 22 Portrait of Saunderson. Engraving of Saunderson from the frontispiece to his *Elements of Algebra*, published posthumously in 1740.

particularly valuable in areas such as mathematics or the recent developments in natural philosophy which the college tutors did not feel equipped to teach. The fact that Newton himself did 'of times... in a manner for want of Hearers, read to the Walls' is, however, an indication that mathematics remained at the margin of the undergraduate curriculum, a situation not improved by the fact that Newton's lectures were often drafts of his *Principia*. This helps to explain the impact of William Whiston, the third Lucasian professor, whose lectures and books provided one of the few means by which Cambridge undergraduates could come to terms with the Newtonian natural philosophy. But at both Oxford and Cambridge (in contrast to the Scottish universities) the tide was flowing away from professorial instruction towards that based on college tutors.[3]

Throughout Cambridge's history there has always been a tussle for dominance between the colleges and the central university. In the Middle Ages the university was largely dominant but from the Reformation onwards the colleges were in the ascendant (though in our own times the balance is beginning to swing back towards the university). In the late seventeenth century the lingering remains of the traditional curriculum with its established round of disputations and other exercises still lent some plausibility to the notion that the central university set a course of studies controlled by the professors. But the eighteenth century was a period that favoured the colleges which, with their wise investments in property, were becoming wealthier. This wealth, coupled with decreasing numbers of students, meant that they could easily take over most undergraduate instruction, thus leaving the professors with little to do.

During Saunderson's tenure, there was still some demand for professorial lectures on natural philosophy, particularly since many of the college tutors did not feel qualified to provide instruction in the field. But, to judge by student guides such as Daniel Waterland's influential *Advice to a Young Student*, college instruction by the 1730s had largely caught up with what the Lucasian professors could offer. This state of affairs helps to account for the faint impression made on the university by the lectures of the fifth Lucasian professor, John Colson (1739–60), for his lack of success in attracting students cannot be entirely put down to his personal inadequacies as a teacher. Nor did Colson's mathematically more distinguished successor, Edward Waring (Lucasian professor, 1760–98), fare much better in attracting students for, as the pedagogue Samuel Parr diplomatically remarked, his 'profound researches were not adapted to any form of communication by lectures'.[4]

The trajectory of the eighteenth-century Lucasian professorship is then something of a downward spiral. At the beginning of the century it could bask in the lustre imparted to it by Barrow and Newton and its impact on the instruction of the university was considerable thanks to the exertions of Whiston and, to a lesser degree, Saunderson. By Colson's time, professorial instruction in mathematics and natural philosophy was both less necessary and less valued as the college tutors assumed control over a curriculum increasingly dominated by mathematics and natural philosophy. To some extent, Colson created a niche for himself by publishing editions or translations of textbooks on mathematics or natural philosophy, while Waring's mathematical publications partially justified his low-key status; but neither had a significant impact within Cambridge. In a university dominated by colleges where the student body (such as it was) was in the hands of tutors, professors often seemed like anachronistic pedants.

But this is to take the long view: when Saunderson was elected following Whiston's expulsion, he had a clear sense of mission. Like Whiston, he sought to win over both the undergraduates and the senior gownsmen to the pure gospel of Newtonianism. He had come to know the 'Great Man' himself and contemporaries attributed his advancement within Cambridge partly to the result of Newton's good offices.

How far Newton involved himself actively in the election for the Lucasian chair is doubtful. Saunderson's chief rival was Christopher Hussey, a Trinity man enjoying the powerful backing of Richard Bentley, the Master of Trinity and a master of academic politicking.

Naturally, Bentley sought to enlist Newton's support: Bentley reckoned that Newton had been indebted to him since 1692 when Bentley
had given his celebrated Boyle lectures. For these lectures, *A Confu-
tation of Atheism*, were among the first major works to publicize the
Principia and to establish it as a resource for defending the existence
of God, as well as the established Anglican order. But Newton was not
to be cajoled, even by the formidable Bentley, and he responded luke-
warmly: 'I have made a resolution not to meddle with this election
of a Mathematick Professor any further then in answering Letters &
have given this answer to some who have desired a certificate
from me'.[5]

None the less, Newton's letter does leave open the possibility
that he did write on behalf of Saunderson, and perhaps did so with
sufficient warmth to sway the electors. The voting record indicates,
however, that Hussey had the support of those in Cambridge most
closely associated with Newton, including Bentley and Sir John Ellys,
the Master of Caius College. Despite this strong support for Hussey,
Saunderson was elected by six votes to four. Bentley's ruthless tactics
had made him many enemies in Cambridge and it is possible that
his identification with Hussey may have hindered, not assisted, his
chosen candidate's prospects. Or, Saunderson's victory might be attributed to his discretion in religious matters: Halley, for example, is
supposed to have remarked that 'Whiston was dismissed for having
too much religion, and Saunderson preferred for having none'.[6]

For the blind and humbly born Saunderson to gain such a post
was a remarkable achievement. Following his birth in Thurlston,
Yorkshire in 1682, the contraction of smallpox at age one cost him
not only his sight but his eyes. Notwithstanding this handicap, his
father, an exciseman, taught him a little arithmetic 'which he took
very readily and made surprizing progress'. His interest in mathematics was also stimulated by a Richard West of Undebank, 'a gentleman
of learning'. Following schooling at the Penistone grammar school he
spent a year or two at the dissenting academy, Christ's College, Attercliffe (Sheffield), despite the fact that the master 'forbad [his pupils] the

Mathematicks, as tending to scepticism & infidelity'. Nevertheless, there must have been some mathematical instruction at Attercliffe since we are told that many students 'made a considerable progress in that branch of Literature', even though Saunderson did not find the traditional curriculum of logic and metaphysics to his taste.[7]

This association of mathematics with infidelity was, however, to prove prophetic since Saunderson appears to have been a Deist with little commitment to revealed religion – in so far as we can gauge his religious views, in an age when, as the Whiston case testified, it was unwise to publicize one's infidelity. That industrious early-nineteenth-century historian of Cambridge, George Dyer, remarked that Saunderson was 'no friend to Divine revelation', though Dyer added that he was said to have taken Holy Communion on his deathbed.[8]

Although probably an exaggeration, Halley's quip concerning the contrasting religious views of Whiston and Saunderson conveys an essential truth about eighteenth-century Cambridge: as an arm of the established Church of England it expected its members not to challenge the central tenets of Anglicanism. The radiance of the Enlightenment, however, was sufficiently strong to stifle attempts at enforcing orthodoxy when unbelief did not take an overtly public form. Saunderson does appear, however, to have tested the limits of Cambridge's tolerance: his entry in the late-eighteenth-century *Biographia Britannica* praises him for his vivacity and honesty, but adds that since he was predominantly remembered for his 'indulgence of women, wine, and profane swearing to such a shocking excess, ... he did more to harm the reputation of mathematics than he did good by his eminent skill in the science'. One wonders, too, what Saunderson's father-in-law, William Dickons, rector of Boxworth, Cambridgeshire, made of the less than devout husband of his daughter, Abigail. Yet, as with Charles Darwin's father, paternal unbelief did not prevent him from placing his son in the Church, for Saunderson's son, John, took orders after graduation. As a clergyman and a fellow of Peterhouse, John also had the leisure to help in the publication of his father's *Algebra*.[9]

After his early introduction to mathematics at school and at Attercliffe, Saunderson became ever more captivated by mathematics through his personal study (which he turned to some practical effect by assisting his father with excise calculations). He aspired to study at Cambridge but this appeared to be a goal denied him by his slender resources. However, in 1707 he made his way there by more indirect means, acting as the personal tutor to Joshua Dunn, the son of a wealthy mercer of Halifax and Attercliffe, who enjoyed the high status of a fellow-commoner at Christ's (which entitled Dunn to dine with the fellows, exemption from undergraduate examinations and higher fees). Thus began a long association by Saunderson with Cambridge University – his high reputation as a private tutor eventually obtaining for him an honorary MA (1711) which, subsequently, was followed by an LL.D. (1728). Christ's College remained his home until 1723 when he moved into a house in Cambridge, where he lived with his wife.

As a private tutor he devoted himself to the enterprise which dominated his Lucasian professorship: winning over the undergraduate body to Newtonianism. In this endeavour he had the support of the then Lucasian professor, William Whiston. Though Saunderson's classes posed some competition to his own, Whiston was ready to embrace a fellow Newtonian proselytizer and, as his willingness to sacrifice his university position indicates, Whiston was not one to put private gain before principle. Predictably, Saunderson's lessons were based on Newton's *Opticks* and *Principia* and his pupils were among those who helped to establish Newtonian natural philosophy as part of the normal fare of undergraduate university disputations: 'We every year heard the Theory of the Tydes, the *Phaenomena* of the Rainbow, the Motions of the whole Planetary System as upheld by Gravity', Richard Davies wrote in an obituary of Saunderson, 'very well defended by such as had profited by his Lectures'.[10]

For Saunderson, being a private tutor meant, inevitably, catering particularly to the wealthier students and especially the fellow-commoners. These gilded youths did not need academic success or even a degree and, consequently, were often less than attentive. Recalling such vexations, Saunderson reportedly exclaimed, 'that if he was to go to hell, his Punishment would be to read lectures in the mathematics to the gentlemen commoners of that university'. Saunderson was not always willing to cast his mathematical pearls before such swine even if it meant pecuniary sacrifice. When the young Horace Walpole became a pupil, Saunderson told him within a fortnight: 'Young man, it is cheating you to take your money: believe me, you never can learn these things; you have no capacity for them'. After a year with another tutor, Walpole acknowledged the justice of Saunderson's assessment and abandoned his struggle with mathematics.[11]

Meanwhile, Saunderson soldiered on with his other pupils and gained an excellent reputation as a tutor. Once elected as Lucasian professor in 1711 he had a more prominent platform on which to continue the work to which he had devoted himself as a private tutor: the promotion of Newtonianism in a university where the increasingly important college tutors were often still mathematically innocent. Around 1730, William Heberden records, 'Newton Euclid and Algebra were only known to those who chose to attend the lectures of Professor Saunderson, for the college lecturers were silent on them'.[12]

As the statutes of the Lucasian chair required, his lectures were deposited in the Cambridge University library and still survive (which, despite the regulation, is not true of those of Colson). They cover most major topics in natural philosophy – hydrostatics, mechanics, optics, sound, the tides, astronomy – all, of course, along Newtonian lines. The number of copies indicates their wide diffusion and they appear to have continued to be used in manuscript well into the late eighteenth century. James Bradley used Saunderson's discussion of hydrostatics in preparing his lectures as Savilian professor of astronomy at Oxford, as did James Wood and Samuel Vince in preparing their late-eighteenth-century natural philosophical textbooks (meant chiefly for Cambridge consumption). Edmund Burke, the celebrated aesthete and opponent of the French Revolution, also commented on the excellence of Saunderson's notes on light and colour.[13]

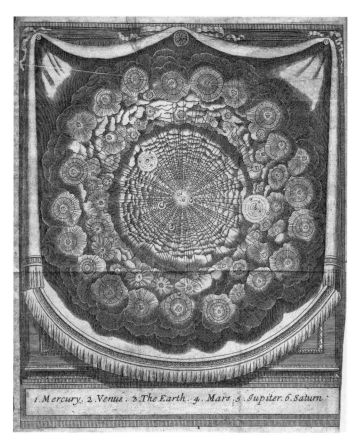

1. Mercury. 2. Venus. 3. The Earth. 4. Mars. 5. Jupiter. 6. Saturn

Figure 23 From *A Week's Conversation on the Plurality of Worlds* (London 1738) Fontenelle's solar system of vortices. While Fontenelle suggested to his audience of aristocratic women that the universe was full of Cartesian vortices, Saunderson and Colson combated such misapprehensions with such works as their translation of Professor Donna Maria Agnesi's *The Plan of the Lady's System of Analyticks*.

As befitted a fellow of the Royal Society (a dignity he held from 1719), Saunderson's enterprises drew extensively on the experimental and mathematical heritage of the *Opticks* and the *Prinicipia*, particularly in his discussion of topics such as 'Of the ascent and suspension of liquors in capillary tubes' or 'Of the circulation of the sap'. Along with his advocacy of Newton, Saunderson was determined to vanquish the 'Great Man's' opponents, particularly the Cartesians. Although in the seventeenth century Descartes's philosophy had enjoyed a strong foothold in Cambridge, it had come to be viewed with increasing disfavour because it was seen to endanger the Anglican conception of the Almighty. The particulars of Cartesian natural philosophy were attacked in order to erode the foundations on which Descartes's theologically suspect world system had been constructed. Thus, for example, it is unsurprising to find that Saunderson's lectures on hydrostatics insisted that the Cartesian notion of fluidity – that is, 'a body whose parts are in Continuall motion' – was 'contrary both to sense & Reason'. On the other hand, Saunderson made certain that he demonstrated to his students that the fundamental principles of the Newtonian cosmos were eminently reasonable: for instance, he ensured that pupils understood that the Copernican system was based on pure reason, which also shows us that one of the central tenets of the Scientific Revolution was not entirely to be taken for granted.

There were also opponents of Newton within the citadel of Cambridge. A year after taking up his Lucasian professorship Saunderson commented dismissively to that industrious purveyor of mathematical gossip, William Jones, that 'There has been nothing publish'd here since my last to you, excepting by one Mr Green Fellow of Clare-Hall of this University'. For Saunderson, the best way of dealing with Greene's work was by treating it with contempt. As he remarked to Jones, 'I can find nothing in it, but ill manners & elaborate nonsense from one end to the other'. The book that Saunderson so unflatteringly described was Robert Greene's *Principles of Natural Philosophy* (1712) which expatiated at length on the inadequacies of the 'Corpuscular System, or the Philosophy of Homogeneous Matter' which he saw as a revival of Epicureanism. Though Greene was cautiously polite in his treatment of Newton, his position inevitably conflicted with Newtonian matter theory. In particular, Greene's work challenged Newton's supposition in *Opticks* that 'it seems probable to me that God in the beginning formed matter in solid, massy, hard, impenetrable, movable particles'. The anti-Newtonianism of Greene became even more evident in his vast *magnum opus*, *The Principles of the Philosophy of the Expansive and Contractive Forces* (1727), which characterized as theologically suspect not only 'the Principles of a Similar and Homogeneous Matter' but also characteristically Newtonian concepts, including absolute space and time and action through a vacuum. Certainly, Newton's disciples had no doubt that Greene's work was intended as an attack on the 'Great Man'. Cotes reported to Newton in 1711 that Greene was publishing a book 'wherein I am informed he undertakes to overthrow the Principles of Your Philosophy'.[14]

Saunderson plainly thought that Greene and other opponents of Newton's philosophy could best be combated by ridicule and vigorous promotion of the self-evident truths of Newtonianism. Given the demands on his time (which must have been compounded by his blindness), Saunderson did this principally through his teaching. However, he made sure that his work was informed by the most recent developments in the Newtonian world. Evidently, he closely followed the progress of Roger Cotes's second edition of the *Principia*, reporting to William Jones in 1712 that 'Sr Is. Newton is much more intent upon his principia than formerly, & writes almost every post about it, so that we are in great hopes to have it in our hands[?] in a very little time'. News that Jones was contemplating an edition of David Gregory's '*Notes upon Sir Isaac Newton's Principia* prompted an enquiry as to whether ''tis an explanation of the whole or only of some parts of the Principia'. But the letter to Jones about this ultimately abortive project also revealed the constant demands on Saunderson's time since he apologized to Jones that 'I have had so much business upon my hands, ever since I had the happiness to be acquainted with you, as has hindered me from carrying on a correspondence with you'.[15]

Thanks in part to Saunderson's exertions during his term of office, the university abandoned the decaying remnants of the traditional scholastic curriculum in favour of mathematics and Newtonian natural philosophy. Such a transformation was not solely the work of the Lucasian professors. Along with the contribution of Whiston and Saunderson was that of the first two Plumian professors of astronomy and experimental philosophy, Roger Cotes (1706–16) and Robert Smith (1716–68), who also promoted Newtonian natural philosophy, especially in its more experimental manifestations. College tutors such as John Rowning of Magdalene, author of the avowedly New-tonian *A Compendious System of Natural Philosophy* (1735, with several reprintings up to 1772) also helped to channel Cambridge's academic energies into the mathematical current for which it became known.

By the 1730s contemporaries remarked on the dominant and, to some, oppressive mathematical atmosphere within the univer-sity. Those of a literary bent like Horace Walpole, whom Saunderson had so quickly and perceptively dismissed, bemoaned Cambridge's mathematical myopia. Hence his lament in 1735 that Cambridge was 'o'errun with rusticity and mathematics' – a view echoed by the poet, Christopher Smart, in his Tripos verses at his graduation in 1742.[16] But such protests did little to stem the mathematical tide which, from the middle of the century, was increasingly institutionalized in the Senate House Examination which replaced the traditional scholastic dispu-tations. Indeed, mathematics so dominated the examination that it eventually took the name 'the Mathematical Tripos'. Thus from 1753 the published Tripos lists were, for the first time, divided into wran-glers and senior and junior optimes – categories which were eventually to develop into the familiar first class and the two subdivisions of the second class.

Beyond the cloistered world of Cambridge, Saunderson's repu-tation was limited by the fact that he published little during his life-time, even if his membership of the Royal Society and the Board of Longitude (on which he served ex officio from 1714) did give him some contact with a larger scientific world. Although manuscripts of his lectures and treatises were circulated widely, he did not follow the common practice of publishing them – perhaps because he was blind. To some extent this deficiency was remedied posthumously. In 1740, the year after his death, his *The Elements of Algebra, in Ten Books: To Which is Prefixed, an Account of the Author's Life and Character, Collected from his Oldest and Most Intimate Acquaintance* was published at the initiative of Richard Davies, physician and fellow of Queens' College. An abridged edition was published in 1756 by the Reverend John Hellins of Trinity College, and in that form four more editions were printed, the last of which appeared in 1792. Hellins paid tribute to 'The Excellence of Professor Saunderson's *Elements of Algebra* [which] is universally acknowledged'. He also added, with a remark that underlines the increasing dominance of the college tutors, that since much of it was intended for the 'Advantage and Amusement to proficients in the general Science of the Mathematics, than of necessary Use to Students in Algebra; some of the principal Tutors in the University of *Cambridge*' were anxious for an abridged, more pedagogically focused work.[17]

Saunderson's work indicates his sympathy for analytical methods which perhaps better suited him because of his blindness. This meant something of a departure from the dominant Newtonian tradition. For, though Newton himself greatly advanced the use of analysis by his actual practice, he retained the traditional veneration for geometry over analysis – hence his remark that 'Algebra is the Analysis of the Bunglers in Mathematicks'. Something of the stir created by Saunderson's approach is evident in the mathematical arguments it prompted between his students and those of the London-based mathematician and military engineer, Benjamin Robins, FRS, 1727 and Copley medallist, 1747. But, for Saunderson, whether algebra or geometry, mathematics in all its forms still retained its mystical aura of the Pythagoreian–Platonic tradition. He regarded mathematics as the one sure guide to truth: 'It is in Mathematics only that truth, that is, rational truth, appears most conspicuous, and shines in her strongest lustre ... whosoever to the utmost of his finite capacity would see truth as it has actually existed in the mind of God from all eternity, he must study Mathematics more than Metaphysics'.[18] It is a passage that underlines the fact that, for all the rumours of his infidelity, Saunderson remained a Deist, if not an orthodox Christian.

Along with the *Algebra*, another posthumous work of Saunderson was his lengthy but revealingly entitled *The Method of Fluxions Applied to a Select Number of Useful Problems: Together with the Demonstration of Mr. Cotes's Forms of Fluents in the Second Part of his Logometria; The Analysis of the Problems in his Scholium Generale and an Explanation of the Principal Propositions of Sir Isaac Newton's Philosophy* (1756). It was a book which served to fuel the growing mathematical blaze within the university, being particularly useful for those aspiring to high honours in the Senate House Examination. Despite this blaze Saunderson was the first Cantabrigian to provide through his lectures a systematic introduction to the fluxional calculus. While the subject of the first part of the book was devoted to the methods of fluxions, the second and third were focused on Cotes's integrals and an analysis of some propositions of the *Principia* (where he remains true to Newton's geometrical approach).[19] Its anonymous editor made the claim that it may 'be reckoned the best for Students in the Universities, of any yet published'. His commentary on Cotes's *Logometria* also testified to 'his intimate Acquaintance with that extraordinary Person'.

Such works, then, enhanced Saunderson's already well-established reputation within Cambridge as a teacher, although they did little to add his name to the larger annals of mathematics. But Saunderson was known throughout Britain, and even beyond, for his extraordinary achievement as an individual in gaining so hotly contested a position as the Lucasian professorship despite his blindness. What particularly fascinated the European philosophers was the extent to which Saunderson could be regarded as a test-case for the Lockean view that all ideas were the outcome of experience. For, as someone who was blind, Saunderson was denied many common experiences (though, of course, Saunderson was blind from infancy, not blind from birth, and may have retained some early experiences – a point not taken up in such discussions). Ten years after Saunderson's

Lettre

```
7   8   5   6   8
8   4   3   5   8
8   9   4   6   4
9   4   0   3   0
```

Il eſt Auteur d'un Ouvrage très-parfait dans ſon genre. Ce ſont des élémens d'algébre où l'on n'apperçoit qu'il étoit aveugle qu'à la ſingularité de certaines démonſtrations qu'un homme qui voit n'eût peut-être pas ren contrées: c'eſt à lui qu'appartient la diviſion du cube en ſix piramides égales qui ont leurs ſom mets au centre du cube, & pour baſes, chacune une de ces faces. On s'en ſert pour démontrer d'une maniere très-ſimple, que toute piramide eſt le tiers d'un priſme de même baſe & de même hauteur.

Il fut entrainé par ſon goût à l'étude des Mathématiques, & déterminé par la médiocrité de ſa fortune & les conſeils de ſes amis, à en faire des leçons publiques. Ils ne douterent point qu'il ne réuſsît au de-là de ſes eſpérances, par la facilité prodigieuſe qu'il avoit à ſe faire entendre. En effet, Saounderſon parloit à ſes Eléves comme s'ils euſſent été privés de la vue ; mais un Aveugle qui s'exprime clairement pour des Aveugles, doit gagner

Lettre Sur Les Aveugles T. 2: pag 36

PL. VI

Figure 24 Saunderson's calculating board. According to Diderot in his *Lettre sur les Avaeugles* (*Letter concerning the Blind*), Saunderson was able to speak to his students as if he had sight.

death Diderot referred to him in his *Letter on the Blind* (1749) when discussing whether a person born blind would be able to grasp geometrical concepts such as the difference between a sphere and a cube. Saunderson's manifest success as a mathematician led Diderot to conclude that, indeed, the blind could grasp such concepts, though an essentially Lockean epistemology was saved by attributing the formation of ideas to experience based on the sense of touch rather than sight.

Diderot even provided a detailed (though somewhat inaccurate) account of the device that Saunderson used to study geometry by means of touch – a device described at length by his successor as Lucasian professor, John Colson, in a section of his posthumous *Elements of Algebra* entitled 'Dr Saunderson's palpable arithmetic decypher'd'. For, as Colson put it, Saunderson devised a form of *'Abacus* or Calculating Table, and with which he could readily perform any Arithmetical Operations, by the Sense of Feeling only'. The device took the form of a sort of wooden grid with a hole at each point of intersection which was capable of accommodating a pin. By using such pins Saunderson could perform complex calculations and 'could place and displace his Pins ... with incredible Nimbleness and Facility, much to the Pleasure and Surprize of all the Beholders'. The device could be used both for arithmetical calculations and to describe 'upon it very neat and perfect Geometrical Figures'. Saunderson also devised his own 'Species of Mathematical Symbols' which could take the place of visible signs by again substituting 'Feeling in the place of Seeing' – symbols he derived from listening to others read mathematical works to him.[20]

Diderot also pointed to other ways in which, in Lockean fashion, Saunderson built up ideas from the building blocks of sensory perception. Saunderson, for example, like other blind people, was particularly sensitive to any slight change in his surroundings, so that he could tell if clouds were passing over the sun and hence when was an appropriate moment for astronomical observations. Saunderson, as Diderot pithily put it, 'saw with his skin'. He also remarked on his finely developed sense of hearing so that he could tell who was approaching by the sound of their step. It was a point also dwelt upon by his Cambridge contemporaries who remarked on his acute ear, evident in his abilities as a flautist. It was related that 'if ever he had walked over a Pavement in Courts, Piazzas, &c, which reflected a Sound, and was afterwards conducted thither again, he could exactly tell whereabouts in the Walk he was placed, merely by the Note it sounded'.[21] By such means Saunderson appears to have constructed a fairly complete picture of the world, enabling him to join in conversation in a way that made contemporaries almost forget about his blindness.

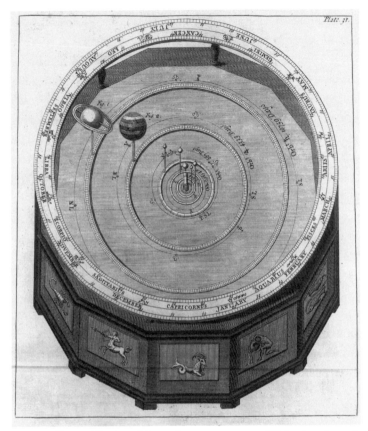

Figure 25   An eighteenth-century orrery. While Saunderson's board helped the visually impaired make complex calculations, other devices, like the orrery, gave eighteenth-century professors and lecturers the ability to enable the sighted to better comprehend the Newtonian universe.

For Diderot, Saunderson was a fascinating case study not only of the way in which the blind could grasp complex mathematical concepts but also as a prosecution witness in Diderot's continuing assault against the claims of religion. For, in Diderot's view, the weaknesses of the argument from design should be particularly evident to someone incapable of seeing the alleged evidences of such design – even if, ironically, the early popularization of Newton's work had been closely associated with clerical popularizers of the design argument. Manufacturing an entirely imaginary conversation, Diderot portrayed Saunderson on his deathbed undermining the arguments advanced by a typical clerical physico-theologian. In the first place, as one that was blind, proofs based on the visible splendours of the universe had no meaning to Saunderson; but even if evidence of design could be conveyed through the sense of touch, this did not necessarily lead to the notion of a divine intelligence. Saunderson's own experience indicated that human beings were too willing to invoke transcendent explanations for what they could not understand:

> People have been drawn to see me from all over England merely because they could not conceive how I was able to do geometry; you must admit that those people did not have very exact notions as to what is or is not possible. We encounter some phenomenon that is, in our opinion, beyond the powers of man, and immediately we say, 'It is the work of God.' Our vanity will admit no lesser explanation.

Saunderson also served as an instance of the problem of evil with his supposed declaration 'Look well at me, Mr. Holmes. I have no eyes. What did we do to God, you and I, that one of us should possess those organs and the other be deprived of them?' Taking the theme of Saunderson's blindness as a instance of the imperfection of the world, Diderot also alluded to the concept of the plurality of worlds – a concept which Saunderson had expounded in his *Elements*. 'How many maimed and botched worlds have melted away', the blind natural philosopher asks, 're-formed themselves, and are perhaps dispersing again at any given moment?'. Despite such scepticism towards divine intelligence, benevolence and design, Diderot portrays Saunderson as finally espousing some form of deistic belief, dying with the cry, 'O God of Clarke and Newton, pity me now'.[22]

In his *A Philosophical Enquiry into the Origin of our Ideas of the Sublime and Beautiful*, Burke argued that Saunderson also served as an important test-case in the relationship between ideas and experience – in particular, ideas relating to colour. What particularly struck Burke was that the blind Saunderson 'gave excellent lectures upon light and colours; and this man taught others the theory of those ideas which they had, and which he himself undoubtedly had not'. Burke then assumed in Lockean fashion that it was impossible to have an idea of something like a colour unless one could build on experience which was not available to a blind person. Saunderson must then, argued Burke, have merely been using the words for colours without possessing the ideas: 'But it is probable, that the words red, blue, green, answered to him as well as the ideas of the colours themselves . . . He did nothing but what we do every day in common discourse'.[23]

Dr Johnson, too, was fascinated by the question of whether Saunderson could discern colours by touch but concluded in a conversation with Boswell, 'that to be sure a difference in the surface makes the difference of colours; but that difference is so fine, that it is not sensible to the touch'.[24] For Johnson, as for Burke or Diderot, Saunderson became an instance of the inseparable link between ideas and experience. All these figures in their different ways used the example of Saunderson to shore up the Lockean tenet that ideas are the product of experience – a tenet that underlay the Enlightenment's confidence that it was possible for humankind to move forward without the inherited burden of innate ideas.

When Saunderson died in 1739 there was no well-established Cambridge candidate for the Lucasian chair, which is itself an indication of the declining status of the professorship. The competition was therefore between the Huguenot refugee, Abraham De Moivre, and John Colson. For all De Moivre's mathematical achievement – for he was the author of pioneering works in the field of probability – he was too old and infirm to be a credible candidate. As the antiquary, William Cole, remarked, De Moivre was 'brought down to Cambridge and created an MA when he was almost as much fit for his coffin; he was a mere skeleton, nothing but skin and bone'. The post therefore went to John Colson, a decision which, according to one contemporary source,

III

Cambridge was later to regret: although he was 'a plain, honest man of great industry ... the university was much disappointed in their expectations of a Professor that was to give credit to it by his lectures'.[25]

As a pedagogue, Colson arrived with good credentials, coming from a family of mathematical instructors. His namesake (fl. 1671–1709) and probable Kinsman ran a mathematical school in Wapping, London, intended particularly for those destined for the sea. Hence Colson's involvement in producing an edition of Sturmy's *Mariner's Magazine* and his role as an examiner in the skills of navigation on behalf of Trinity House. John Colson senior also took a close interest in astronomy, contributing to the Royal Society an observation of the solar eclipse of 1 June 1676. Such connections led to his friendship with Halley and a rather tense relationship with Flamsteed, who took him to task for spreading the story that Newton's account of the moon's motions was based on Halley's observations. But Flamsteed evidently continued to hold a high opinion of Colson's pedagogical abilities since he referred on to him Jacob Bruce who came to London in 1698–9 as part of Peter the Great's entourage. Bruce studied under Colson for six months and returned to Russia with a collection of scientific works which included accounts of the Newtonian system. Subsequently Bruce was involved in the foundation of a Russian navigation school and the St Petersburg Academy of Sciences.[26]

Another mathematical member of the Colson clan was Nathaniel (fl. 1674), whose *Mariner's New Kalendar* included an advertisement for John Colson's school as a place where 'are taught these Mathematical Sciences, viz: Arithmetic, Geometry, Algebra, Trigonometry, Navigation, Astronomy, Dialling, Surveying, Gauging, Fortification and Gunnery, the Use of the Globes and other parts of the Mathematics'. Yet another mathematical Colson was Lancelot (fl. 1668–87), whose astronomical observations appear in the correspondence of Newton and Flamsteed.[27]

In contrast to these London-based Colsons, John Colson, the future Lucasian professor, was born at Lichfield, the son of the vicar-choral of the cathedral. He then matriculated at Christ Church, Oxford in 1699, though he left without taking a degree. At this time, Oxford was a centre for the dissemination of Newton's work, thanks to the energetic activity of the two Episcopalian exiles from Scotland, David Gregory and John Keill, and it is possible that they may have helped to stimulate further the young Colson's interest in the mathematical sciences. It is also possible that after Oxford Colson taught at his namesake's school in London. If he did so, this experience, together with his family connections, help to explain why he was appointed in 1709 as the first master of the new mathematical school founded at Rochester by Sir Joseph Williamson – a post that brought a house and a salary of one hundred pounds a year. Certainly, the goals of this new school – to educate boys 'towards the Mathematicks and all other Things which may fitt and encourage them for the Sea Service' – were very similar to those of the school run by John Colson senior in London.[28]

In a society offering few opportunities for mathematical careers, the Mastership was a much sought-after appointment. Among his rivals was the hapless Christopher Hussey, who was subsequently defeated by Saunderson in the election for the Lucasian chair. As a fellow of Trinity, Hussey was able to enlist Newton's aid through the good offices of the college's Master, the redoubtable Richard Bentley, who intended to 'get of him [Newton] a Character of Mr. Hussey'. Consequently, after interviewing Hussey Newton wrote a testimonial about his mathematical abilities, describing Hussey as 'very well qualified'. But, as Bentley had feared, the post went to Colson – something which, predictably, Bentley attributed to the fact 'y^t Interest rather than Merit will prevail in ye Election, & one Coleson has ye best friends'. Bentley believed that these 'friends' of Colson swayed the seventeen electors, who were local dignitaries such as the mayor, the recorder and the 'eldest Resident Prebendary' (who may have had some clerical connection with Colson's cathedral-based father or with Colson's godfather, the ecclesiastical historian, John Strype). Perhaps, too, the Royal Society had some say in Colson's appointment since the Rochester school was patronized by Sir Joseph Williamson, the former President of the Royal Society from 1677 to 1680. Certainly it did not hurt that Colson's Kinsman had been friends with the likes of Halley and Flamsteed. True to family tradition, once Colson was appointed he shared his good fortune with his kinsfolk: he appointed his sister, Mary, as his housekeeper and his brother, Francis, as an usher.[29]

Colson must at some stage have taken orders, for after 1724 he was able to augment his salary as Master of the school with the living of Chalk near Gravesend and when he died in 1760 he was rector of Lockington, Yorkshire. His theological interests are evident, too, in his translation (along with Samuel D'Oyley) from the French of the massive *An Historical, Critical, Geographical, Chronological, and Etymological Dictionary of the Holy Bible, in Three Volumes.*

Even before assuming his position at Rochester, Colson established his credentials at the Royal Society with a mathematical paper in the *Philosophical Transactions*, 'The universal resolution of cubic and biquadratic equations, as well analytical as geometrical and mechanical' (1707), and he was duly elected a fellow in 1713. The only other original mathematical papers that he published were an 'Account of negativo-affirmative arithmetick' (1726) and 'The construction and use of spherical maps' (1736), which also appeared in the *Philosophical Transactions*.

On 23 April 1728 Colson was formally entered as student at Emmanuel College, Cambridge, and shortly afterwards he was one of the seventy-one persons granted a degree when George II visited the university. At this stage Colson's links with Cambridge appear to have been rather tenuous. His place on the King's list probably derived from the good offices of the mathematician, Robert Smith, a fellow of Trinity and from 1716 Plumian professor of astronomy and experimental philosophy. Certainly, Smith had good connections at court, being master of mechanics to George II, as well as being mathematical preceptor to William, Duke of Cumberland. It was largely thanks to Smith's influence that Colson eventually obtained the Lucasian chair.

Colson continued to hold the post of Master at the Rochester school until a year after he was made Lucasian professor in 1739. By

Figure 26 Portrait of John Colson. From the collection of the Cambridge antiquarian, William Cole.

1737, however, he was resident in London, for it was there that the young David Garrick (who, like Colson, was from Lichfield) boarded with him while being instructed in 'the mathematics, philosophy and human learning'. On his trip to London Garrick was accompanied by

Samuel Johnson, who is said to have modelled his character Gelidus in the *Rambler* (no. 24) on Colson.[30]

It was in this period that Colson published the work for which he is chiefly remembered and which probably did much to secure the Lucasian chair: *Sir Isaac Newton's Method of Fluxions; Translated from the Author's Latin Original not yet Made Publick. To Which is Subjoined a Perpetual Comment on the Whole Work* (London, 1736 and 1737). Its dedication to William Jones made apparent Colson's debt to the mathematician, friend of Newton, and vice-president of the Royal Society: he profusely thanked Jones for 'the free access you have always allow'd me, to your Copious Collection of whatever is choice and excellent in the Mathematicks'. The gratitude was merited, since Colson's edition was based on Jones's transcript of Newton's tract.[31] The work was, however, considerably more than a simple translation, for Colson added to it some two hundred pages of explanatory notes. He also endeavoured to use the work to place the calculus of fluxions on a sounder mathematical (and, in particular, geometrical) footing.

The footings of the Newtonian fluxional calculus had been undermined not only from continental supporters of the Leibnizian differential calculus, but also from within the British Isles, notably from Bishop Berkeley in his *The Analyst* (1734) and *A Defence of Freethinking in Mathematics* (1735). For Berkeley, the analytical foundations of fluxional calculus, with its reliance on the concept of infinitesimals, was obscure. It was philosophically and even theologically corrupt: 'if we remove the veil and look underneath, if, laying aside the expressions', Berkeley grumbled, 'we set ourselves attentively to consider the things themselves which are supposed to be expressed or marked thereby, we shall discover much emptiness, darkness and confusion; nay, if I mistake not, direct impossibilities and contradictions'.[32]

Within Cambridge there were those who regarded such a critique as an attack on the whole Newtonian heritage with which the university was now identified. The sense of outrage in Cambridge – a

Figure 27 In Newton's *Method of Fluxions*, John Colson's 'perpetual commentary' responded to Berkeley's vehement attack on 'infidel mathematicians' by showing that the calculus was a rational and tangible means of expressing real motions in space. To explain the second derivative, Colson has the reader, presumably a Cambridge undergraduate, imagine an attempt to pot two ducks with one shot while the Greek masters look upon the action. Colson presumes that this '*sensibiles sensibilium velocitatum mensure*' becomes 'not only the object of the Understanding, and of the Imagination ... but even of sense too'.

university which had prided itself on welding the defence of Christianity to the promotion of Newtonian natural philosophy – was all the more violent since the bishop had linked infinitesimals and heterodoxy. This Cantabrigian outrage is encapsulated in the title of one major response to Berkeley: James Jurin's *Geometry no Friend to Infidelity: Or, a Defence of Sir Isaac Newton and the British Mathematicians* (1734), who wrote under the revealing pseudonym, *Philalethes Cantabrigensis*. A year later Jurin's defence was followed by his *The Minute Mathematician: Or, The Free-thinker no Justthinker*. Jurin, a former fellow of Trinity and, at Bentley's insistence, an early popularizer of Newton's work, had been, like Colson, a mathematical educator, having obtained a position at Christ's Hospital School with the support of Bentley before becoming master of Newcastle grammar school.

For Jurin, Berkeley's work was unpatriotic both because it was a critique of the 'British mathematicians' and because it undermined the mathematical achievement of Britain's great cultural hero, Isaac Newton. For, as he wrote in his first pamphlet, Berkeley's 'avowed design' is 'to lessen the reputation and authority of Sir *Isaac Newton* and his followers, by showing that they are not such *great masters of reason, as they are generally presumed to be'*. More dangerously, in Jurin's estimation, Berkeley implied that science could lead to infidelity when the energies of many within the Established Church and (particularly within Cambridge) had been devoted to demonstrating the opposite: 'I am afraid,' Jurin asserted, 'it would be a great stumbling block to men of weak heads, if they were made to believe, that the justest and closest reasoners were generally infidels.' There is even a whiff of anticlericalism in Jurin's response to Berkeley's assault, as when, in the second pamphlet, he contrasted the intellectual freedom of the mathematicians with the clergy who are bound by creeds. Discussing his hostility to Berkeley's clerical supporters, he remarked that, 'I am a Layman. If the Clergy obtain more power, I shall have less liberty: if they will have more wealth, I am one of those must pay to it.'[33]

By contrast, Colson was less hostile towards 'the very learned and ingenious Author of the Discourse call'd *The Analyst*'. He sought to counter Berkeley's critique by placing the fluxional calculus more securely on a geometrical rather than an analytical foundation. By doing so, he helped to provide that form of calculus with a more kinematical approach which corresponded more directly with the behaviour of a body in motion. As he himself put it, it was his aim to use moving diagrams to exhibit 'Fluxions and Fluents Geometrically and Mechanically...so as to make them the objects of Sense and ocular Demonstration'. Thanks to Colson, then, fluxional calculus thereafter had a much more overt geometrical foundation which served to distance it from the use of infinitesimals and thus distinguish it more clearly than in the past from the differential calculus. But, for Colson, such an enterprise was not an innovation but rather the recovery of the true Newtonian heritage: 'For it was now become highly necessary, that at last the great Sir *Isaac* himself should interpose, should produce his genuine Method of Fluxions, and bring it to the test of all impartial and considered Mathematicians'.[34]

When Colson was duly rewarded with the Lucasian professorship for his defence of the Newtonian mathematical heritage, he also gained with it the first Taylor lectureship in mathematics at Sidney Sussex College. As Lucasian professor, Colson, who appears to have had considerable facility with languages, chiefly contributed to the scholarly world by publishing scientific translations – for his lectures made little impact on the university. But, as we have seen, the need for such lectures had greatly diminished now that Newtonian natural philosophy and mathematics had become part of the staple of college instruction. Fittingly, Colson's first professorial publication paid tribute to Saunderson, his predecessor as Lucasian professor, for prefixed to the posthumous 1740 edition of Saunderson's *Elements of Algebra* was Colson's dissertation on 'Dr Saunderson's palpable arithmetic decypherd'. Evidently, too, his good opinion of Saunderson was reciprocated, since in his *Algebra* Saunderson refers to 'The learned Mr. John Colson, a gentleman whose great genius and known abilities in these sciences I shall always have in the highest admiration and esteem'.[35]

In the often parochial world of English and, *a fortiori*, Cambridge science, Colson's translations exposed students to continental natural philosophy. For, in 1744, he published a translation from the Latin of the great Dutch experimentalist and Newtonian popularizer, Peter van Musschenbroek's *The Elements of Natural Philosophy Chiefly Intended for the Use of Students in Universities* (though with a preface which, in good English Newtonian style, dissented from the Dutchman's reliance on Leibniz). This was followed in 1752 by a translation from the French of the *Lectures in Experimental Philosophy* by Abbé Nollet, who was particularly well-known for his work on electricity. This in some ways was a brave publication in Newtonian Cambridge since Nollet, a thorough-going experimentalist, was agnostic in his allegiance to any particular system of natural philosophy. Hence Nollet's claim in the preface to the work that 'I do not list myself here, under the standard of any of the celebrated philosophers. 'Tis not the Physicks of *Des Cartes*, nor of *Newton*, nor of *Leibniz*, that I propose particularly to follow . . . I take nothing upon their bare Word, except what I find to be confirmed by Experience.'[36]

In old age Colson even set to work to learn Italian to translate the mathematical work of Donna Maria Agnesi, that rare phenomenon in the eighteenth century: a female professor of mathematics (at the university of Bologna). One of his intentions in doing so was to make the study of mathematics more accessible to women, a goal reflected in the title of his manuscript draft, *The Plan of the Lady's System of Analyticks*, and the detailed exposition of contents it included. Such a goal of ensuring that English women did not fall behind those of Italy in the study of mathematics was an unusual one in male-dominated Cambridge and perhaps reflects the fact that Colson brought to the post of Lucasian professor a long family tradition of mathematical instruction which could go beyond the immediate needs of the university. More predictably, it was also a work which could appeal to Colson as a demonstration of the value of fluxions.[37] However, he died before it could appear and it was eventually published posthumously in 1801 by the Reverend John Hellins, FRS (the editor of Saunderson's abridged *Algebra*) at the expense of Baron Maseres.

Such was the contribution of Colson to the consolidation of natural philosophy and mathematics at Cambridge. But, in many ways, the task had already been achieved by the time of Colson's election and the role of the Lucasian professor had become somewhat superfluous as such studies had become institutionalized in the college teaching and university examinations. Much of the period when Colson held office (1739–60) was one when the great initial wave of enthusiasm for Newton's work had subsided into a dutiful orthodoxy as Cambridge drilled its students in the accepted canon of Newtonianism. It is significant, for example, as Geoffrey Cantor points out, that only two new textbooks of natural philosophy were published at Cambridge between 1750 and 1780 as compared with eleven in the period 1720 and 1750.[38]

When Colson died in 1760, then, the role of the Lucasian professor had shrunk from one of advocacy and catalysis for mathematical (and especially Newtonian) studies to that of echoing the instruction that undergraduates routinely received from the tutors. The revival of the chair as an active focus of mathematical activity required a more pluralist intellectual climate in which the confidence that Isaac Newton, Cambridge's most famous son, had revealed all – or very nearly all – was challenged. It also required a re-evaluation of the role of professors whereby their role as educators could be better integrated into the round of undergraduate teaching and their role in promoting their subject through research could receive greater recognition. Such changes, however, were to be the work of the next century when the Lucasian chair and the university as a whole were energized by the reforming impulse of the age.

Notes

1 Anecdote of Martin Folkes in King's College, Cambridge, Keynes MS 130.
2 Anecdote of John Durham in King's College, Cambridge, Keynes MS 130.

3 Anecdote of Humphrey Newton in King's College, Cambridge, Keynes MS 130.

4 Samuel Parr, *A Spital Sermon ... to Which are Added Notes* (London, 1801), p. 121.

5 Newton to Bentley, 20 October 1709, *The Correspondence of Isaac Newton*, H. W. Turnbull, J. F. Scott, A. R. Hall and Laura Tilling (eds) (7 vols, Cambridge, 1959–77), vol. vii, p. 479.

6 J. Tattersall, 'Nicholas Saunderson: "The blind Lucasian professor"', *Historia Mathematica*, 19 (1992), 356–70 on 366.

7 Saunderson's early arithmetic in Add. MS 4223, f. 194 (biographical anecdotes from the Birch collection); maths in Sheffield in J. W. Ashley Smith, *The Birth of Modern Education: The Contribution of the Dissenting Academies 1660–1800* (London, 1954), p. 109; Saunderson's logic in Add. MS 4223, f. 194v.

8 G. Dyer, *The Privileges of the University of Cambridge* (London, 1824), vol. ii, pp. 142–3.

9 Saunderson's debauchery cited in Tattersall, 'Nicholas Saunderson', p. 362; John Saunderson's work in J. Peile, *Biographical Register of Christ's College, 1505–1905* (2 vols, Cambridge, 1910–13), vol. ii, p. 243.

10 Whiston's support in *Gentleman's Magazine*, 24 (1754), 373. Davies' reminiscence in N. Saunderson, *The Elements of Algebra, in Ten Books: To Which is Prefixed, an Account of the Author's Life and Character, Collected from his Oldest and most Intimate Acquaintance* (Cambridge, 1740), p. vi.

11 Saunderson on gentleman commoners in *Gentleman's Magazine*, 24 (1754), 373; Walpole's study in 'The life and works of Thomas Gray', *Quarterly Review*, 94 (1853–4), 1–48 on 8.

12 C. Wordsworth, *Scholæ Academicæ* (Cambridge, 1877) p. 66.

13 Saunderson's lectures are held in Cambridge University Library, Add. 589. Other copies are Add. 6312, 2977. (There is also a set of student notes from his lectures by John Mickleburgh (professor of chemistry, 1718–56) in Gonville and Caius College Library, MS 723/74a.) Outside Cambridge there are copies at Bodleian, MS Rigaud 3-4, British Library, MS Add. Eg. 834 and University College, London, MS Add. 243; Burke on colour in Tattersall, 'Nicholas Saunderson', 363.

14 Saunderson's response to Greene in Newton, *Correspondence*, vol. v, p. 247, Saunderson to Jones, 16 March [1711/12]; Newton on matter in H. Thayer, *Newton's Philosophy of Nature. Selections from his Writings*

(New York, 1974), p. 175; Cotes on Greene in Newton, *Correspondence*, vol. V, p. 166; Greene on Newton in R. Greene, *The Principles of the Philosophy of the Expansive and Contractive Forces* (Cambridge, 1727), preface, pp. 17, 41–7.

15 Saunderson to Jones, 16 March [1711/12], Newton, *Correspondence*, vol. v, p. 247; Saunderson to Jones, 4 February 1713–4, S. Rigaud, *Correspondence of Scientific Men of the Seventeenth Century* (2 vols, Oxford, 1841), vol. i, p. *264.

16 W. S. Lewis (ed.) *The Correspondence of Horace Walpole*, vols, 13–14 (New Haven, CT., 1948), p. 94 and A. Sherbo, *Christopher Smart: Scholar of the University* (East Lansing, MI., 1967), p. 27.

17 N. Saunderson, *Select Parts of Saunderson's Elements of Algebra for the Use of Students at the Universities*, 5th edn. Revised and corrected by the Rev. John Hellins of Trinity College, Cambridge, p. iii.

18 G. Hiscock, *David Gregory, Isaac Newton and their Circle* (Oxford, 1937), p. 42; Saunderson's approach to algebra in N. Guicciardini, *The Development of Newtonian Calculus in Britain 1700–1800* (Cambridge, 1989), p. 24; Saunderson, *Elements of Algebra*, vol. ii, p. 740.

19 Guicciardini, *The Development*, p. 24.

20 *Lettre sur les Aveugles, à l'Usage de ceux qui Voient* in *Oeuvres Philosophiques de Mr. D[iderot]* (Amsterdam, 1772), vol. ii, pp. 73, 29–36. On Diderot's inaccuracies see M. Kessler, 'A puzzle concerning Diderot's presentation of Saunderson's Palpable arithmetic', *Diderot Studies*, 1981 (20), 159–73; Saunderson, *Elements of Algebra*, pp. xx–xxv.

21 Diderot, *Lettre*, p. 46; Saunderson, *Elements of Algebra*, p. xiii.

22 Diderot, *Lettre*, 48–55. Translation from L. Crocker (ed.), *Diderot's Selected Writings* (New York, 1966), 20–4.

23 E. Burke, *A Philosophical Enquiry into the Origin of our Ideas of the Sublime and Beautiful*, J. T. Boulton (ed.) (London, 1958), p. 169.

24 J. Boswell, *Life of Johnson* cited in K. MacLean, *John Locke and English Literature of the Eighteenth Century* (New Haven, 1936), p. 107.

25 Cole on De Moivre, BL, Add. 5866, 200. My thanks to Dr A. Capern for transcriptions of this. Colson's achievements in *History and Antiquities of Rochester* (1772), vol. viii, p. 220 cited in BL, Add. 5866, f. 3.

26 Colson on navigation in E. G. R. Taylor, *The Mathematical Practitioners of Tudor and Stuart England* (Cambridge, 1954), p. 264; Flamsteed to Colson, 10 October 1698, Newton, *Correspondence*, vol. iv, pp. 284–5;

Jacob Bruce in V. Boss, 'Russia's first Newtonian', *Archives Internationales de l'Histore des Sciences*, 15 (1962), 233–65.

27 John Colson's curriculum in Taylor, *Mathematical Practitioners*, p. 264; Flamsteed to Halley, 17 February 1680/1; Newton to Flamsteed, 16 April 1681, [12 January 1684/5], Newton *Correspondence*, vol. II, pp. 339–40, 365.

28 R. V. and J. Wallis, *Biobibliography of British Mathematics and its Applications. Part II, 1701–1760* (Newcastle-upon-Tyne, 1986), p. 29; Colson's school in Taylor, *Mathematical Practitioners*, 530.

29 Bentley on Hussey in Bentley to Cotes [21 May 1709], J. Edleston, *Correspondence of Sir Isaac Newton and Professor Cotes* (London, 1850), p. 1; Newton on Hussey in Richard Westfall, *Never at Rest. A Biography of Isaac Newton* (Cambridge, 1980), p. 700. Westfall surmises that the interview was for a post at the Christ's Hospital School but Cotes in a letter to Newton refers to the occasion on which Hussey 'formerly waited upon You for Your recommendation of him to a Mathematical School at Rochester'. Cotes to Newton 25 October 1711, Newton, *Correspondence*, vol. V, p. 202; Bentley on election in Edleston, *Correspondence*, p. 2; Colson's family in Add. 5866, f. 1.

30 T. Davies, *Memoirs of the Life of David Garrick* (2 vols, Dublin, 1780), vol. I, p. 9; Samuel Johnson on Colson in Taylor, *Mathematical Practitioners*, 115.

31 J. Colson, *Sir Isaac Newton's Method of Fluxions* (London, 1736–7), p. iv; D. Whiteside (ed.), *The Mathematical Papers of Isaac Newton* (Cambridge, 1967–81), vol. VIII, p. xxiii.

32 H. Pycior, *Symbols, Impossible Numbers, and Geometric Entanglements. British Algebra Through the Commentaries on Newton's Universal Arithmetick* (Cambridge, 1997), p. 233.

33 [J. Jurin], *Geometry no Friend to Infidelity* (London, 1734), pp. 9–10; [J. Jurin], *The Minute Mathematician* (London, 1735), pp. 2–3, 9.

34 Colson on Berkeley in Pycior, *Symbols*, p. 301; Colson, *Method of Fluxions*, vol. X, p. 234: cited in Guicciardini, *The Development*, p. 57; N. Guicciardini, 'Flowing ducks and vanishing quantities' in S. Rossi (ed.), *Science and Imagination in XVIII-century British Culture* (Milan, 1987), pp. 231–5 on 231.

35 Saunderson, *Elements of Algebra*, vol. II, p. 720.

36 Jean Antoine Nollet, *Lectures in Experimental Philosophy*, translated by
   J. Colson, p. xvi, (London, 1752).
37 Colson's draft manuscript in Cambridge University Library, MSS
   Ee.2.36-38; W. Johnson, 'Contributions to improving the teaching of
   calculus in early nineteenth century England', *Notes and Records of the
   Royal Society*, 49 (1995), 93–103 on 100.
38 G. Cantor, *Optics after Newton. Theories of Light in Britain and Ireland,
   1704–1840* (Manchester, 1983), p. 5.

Illustrations

IV

SCIENCE, RELIGION AND THE FOUNDATIONS OF MORALITY IN ENLIGHTENMENT BRITAIN*

In an age which has long been schooled in the need to avoid the intellectual promiscuity of confusing 'is' and 'ought' it is natural for us to regard science and ethics as inhabiting distinctly different realms. It is difficult, then, to recapture the view that found its clearest statement in the work of Aristotle and his numerous Western scholastic commentators up to the end of the seventeenth century that all fields of philosophical endeavour - whether metaphysics, natural philosophy or ethics - were indissolubly linked. That, in Aristotelian language, all phenomena, whether that of the human or the inanimate world, were informed by final causes which ultimately served to bring the True and the Good into fundamental harmony. But it was one of the avowed aims of the Scientific Revolution which overthrew the Aristotelian-based cosmology to cut itself loose from such final causes. For they were viewed as obstacles in the way of a properly focussed exploration of the way in which Nature actually was rather than the way it ought to be. Thus for Francis Bacon 'The Inquisition of Final Causes is barren and like a virgin consecrated to God produces Nothing'.[1]

But the habits of mind which sought some fundamental unity in all fields of human understanding died hard and the impulse to link one's understanding of the natural world with one's conception of the Good was to continue into the eighteenth century - and, indeed, shows some signs of coming to life again in our own times.[2] For

---

* My thanks to Alan Tapper, my colleague, Jim Franklin, and my brother, Robert Gascoigne, for their comments on an earlier draft of this paper
[1] J G Crowther, *Francis Bacon. The first statesman of science* (London, 1960), 70.
[2] As for example in recent writings about systems of virtue ethics, ecological theorising about the relations between humanity and nature or in the revival of forms of natural theology. Philosophers have also become increasingly wary of attributing apodictic factual status to any

eighteenth-century Britain the achievements of modern science were encapsulated in the work of Newton and so the search to find some bridge between science and ethics was largely linked to the larger issue of the reception of Newton's work. The object of this paper is, then, to examine the ways in which a number of notable eighteenth-century moralists attempted to enlist the vast prestige of Newton in constructing systems of ethics which, they argued, could marry the Good and the True.

I

In assessing Newton's philosophical impact it is natural to begin with Newton's virtual philosophical spokesman, Samuel Clarke (1675-1729), who fought the good fight on behalf of the Master when dealing with such foreign philosophical nuisances as Leibniz.[3] In many ways Clarke the polymath - theologian, natural philosopher and classicist - kept alive in the increasingly uncongenial intellectual environment of the eighteenth century the more traditional view of knowledge as a seamless web which could not be readily hacked up into separate disciplines. This concern with maintaining the unity of knowledge revealed itself most obviously in his influential endeavours to demonstrate the consonance between Newtonianism and Christianity rightly understood. One aspect of this larger intellectual mission was the attempt to yoke together a Christian-based ethics with the structures of thought he derived from his close study of Newtonianism.

The central philosophical problem which Clarke faced in undertaking such a mission was to reconcile the tendency of the mechanical philosophy, even in its Newtonian guise, towards philosophical determinism with some conception of human liberty without which morality or, at least, Christian morality, was impossible.[4] Appropriately one of the two propositions that he

---

field - for a discussion of this latter issue see Peter Railton, 'Facts and values', *Philosophical Topics*, xiv (1986), 6.
[3] In this section I am drawing on some material from my forthcoming entry on Samuel Clarke for the *New Dictionary of National Biography*.
[4] John Gay, 'Matter and freedom in the thought of Samuel Clarke', *Journal of the History of Ideas*, xxiv (1963), 85-105.

defended when he took his Doctor of Divinity degree at Cambridge was 'Without the Liberty of Humane Actions there can be No Religion'. The other indicated a similarly central strand in his overall *oeuvre:* 'No Article of the Christian Faith, delivered in the Holy Scriptures, is Disagreeable to Right Reason'.[5] The defence of liberty was the thread that ran through all his philosophical works just as the defence of political and religious liberty - which he equated with the Whig Protestant Hanoverian establishment - was a central preoccupation of much of his theological writings.

'This Liberty, or Moral Agency', as his admiring first biographer, his fellow Low Churchman, Benjamin Hoadly, wrote 'was a Darling Point to Him'.[6] He first addressed it directly in his first set of Boyle lectures which was revealingly entitled - *A demonstration of the being and attributes of God. More particularly in answer to Mr Hobbes, Spinoza and their followers. Wherein the notion of liberty is stated, and the possibility and certainty of it proved, in opposition to necessity and fate* (1704) - a work which was intended to deal with the foundations of natural religion by providing *a priori* philosophical reasoning to substantiate a belief in a benevolent Deity.

As historians such as Metzger[7] and Jacob[8] have shown, it was a work which played a major part in the early dissemination of some of the chief conclusions of the *Principia,* for Clarke's *Demonstration of the being and attributes of God* bears strong marks of his training in Newtonian natural philosophy. In the first place, Clarke, as befits a student of the *Principia mathematica,* sought to apply to metaphysics a mathematical, *a priori,* style of reasoning. As he wrote in the preface, the argument of the work is 'as near as Mathematical as the Nature of such a Discourse would allow'(p.i). Along with his mathematical method, Clarke also exhibited a more direct debt to Newtonianism with his attempt to

---

[5]   Benjamin Hoadly ed., *The works of Samuel Clarke* (4 vols, London, 1738), I, xvi.
[6]   *Ibid.,* I, ix.
[7]   Hélène Metzger, *Attraction universelle et religion naturelle chez quelques commentateurs anglais de Newton* (Paris, 1938).
8   M C Jacob, *The Newtonians and the English Revolution 1689-1720* (Hassocks, Sussex, 1976).

demonstrate that Newtonian science served to demonstrate the existence of an Intelligent Designer who both created and sustains the universe. He pointed, for example, to the demonstration in the *Principia* of the *'inexpressible Nicety'* of the balance between the annual motion of the planets and the force of gravitation towards the sun as an instance of a directing Providence (pp.229-30).

As well as providing an instance of the shaping hand of the Creator at work in the universe the Newtonian conception of gravitation also served as part of his attack on philosophical systems which weakened the concept of human liberty which, as we have seen, underlay his conception of the whole nature of Christian morality. Thus he attacked the determinism which lay at the base of Spinoza's claim that matter was a necessary being without need of a Creator by arguing that 'if Gravitation be an Universal Quality or Affection of *All Matter*; then there is a Vacuum; (as is abundantly demonstrated by Mr. *Newton*:) and if there be a Vacuum, then Matter is not a Necessary Being' (p.49).

He returned to such issues a few years later in his controversy with the Deist, Anthony Collins, again invoking Newton in defence of his view that matter alone could not explain all aspects of Nature. For had not Newton demonstrated that gravity had no material cause thus underlining the folly of those *'Materialists* [who] would undertake to explain the phaenomena of Nature *Mechanically,* by the mere Powers of Matter and Motion'.[9] Such an argument supported his more general thesis that matter alone could not explain such phenomena as human thought or, indeed, human free will and the moral choice that it made possible. Hence his attack on Collins's materialistic views of human nature as being 'totally destructive of Religion'.[10]

Having established the philosophical bases of human liberty, and hence of moral choice, in his first set of Boyle lectures in the second - *A discourse concerning the unchangeable obligation of natural religion, and the truth and certainty of the Christian revelation* (1705) - Clarke proceeded to examine in more detail the nature of morality and its relations with Christian revelation. True

---

[9] Hoadly ed., *The works of Samuel Clarke*, III, 849.
[10] *Ibid.,* 906.

to the traditions of natural law, Clarke was emphatic that morality accorded with the ultimate ordering of Creation rather than being the result of the arbitrary decree of God. Such a rationalistic conception of morality in some ways lent comfort to Clarke's Deistic opponents with their view that morality could be formulated without recourse to a notion of Revelation but, needless to say, Clarke dissented from such a position. Human beings, he urged, were too swayed by their passions or simply too preoccupied to arrive at an acceptable conception of morality through their own reasoning. Moreover, Revelation provided an essential element lacking in any system of ethics derived from reason: the assurance of reward or punishment in the next life.

In constructing the case for the truth of Christian Revelation, Clarke acknowledged that, in contrast to the arguments about the bases of natural religion in his first set of Boyle lectures, 'the same *demonstrative* force of reasoning, and even *Mathematical* certainty, which in the main Argument was there easy to be obtained, ought not here to be expected; but that such *moral* Evidence, or mixt Proofs from Circumstances and Testimony' (p.14). Clarke's preference, then, was plainly to defend both Christianity and morality by the deductive methods which he regarded as one of the glories of Newton's system. However, where necessary, he would also employ the empirical evidence which he derived from history and human experience. So, too, did Newton combine mathematical deduction with experimental data in the *Principia* while in the *Opticks* (which Clarke translated into Latin) he almost entirely based his work on an *a posteriori* resort to experience.

But, where possible, Clarke took the high ground of deductive certainty. Thus in his second Boyle lectures he closely linked the foundations of Christian morality with the basic premises about the nature of the Deity which he had established by *a priori* methods in the first Boyle lectures arguing 'That the Practical Duties which the Christian Religion injoyns, are all such, as are most agreeable to our natural Notions of God' (p.13). Indeed, he went so far as to draw a parallel between the processes of moral and mathematical reasoning arguing that 'The Reason which Obliges every Man in Practise, so to deal always with another, as he would reasonably expect that Other should in like circumstances deal with Him; is the very same, as that which forces him in Speculation to affirm that if one Line or Number be equal to another, That other is reciprocally equal to it' (p.86).

If, as Clarke kept insisting, moral principles were inseparably intertwined with the overall ordering of Nature it followed that morality was something that was intrinsic to God's Creation rather than being imposed upon it by God. True to the anti-Calvinist tenor of his thought Clarke was strongly opposed to any voluntarist conception of morality as something imposed by the Will of God rather than being rooted in God's overall ordering of things. In his sermon 'Of the Goodness of God', he urged that 'Nothing therefore can be more absurd, than the doctrine which has sometimes been advanced; that *Goodness* in *God*, is not the same thing as Goodness *in Men*'.[11] Significantly, as a child this youthful prodigy was much exercised in wrestling with the proposition of whether or not God could lie.[12] For the mature Clarke the answer would be a ringing 'no' as evidenced by his argument in his second Boyle lectures 'that which is Holy and Good ... is not therefore Holy and Good, because it is commanded to be done; but is therefore commanded by God, because it is Holy and Good'. More simply he made the same point by declaring 'that Virtue and Vice are eternally and necessarily different' (pp.110, 116).

Clarke's conception of religion as embodying an eternal and unchanging morality implicit in the very laws of Nature was reinforced not only by his rejection of Calvinist voluntarianism but also his resolute Protestant anti-sacerdotalism. Religion for Clarke was about morality rather than sacred rites. As he urged in his sermon on 'How to judge of Moral Actions': 'The *End* and *Design* of all Religion... *the ultimate View and Fundamental Intention of all religious Truths,* implanted in men either by Nature or Teaching; is the Practice of Virtue.'[13] In Clarke's view, then, one of the marks of the divine origin of Christianity was that it had strengthened the significance of basic moral tenets by downplaying

---

[11] Hoadly ed., *The works of Samuel Clarke*, I, 91.

[12] J.P.Ferguson, *An eighteenth century heretic. Dr. Samuel Clarke* (Kineton, Warwick, 1976), 2.

[13] Hoadly ed., *The works of Samuel Clarke*, I, 250.

the importance of the ceremonial aspects of religion. Thus in one of his sermons he proclaimed that 'All Positive and ritual injunctions whatsoever, can be but subordinate to the Practice of moral Virtues.'[14]

Like true natural philosophy, then, the path of true religion lay in discerning and following the laws which God had encoded into the very foundations of the universe. The true Newtonian Christian derived his religion from the Word of God as proclaimed in Scripture and the Mind of God as revealed in the ordering of the moral and natural world. Ironically, however, such a position was, as his Cambridge Whig colleague, Daniel Waterland, pointed out one likely to give intellectual sustenance to their Deistic counterparts. For if the positive duties of religion (such as the obligation to attend Divine Service) were as inferior as Clarke suggested to moral obligations what necessity was there for a religious establishment and, with it, a clerical estate enjoying the privileges of which Clarke himself was a beneficiary? Could not one simply derive moral principles from a study of the natural and human world without recourse to either Scripture or the teaching of the Church?

Waterland was prompted to make these very pertinent points in response to Clarke's posthumously published *An exposition of the church-catechism* (1729) in which Clarke returned to the theme that had run throughout all his theological writings: the subordination of church-imposed 'positive' duties to the moral duties ordained by the will of God as made manifest in Scripture and natural law. 'Tis the peculiar Excellency and Advantage of the Christian Religion', he wrote in this work, 'that it is not burdened, as the *Jewish* was, with a multitude of external *Rites* and *Ceremonies*.'[15] For Waterland such a position challenged the clergy's claim to a divinely-sanctioned authority. By thus 'lay[ing] it down for a rule and a principle, that *positive* precepts or duties are never to be compared with *moral*', argued Waterland, Clarke had unintentionally given support to the rise of infidelity for 'Deism has sprung

---

[14] *Ibid.*, I, 708.
[15] Hoadly ed., *The works of Samuel Clarke*, III, 708.

up out of the same doctrine about *moral* and *positive* institutions'.[16]

But Clarke was dead when Waterland raised such objections and, in any case, Clarke would probably have been unmoved by objections to positions which lay at the very foundations about his views about the nature of religion and morality. In Clarke's scheme of things it was axiomatic that true religion was about moral rather than positive duties and confirmation for such a position could be found in the fact that God Himself had made known His Will through the laws of nature which governed both the realms of natural philosophy and morality. It was a parallel he spelled out in his revealingly entitled sermon, 'The Practice of Morality leads to the Practice of the Gospel', which took as its point of departure the familiar Clarkian tenet that 'The *Duty* and *Happiness* of natural Creatures, is the Practice of Righteousness and true Virtue, founded upon a Belief of the Being and Government of God'. By doing so human beings were partners in the divine plan 'by preserving the Harmony of the *Moral* World, in like manner as the Wisdom of his Government over *the Natural and Material* World is shown forth in the regularity of all its Motions'. By contrast, 'By Sin, Moral Agents oppose and bring disorder into this kingdom of God'[17]. Moreover, as he argued in another sermon: 'The Moral World, is of infinitely greater Importance: It is That, for the sake of which the material World was created; and without which, this beautiful and stupendous Fabrick of the inanimate Universe is Nothing.'[18]

For Clarke, then, something of the traditional parallel between the microcosm of man and the macrocosm of the universe had survived the corrosive effects of the Newtonian world-view in merging heavenly and terrestrial motion. Humankind could discern the Divine Will not only in the laws of the universe as a whole but also in the moral laws which ought to shape their ends. If human beings did not conform to such moral laws they were subverting the Divine economy not only within their immediate social sphere but

---

[16] W. Van Mildert ed., *The works of Daniel Waterland* (11 vols. in 12, Oxford, 1823), IV, 99.

[17] Hoadly ed., *The works of Samuel Clarke*, II, 231.

[18] *Ibid.*, II, 28.

also in the universe more generally. Both natural and moral laws could be deduced from the evident will of God as revealed in both Scripture and the Book of Nature. The advent of the Newtonian philosophy was, in a fundamental sense, part of God's Revelation by making plain the Divine Presence in shaping the order and regularity of the universe thus exemplifying Clarke's basic premise that the moral and natural laws worked in harmony. To adapt the peroration of *The origin of species* of Charles Darwin who was reared on the traditions of natural theology that found their philosophical basis in Clarke: there is a grandeur in this view of nature with its confidence that whilst this planet has gone cycling on according to the fixed law of gravity it also conformed to larger designs which were embodied in the laws of morality. In short, Clarke still leaves the door open to some residual notion of final causes. So strong was the weight of this deeply rooted tradition of linking the Good and the True that it can also be found in other notable moralists of the English and Scottish Enlightenment even though they departed significantly from Clarke's attempt to deduce a system of Christian morality by the use of reason.

II

Not surprisingly, the problem of how to combine the Newtonian world-view with some conception of a universe that conformed to moral laws was a particular preoccupation of Newton's own university of Cambridge thanks in large part to the theological and scientific influence of Clarke. The general tenor of this continuing quest to link Newtonian natural philosophy and Christian ethics can be roughly and perhaps rather simplistically characterised by adopting two strategies: firstly, the Clarkian deductive methods which he had so admired in the *Principia* and, secondly, a more empirically based approach which was to draw more on the *Opticks* than the *Principia* and which also owed much to Lockean epistemology. The first approach meshed naturally with the natural law theorists such as Grotius and Pufendorf who were adapted to the needs of the Cambridge curriculum by figures such as Thomas

Rutherforth.[19] In doing so, however, such Cambridge moralists, like their Continental models, widened the scope of their discussion to include an account of how such an underlying set of natural laws manifested itself in the complexities of social and political reality. This was an exercise that involved a good deal of appeal to *a posteriori* reasoning from experience - an indication that the deductive and the experimental approaches were as often as intertwined in the study of ethics as they were in the study of nature. The second approach, which was more overtly linked to experience and experiment, took a more original path as instanced by the work of David Hartley (fellow of Jesus College, Cambridge, 1727-30) in attempting to move the basis of discussion away from the cosmic stage of Clarke and to focus more closely on the way in which the human mind worked.

Though Hartley painted on a smaller canvas than Clarke his intention was the same: to demonstrate the existence of a Divine Intelligence through revealing the laws by which it operated. Whereas Clarke attempted to discern such laws in the movements of the heavens and the moral ordering of human society Hartley took as his terrain the operation of the human mind. While Clarke largely looked to the principle of gravitation to justify the ways of God to Man, Hartley turned to the concept of association to reveal the order and simplicity of the way in which the human mind worked and hence its status as a divine construct.[20]

In constructing his theory of association Hartley was plainly indebted to Locke's theory of the way in which human mind built up its ideas from the raw material provided by sensation and experience. But in helping to provide the physical mechanism by which Hartley linked external experiences with their effect on the human mind he turned to Newton's exposition of the effect of

---

[19] John Gascoigne, *Cambridge in the age of the Enlightenment. Science, religion and politics from the Restoration to the French Revolution* (Cambridge, 1989), 127.
[20] Barbara Oberg, 'David Hartley and the association of ideas', *Journal of the History of Ideas*, xxxvii (1976), 441-54 and R Oldfield and Kathleen Oldfield, 'Hartley's "Observations on Man"', *Annals of Science*, i (1951), 371-81.

vibrations on the human eye as outlined in Query 23 of the *Opticks*.[21]

A lesser known influence was the obscure Cambridge moralist, John Gay (1600-1745), who, like Hartley, linked associationism in psychology with utilitarianism in morals. As Hartley acknowledged it was Gay's *Preliminary discourse concerning the fundamental principles of virtue and morality* (1731) which 'put me upon considering the Power of Association'.[22] For both Gay and Hartley the link between associationism and what was later called utilitarianism was, as Gay put it, that 'as pleasure and pain are not indifferent to him [the generic man], nor out of his power, he pursues the former and avoids the latter' and, consequently, 'that which he apprehends to be apt to produce pleasure, he calls *good*'.[23]

It followed that actions could be categorised by the extent to which they contributed to the sum of human happiness thus laying the foundation for a system of utilitarian ethics. But though such a system of ethics eventually became most closely associated with the atheistic Bentham, for Gay and Hartley, as for their Cambridge disciple, William Paley, it was possible to place such a moral code in a theistic and indeed Christian framework. For Hartley and Gay the beauty of their *schema* was that it illustrated that the laws of natural philosophy, as instanced in the theory of association and its physical basis in vibration, could be shown to be closely intermeshed with a system of morality [24] - that, as Clarke had attempted to argue in a different manner, the True and the Good could be shown to be two sides of the same divinely-ordained coin. Utilitarianism provided a means by which to discern the direction in which God, the ultimate Good, wanted human beings to go.

[21] H Guerlac, *Essays and papers in the history of modern science* (Baltimore, 1977), 161, and C U M Smith, 'David Hartley's Newtonian neuropsychology', *Journal of the History of the Behavioural Sciences,* xxiii (1987), 123-36.
[22] M Webb, 'A new history of Hartley's "Observations on Man"', *Journal of the History of the Behavioural Sciences,* xxiv (1988), 203.
[23] E A Burtt ed., *The English philosophers from Bacon to Mill* (New York, 1939), 766-67.
[24] R Marsh, 'The second part of Hartley's system', *Journal of the History of Ideas,* xx (1959), 264-73.

Hartley even hinted that the theory of association and its concomitant utilitarian morality provided a path back to humankind's first innocence. 'Association', he wrote in his magnum opus, *Observations on man* (1749), 'has a Tendency to reduce the State of those who have eaten of the Tree of Knowledge of Good and Evil, back again to a paradisiacal one' (p.83).

For association, with its progression from simple reactions to complex ones, provided a means of moral education as we are led from a childlike association of pleasure from immediate gratification to associate pleasure with altruistic actions. As the devout Hartley wrote: 'the great Business and Purport of the present Life, [is] the transformation of sensuality into Spirituality, by associating the sensible Pleasure, and their Traces, with proper foreign Objects' (p.214). By such means 'the sensible Pleasures and Pains must be transferred by Association more and more every Day, upon things that afford neither sensible Pleasure nor sensible Pain in themselves, and so beget the intellectual Pleasure and Pains' (p.82). In short, concluded Hartley using the language of Aristotelian causality, 'all the Things which have evident final Cause, are plainly brought about by mechanical means' (p.338).

III

In their different ways, then, both Clarke's system of Newtonian theology and Gay's and Hartley's theory of association sought to give an ethical and, indeed, Christian colouring to an essentially mechanistic conception of (in Clarke's case) the workings of the universe or (in Hartley's case) the working of the human mind. This same impulse to draw together the realms of natural and moral philosophy can be found, too, in the Scottish universities and, above all, at Glasgow, the home *par excellence* of moral philosophy in the first half of the eighteenth century. Indeed, it was to be one of the most striking features of the Scottish universities that, despite the dissolution of the traditional scholastic curriculum at the end of the seventeenth century, they maintained, in the changed intellectual conditions of the Enlightenment, something of that philosophical preponderance which had been a traditional feature of European universities since the High Middle Ages.[25] One indication of the continuing philosophical vitality of the Scottish universities was the impulse to attempt to link the different branches of philosophy - most notably natural and moral philosophy - which elsewhere were allowed more and more to drift into separate intellectual abodes. As in England, the institution of an established church, with its close links with the universities, provided an intellectual environment in which there was a natural tendency to emphasise the consonance between Christianity and the fruits of the Enlightenment. In particular, it was the task of a professor of moral philosophy in the Scottish universities to defend the rationality of the ethical teachings of Christianity.[26] However, in Scotland where clerical control of the universities was weaker than in England this Christianising tendency was sometimes more lightly worn or, as in the case of Adam Smith, maintained only in a vestigial Deistic form.

Of the Scottish moralists by far the most influential was Francis Hutcheson, professor of moral philosophy at Glasgow from 1729 to 1746, who, by arguing for a fundamental moral sense, put British moral philosophy on a new footing. He also largely established the parameters within which the ethical debates of the Scottish Enlightenment were conducted.[27] True, Hutcheson owed some debt to Shaftesbury with his conception of an innate moral sense but Hutcheson refined and developed such a view arguing for an ethical consciousness which transcended the self-interest of the agent.[28] Thus Hutcheson talks of 'that *moral administration* we

---

[25] J Veitch, 'Philosophy in the Scottish universities', *Mind*, ii (1877), 74-91, 207-34.

[26] Alasdair MacIntyre, *Whose justice? Whose rationality?* (Notre Dame, 1988), 248-251.

[27] Richard Sher, 'Professors of virtue: the social history of the Edinburgh moral philosophy chair in the eighteenth century' in M A Stewart ed., *Studies in the philosophy of the Scottish Enlightenment* (Oxford, 1990), 96, and T Campbell, 'Francis Hutcheson: "Father" of the Scottish Enlightenment' in R.H. Campbell and A.S.Skinner eds., *The origins and nature of the Scottish Enlightenment* (Edinburgh, 1982), 167-85.

[28] Stephen Darwall, *The British moralists and the internal 'ought'* (Cambridge, 1995), 208-9.

feel within ourselves, that structure of our souls destined to recommend all those kind and generous affections which resemble the moral perfections of God'.[29]

Though Hutcheson argued that 'We are not to imagine, that this moral Sense, more than the other Senses, supposes any innate ideas, Knowledge, or practical Proposition: We mean by it only a Determination of our Minds to receive amiable or disagreeable Ideas of Actions'[30] the tenor of his system of ethics departed from the tradition of moral philosophy deriving from the Lockean rejection of innate ideas. More markedly it diverged from the two other dominant moral philosophical paradigms which prevailed within Britain at the time: that of the hedonists represented by Hobbes and, more recently, by Mandeville - a view which Hutcheson dismissed as one that 'can never account for the principal actions of human life'[31] - and that of the natural law theorists with their belief that God had impressed a moral order on the universe which humankind could discover. Along with his rejection of the natural law theorists went Hutcheson's dismissal of Clarke's rationalist attempt to deduce moral laws.[32] Indeed, as a student at Glasgow in 1717, Hutcheson sent Clarke a letter criticising the centre-piece of Clarke's deductive system of religion and morality: the *a priori* proof for the existence of God.[33] By contrast, Hutcheson argued that the origin of moral behaviour is not external to humankind waiting to be discovered but rather internal deriving from a fundamental moral sentiment which precedes the

---

[29] Francis Hutcheson, *A system of moral philosophy, in three books* (2 vols, London, 1755), I, 114-15.

[30] Francis Hutcheson, *An inquiry concerning the original of our ideas of virtue or moral good*, in L A Selby-Bigge ed., *British moralists being selections from writers principally of the eighteenth century* (2 vols., Oxford, 1897), I, 83.

[31] Francis Hutcheson, *Illustrations on the moral sense* ed. Bernard Peach (Cambridge, Mass., 1971), 117.

[32] Henning Jensen, *Motivation and the moral sense in Francis Hutcheson's ethical theory* (The Hague, 1971), 68.

[33] William Scott, *Francis Hutcheson. His life, teaching and position in the history of philosophy* (Cambridge, 1900), 15-16.

exercise of reason.[34] A consequence of this position was that Hutcheson and his disciples were less inclined than Clarke or the natural law theorists to argue for a fixed and unchanging moral code and were more inclined to recognise that morality could be shaped by individual and cultural differences - thus providing an intellectual foundation for the social theorising that was so notable a feature of the Scottish Enlightenment.[35] Though Hutcheson based his moral philosophy on different premises than those which had hitherto largely prevailed within Britain, he and his fellow Scots retained the impulse to link moral and natural philosophy as sharing in a fundamental unity. In accord with the philosophical traditions of the Scottish universities Hutcheson attempted to link these two branches of philosophy by bringing to the study of moral philosophy the same attention to empirical evidence that had transformed natural philosophy,[36] devoting himself to what his Glasgow colleague, William Leachman, called 'the science of human nature and morality'.[37] For Hutcheson, too, drew his system of moral philosophy under the protecting umbrella of Newtonianism. Thus in his *An inquiry concerning the original of our ideas of virtue or moral good* he likens the principle of 'universal Benevolence toward all Men' to 'that Principle of Gravitation, which perhaps extends to all Bodys in the Universe; but, like the Love of Benevolence, increases as the Distance is diminish'd, and is strongest when Bodys come to touch each other'.[38] As with the English moral philosophers we have examined, the intellectual matrix which linked the two branches of philosophy was natural theology. For Hutcheson the moral sense was 'fixed for us by the AUTHOR of our Nature, subservient to the

---

[34] W Blackstone, *Francis Hutcheson and contemporary ethical theory* (Athens, Georgia, 1965), 41 and William Frankena, 'Hutcheson's moral sense theory', *Journal of the History of Ideas,* xvi (1955), 360.
[35] A L Donovan, *Philosophical chemistry in the Scottish Enlightenment* (Edinburgh, 1975), 54.
[36] J Rendall, *The origins of the Scottish Enlightenment* (London, 1978), 75.
[37] Darwall, *The British moralists and the internal 'ought'*, 243.
[38] Selby-Bigge, *British moralists* , I, 130.

Interest of the *System*[39] in the same way that Newton and his followers regarded gravitation as being implanted by the Creator to give direction to His Creation.

This assumption that moral and natural philosophy shared a fundamental unity was a well-worn theme within the philosophy schools of the eighteenth-century Scottish universities. Hutcheson's predecessor, Gershom Carmichael (Glasgow's inaugural professor of moral philosophy, 1727-29),[40] began his edition of Samuel Pufendorf's textbook on natural law with an assertion of the way in which the science of morality should keep pace with advances in the science of nature.[41] Such a view that moral and natural philosophy were bound together was maintained not only by professors of moral philosophy like Hutcheson and Carmichael but also by natural philosophers like Colin MacLaurin (professor of mathematics at Edinburgh, 1725-46). For in his *An account of Sir Isaac Newton's philosophical discoveries* (1750) MacLaurin viewed philosophy as a seamless web of which the study of nature was a part — a part 'chiefly to be valued as it lays a sure foundation for natural religion and moral philosophy' (p.3).

At Marischal College, Aberdeen, the long continuance of the regenting system appears to have further reinforced the impulse to demonstrate the close integration of moral and natural philosophy. For the traditional regenting system involved one individual teaching all branches of philosophy - a system not abandoned there until 1753 (and not at King's College, Aberdeen until 1799) whereas it had been replaced by professorial teaching at Edinburgh as early as 1708. As Wood has shown, at Aberdeen it became

---

[39] Francis Hutcheson, *An essay on the nature and conduct of the passions and affections. With illustrations on the moral sense* (London, 1728), 117.

[40] On whom see J Moore and M Silverthorne, 'Gershom Carmichael and the natural jurisprudence tradition in eighteenth-century Scotland' in I Hont and M Ignatieff eds, *Wealth and virtue. The shaping of political economy in the Scottish Enlightenment* (Cambridge, 1983), 73-87 and J Moore and M Silverthorne, 'Natural sociability and natural rights in the moral philosophy of Gershom Carmichael' in V Hope ed., *Philosophers of the Scottish Enlightenment* (Edinburgh, 1984), 1-12.

[41] Gershom Carmichael, *S. Puffendorfii de officio hominis et civis, juxta legem naturalem, libri duo* (Edinburgh, 1724), v.

almost a truism of moral philosophy teachers that it shared the methods and assumptions of natural philosophy.[42]

In his 1720 graduation theses George Turnbull dwelt on this point invoking in support no less an authority than Newton himself who, in the 31st query of the *Opticks*, had written of the method of analysis and synthesis that 'if natural Philosophy in all its Parts, by pursuing this Method, shall at length be perfected, the Bounds of Moral Philosophy will be also enlarged'.[43] Later, in the 'Epistle Dedicatory' to his *Principles of moral and Christian philosophy* (1740), Turnbull once again affirmed that the Newtonian enterprise would contribute to the advancement of moral as well as natural philosophy describing his object 'to account for MORAL, as the great *Newton* has taught us to explain NATURAL Appearances'. For the order revealed by Newton in the natural world prompted the conclusion 'that the same admirable order shall prevail for ever, and consequently that due care will be taken of virtue, in all its different stages, to all eternity' (p.[i]). Newton himself, he pointed out in his preface, thought that the establishment of the true system of natural philosophy would bring with it 'the enlargement [of] Moral Philosophy' (p.iii).

In pursuing such a philosophical path Turnbull saw himself as following the lead of Hutcheson in viewing both natural and moral philosophy as being based on 'the same first principles' and being 'carried on in the same method of investigation, induction, and reasoning' (p.2). With a gesture towards the teleological assumptions that underlay the traditional conception of final causes Turnbull acknowledged the unity of philosophy arguing that 'when natural philosophy is carried so far as to reduce phenomena to good general laws, it becomes moral philosophy' (p.8). This same theme was taken up by his Aberdonian colleague, David Fordyce, who argued that moral and natural philosophy were linked not only by their ultimate goals but also by their methods. For, wrote Fordyce, '*Moral Philosophy* has this in common with *Natural Philosophy,* that it appeals to *Nature* or *Fact*; depends on Observation, and

---

[42] Paul Wood, 'Science and the pursuit of virtue in the Aberdeen Enlightenment', in Stewart, *Studies*, 127-49.
[43] *Ibid.,* 131.

builds its Reasonings on plain uncontroverted Experiments, or upon the fullest Induction of Particulars of which The Subject will admit'.[44]

Even that Scottish moralist, David Hume, from whom the famous distinction between fact and value - sometimes called 'Hume's law' - derives, was less inclined to keep these two spheres as rigidly distinct as some of his modern commentators suggest. True, in his *A treatise of human nature* there appears the classic passage:

> In every system of morality, which I have hitherto met with, I have always remark'd, that the author proceeds for some time in the ordinary way of reasoning, and establishes the being of a God, or makes observations concerning human affairs; when of a sudden I am surpriz'd to find, that instead of the usual copulations of propositions, *is*, and *is not*, I meet with no proposition that is not connected with an *ought*, or an *ought not*. This change is imperceptible; but is, however, of the last consequence. For as this *ought*, or *ought not*, expresses some new relation or affirmation, 'tis necessary that it shou'd be observ'd and explain'd; and at the same time that a reason should be given, for what seems altogether inconceivable, how this new relation can be a deduction from others, which are entirely different from it.[45]

But it has been argued that Hume was here attacking the attempts of moralists like Samuel Clarke or William Wollaston to construct a system of ethics on a rationalistic basis rather than attempting totally to distinguish the realm of moral philosophy from that of fact and observation.[46] Such an interpretation is consistent with his critique elsewhere in the *Treatise* of rationalistic ethics as when he writes that, ''tis impossible, that the distinction betwixt moral good

---

[44] David Fordyce, *Elements of moral philosophy: in three books* (1754) with a new introduction by John Price (Bristol, 1960), 7.

[45] David Hume, *A treatise of human nature*, with text revised and notes by P.H. Nidditch (Oxford, 1978), 469.

[46] W D Hudson, 'Hume on *Is* and *Ought*', in Chappell, *Hume*, 297, and Marie Martin, 'Hutcheson and Hume on explaining the nature of morality, Why it is mistaken to suppose Hume ever raised the "is-ought" question', *History of Philosophy Quarterly*, viii, no. 3 (1991), 277-89.

and evil, can be made by reason; since that distinction has an influence upon our actions, of which reason alone is incapable'.[47]

Hume was in sympathy with Hutcheson in attempting to construct a system of ethics based not on *a priori* assumptions but on the facts of human nature and society - an area in which such scientific activities as observation and even experiment played an important role.[48] In this sense Hume himself breached the fact/value divide by drawing on empirical data about human behaviour to inform his discussion of the nature of morality.[49] Moreover, as one educated in the Scottish philosophical tradition, Hume naturally regarded his studies in natural philosophy as complementing his work in moral philosophy. He implies such a connection for example when he wrote in his 'A kind of history of my life' that 'I found that the moral Philosophy transmitted to us by Antiquity, labor'd under the same Inconvenience that has been found in their natural Philosophy, of being entirely Hypothetical, & depending more upon Invention than Experience'.[50]

Hume, then, regarded the study of natural and moral philosophy as being intertwined though there remains the more fundamental issue of whether he considered morality as entirely a matter of human sentiment and convention and as therefore having no connection with any ultimate reality[51] - though one could also make a case for Hume being sceptical of any such reality in any connection whether it be in the realm of ethics or natural philosophy. In short, did Hume not so much distinguish between 'is' and 'ought' as dismiss the category 'ought' altogether? As an admirer and follower of Hutcheson, Hume appears to attribute such

---

[47] Hume, *A treatise*, 462.

[48] David Norton, 'Hume, human nature, and the foundations of morality' in David Norton ed., *The Cambridge companion to Hume* (Cambridge, 1993), 149.

[49] Alasdair MacIntyre, 'Hume on "is" and "ought"' in V C Chappell ed., *Hume* (New York, 1966), 247.

[51] Norton, *The Cambridge companion to Hume*, 348.

[51] Such for example is the view taken by Anthony Flew in 'On the interpretation of Hume', in Chappell, *Hume,* 283, or N Capaldi in 'Humes's rejection of "ought" as a moral category', *The Journal of Philosophy,* lxvi (1966), 126-37.

a view to Hutcheson as well as himself as when he wrote to him in 1740: 'I wish from my heart I could avoid concluding that since morality, according to your opinion as well as mine, is determined merely by sentiment, it regards only human nature and human life. This has often been urged against you, and the consequences are very momentous'.[52] Such an entirely subjectivist view of morality has also been attributed to both Hutcheson and Hume by modern commentators such as Stafford but, although it has some plausibility for the arch-sceptic Hume, it is difficult to reconcile with the firmly theistic and even teleological framework within which Hutcheson constructs his moral system.[53]

As in so many other ways so, too, in his ethical theorising Hume is both elusive and atypical in an age still saturated by the assumption that there was an ultimate reality which could be discovered by the methods of the natural philosopher and which had implications for the understanding of moral philosophy. In a period when Christian orthodoxy was under question but when theistic assumptions continued to shape the whole conduct of intellectual life it naturally followed that the study of Creation would lead to a better knowledge of the mind of the Creator and hence the nature of the Good as well as the True. In both eighteenth-century England and Scotland it was part of the mental furniture of Enlightenment culture that, in some sense, the rational processes that had produced Newton's great monument to the capacities of the human mind could and, indeed, should be used as a means of arriving at an understanding of that ancient problem: the nature of the Good Life. As the poet, Alexander Pope, put it in Epistle One of his *Essay on Man*: 'Account for Moral, as for Natural Things'.

It was a sentiment that had the blessing of the Master himself who, at the end of the *Opticks,* had urged natural and moral philosophers to follow the same methods for their ultimate goal was the same. 'For', wrote Newton, 'so far as we can know by natural philosophy what is the first cause, what power he has over

---

[52] J Y T Greig ed., *The letters of David Hume* (2 vols., Oxford, 1932), I, 40.
[53] Martin Stafford, 'Hutcheson, Hume and the ontology of morals', *Journal of Value Inquiry*, xix (1985), 133.

us, and what benefits we receive from him, so far our duty toward him, as well as that toward one another, will appear to us by the light of nature'.[70] Despite the assault on final causes, then, there still remained a belief that in an ultimate sense natural and moral philosophy were both directed towards the discovery of the same reality. Thus the Enlightenment in its English and Scottish forms still regarded the Good and the True as belonging to the same intellectual universe nor did it construct quite as formidable a *cordon sanitaire* between Fact and Value as has become so engrained a part of the philosophical furniture of the twentieth century.

---

[70] H S Thayer, *Newton's philosophy of nature: selections from his writings* (New York, 1953), 179.

V

The Teaching of Philosophy within the British Universities and Learned Societies of the Eighteenth Century

England

Oxford and Cambridge

Both eighteenth-century Oxford and Cambridge were heirs to a similar philosophical heritage. Like most universities in Europe they were the beneficiaries of long centuries of close philosophical analysis and debate as the Aristotelian canon which had been absorbed by Western Europe in the High Middle Ages had been sifted and refined by generations of teachers and students. Part of the intellectual and pedagogical attractiveness of this scholastic philosophical system – the roots of which lay in Aristotle's work but with branches that spread out to absorb a range of philosophical positions (Schmitt 1973 [*1]) – was that it represented a clearly integrated approach to most branches of knowledge. From the basic metaphysical premises the student proceeded to the study of both physics (or natural philosophy) and ethics while the study of these different branches of philosophy was closely regulated by that most distinctive and influential of Aristotle's bequests to posterity: classical logic. The scholastic order represented not only a closely integrated philosophical system but also one which provided a clear pedagogical framework. For the study of logic served not only as the basis for the student's studies but also as the chief weapon in the public disputations which served both as a means of training and examining students as they demonstrated publicly their competence in both logic and Latin by defending a thesis.

By the late seventeenth century, however, the closely woven fabric of the traditional scholastic curriculum in which a set of common philosophical assumptions had underlain and drawn together the different elements of university instruction had begun to unravel (Schobinger 1988 [*3a:6–11]). This was chiefly the result of the impact of the 'new science' which undermined the traditional teaching of natural philosophy, an area of knowledge which had always been a major component of the undergraduate curriculum. Moreover, the methods and conclusions of the new science challenged some of the basic Aristotelian premises which had given unity and direction to the scholastic curriculum. In particular, it challenged the assumptions 'that the material and spiritual worlds were not clearly distinguishable and that the syllogism was the key to knowledge' (Brockliss 1981 [*2:153]). For the close integration of scholasticism was a source of weakness as well as strength: if one major area of the scholastic order, such as natural philosophy, was withdrawn from the controlling influence of fundamental Aristotelian premises then the scholastic structure as a whole was gravely weakened. Confidence in the explanatory utility of Aristotelian-based metaphysics or logic began to wilt in the face of the evident success of the new science in explaining natural phenomena in terms of fundamentally different philosophical premises and by using methods, such as experiment and mathematics, which were largely foreign to scholastic natural philosophy.

But Oxford and Cambridge were to diverge considerably in the extent to which they were to dismantle their traditional scholastic curriculum as a consequence of such a challenge. At Cambridge, the study of modern natural philosophy and its close twin, mathematics, came to be accorded such a degree of attention that not only was scholastic natural philosophy forsaken but, along with it, most other branches of the scholastic order including metaphysics, logic and Aristotelian ethics. At Oxford, by contrast, although scholastic natural philosophy faded from view, natural philosophy more generally did not achieve such a dominant position. As a consequence, other branches of the traditional scholastic order survived and remained a part of the Oxford curriculum until the nineteenth century. This was most notably the case with logic which, in a partially modernised but fundamentally Aristotelian form, survived the dissolution of scholasticism to remain a major part of an Oxford undergraduate's studies. Ethics, too, retained more of an Aristotelian stamp than at Cambridge. And, at Oxford, the vacuum created by the demise of scholastic natural philosophy was largely filled by increased attention to the study of the classics – a study which, as the Renaissance humanists had so frequently complained, largely fell outside the scholastic order but which did not pose the same philosophical challenge to the residues of scholasticism that the ever-increasing attention to modern natural philosophy was to do at Cambridge.

The reasons for this divergence between the two universities, which in the late seventeenth century were so intellectually similar, appear to have derived from their different political and religious sympathies – or so at least I have argued at length elsewhere (Gascoigne 1984 [*79]; 1989 [80]). Oxford remained only very partially reconciled to the political and religious order ushered in by the Glorious Revolution of 1688 and the coming of the Hanoverians in 1714 (Ward 1958 [*36]) – an order that involved a considerable desacrilising of the role of both Church and Crown and a break with many aspects of tradition. The political and religious tenor of Cambridge, by contrast, was transformed – albeit with much reluctance and resistance – by the use of patronage on the part of bishops and politicians sympathetic to the new order. Their influence fostered the growth of a form of theology that emphasised natural theology and the role of reason. For the clerical supporters of the post-revolutionary settlement in Church and State tended to be more sympathetic to styles of theology which emphasised the consonance of reason and religion. Similarly, the apologists for the new political order, such as Locke, claimed that it could be justified by rational principles such as the notion of the social contract – in contrast to the mystifications of tradition employed by the defenders of the Stuarts.

Consistent with a latitudinarian theology in which the claims of reason loomed large the clerical defenders of the new order – who, more and more, came to be the dominant voices within Cambridge – urged the apologetic importance of demonstrating the consonance between Christianity and the new science and, in particular, the great cosmic synthesis so recently achieved by Newton. As a consequence, at Cambridge natural philosophy was invested with a particular status and dignity as the natural ally of the Argument from Design. Moreover, once thus established institutional imperatives served to magnify the attention accorded to the mathematical sciences. For, in the ever more competitive Senate House Examination which over the course of the eighteenth century largely supplanted the traditional disputations as a means of examining students, mathematics was more and more used as a means of minutely discriminating between students (Gascoigne 1984 [*48]). At Oxford, by contrast, natural philosophy never enjoyed quite the same theological attention and so traditional scholastic-based subjects such as logic and ethics still retained a considerable presence in the undergraduate curriculum. Moreover, at Oxford, where a nostalgia for the forms of Church and Crown which had prevailed under the Stuarts remained a part of university culture until well into the eighteenth century, the hold of tradition – including academic tradition – was never as effectively challenged as at the more whiggish, low-church Cambridge.

A widely differing emphasis on the study of natural philosophy was, then, the first and most fundamental difference between the two universities' teaching of philosophy in the eighteenth century. Natural philosophy in eighteenth-

V

century Cambridge meant, of course, first and foremost the study of Newton, the university's most famous son. Cartesianism which at Cambridge, as at other European universities, had played an important role in undermining the traditional scholastic curriculum (Gascoigne 1990 [*3:215–20]) left only a fleeting mark on the undergraduate curriculum at the beginning of the eighteenth century and these few Cartesian remnants were soon expunged by the rising tide of Newtonianism. When the century began Rohault's Cartesian-based textbook, *Traité de Physique* [*387], was still used in the Latin translation by Samuel Clarke but primarily for its notes. These became ever more Newtonian and therefore ever more at variance with the primary text as it passed from the first Cambridge (1696), still largely Cartesian, edition, to the third (1710) openly Newtonian edition under the guidance of Samuel Clarke, an early convert to Newtonianism while a student at Cambridge in the 1690s (Hoskin 1961 [*75]). 'By this means', wrote Bishop Hoadly in his introduction to Clarke's works, 'the True Philosophy [i.e.Newtonianism] has without any Noise prevailed ... and his Notes, the first Direction to those who are willing to receive the Reality and Truth of Things in the place of Invention and Romance [i.e.Cartesianism]' (Hoadly 1738 [*59:I, ii]).

Newton reportedly had 'made the *Principia* abstruse to avoid being baited by the little smatterers in mathematics' (Derham [*53]); not surprisingly, then, it was not suited to serve as an undergraduate textbook of natural philosophy. But, along with the protean Rohault, Cambridge was to produce an increasing number of popularisations and simplifications of the Newtonian system which served both to establish his natural philosophy as a part of the curriculum as well as helping to make Newton's general conclusions part of the culture of elite English society more generally. William Whiston, Newton's successor as Lucasian professor of mathematics, led the way by publishing two sets of Newtonian-based lectures in 1707 [*391] and 1710 [*392]. This was later followed by other Cambridge Newtonian texts such as those by John Clarke (1730 [*495]), John Rowning (1735–42 [*396]) and Thomas Rutherforth (1748 [*363]).

At Cambridge, the Newtonian heritage was chiefly a mathematical one and derived principally from the *Principia*. But the more overtly experimental aspects of the Newtonian tradition, as exemplified in the *Opticks*, also left their mark on the undergraduate curriculum. Newton's first editor, Roger Cotes of Trinity College, amply fulfilled the twofold requirements of his position as the foundation Plumian Professor of Astronomy and Experimental Philosophy by combining his astronomical observations with 'a course in philosophical experiments'. These experiments, wrote Robert Smith in the preface to his edition of Cotes's *Hydrostatical and Pneumatical Lectures* (1738 [*397: Preface]), were 'performed before large assemblies at the Observatory in Trinity College Cambridge; first by the Author in conjunction with Mr. Whiston ... [and] then by the Author alone, and after his decease by my self'. Smith, then, helped to institutionalise the tradition of giving lectures in experimental philosophy which had been begun by Cotes and was to be a feature of Cambridge throughout the eighteenth century. Appropriately, Smith's two major scientific works on optics (1738 [*398]) and harmonics (1749 [*400]) drew heavily on experimental data. The former work was much the more influential since it helped to establish the view that the corpuscular theory of light represented the true Newtonian orthodoxy. For it was the aim of Smith's work to demonstrate the coherence of Newton's total *oeuvre* by attempting to link together the queries of Newton's *Opticks* and the laws of motion of the *Principia* through a treatment of light based on thoroughgoing corpuscular principles (Steffens 1977 [*77:27f.]; Cantor 1983 [*18:34–39]).

Though the teaching of experimental philosophy was kept alive in Cambridge throughout the eighteenth century it was more and more overshadowed as the century progressed by an increasing emphasis on the more overtly mathematical sciences. Such a development reflects the increasing importance of the Senate House Examination which, after 1753 – when students were first divided into classes of honours (wranglers and senior and junior optimes) – more and more determined the award of college fellowships and hence, ultimately, access to college clerical livings. Though the Senate House Examination included some questions on non-scientific matters such as moral philosophy and theology it was chiefly an examination in natural philosophy and more particularly those branches of natural philosophy which were most amenable to mathematical treatment. For mathematics lent itself to the fine discriminations between candidates that the increasingly competitive character of the examination demanded. Moreover, mathematics – which largely meant geometry – could be regarded as a substitute for logic the study of which was still prescribed by the Elizabethan Statutes of 1570 as well as long centuries of academic practice.

Ironically, Oxford had been in some ways quicker to assimilate the *Principia* than Cambridge but, nonetheless, this early interest in Newton's work did not take root as successfully as at Cambridge – largely because the political and religious climate at Oxford provided a less fruitful environment for Newtonianism to flourish. This early interest in Newton at Oxford largely derived from Scottish rather than English roots since, when David Gregory was translated from his chair of mathematics at Edinburgh to the Savilian professorship at Oxford in 1691, he introduced the teaching of Newtonianism – Newton having attracted more early disciples in Scotland than England. Gregory also strengthened such links by an extended visit to Cambridge in 1694 in order to consult Newton and to delve into his rich store of scientific manuscripts.

At Oxford, Gregory's astronomical lectures (which were published in 1702 [*389]) provided an introduction to parts of the *Principia* and especially those concerned with astronomy. A more comprehensive introduction to Newtonian natural philosophy was provided by Gregory's former student from Edinburgh, John Keill, who followed Gregory to Oxford in 1691 and there (as John Desaguiliers put it) was the 'first who publickly taught *Natural Philosophy* by *Experiments* in a mathematical manner' (Strong 1957 [*73:53]). His lectures were printed in 1702 [*390] and provided the first popular exposition of Newtonian natural philosophy.

But, after this promising beginning, the study of natural philosophy lost momentum at Oxford. Early in the nineteenth century Edward Tatham, a would-be university reformer, could ask of Oxford's examination system, 'Why ... is NATURAL PHILOSOPHY the Queen of all Theoretic Science, one of the fittest subjects of academical Education, *totally omitted*?' (Sutherland and Mitchell 1986 [*42:631]). But though at Oxford natural philosophy did not receive the institutional reinforcement of being integrated into a formalised system of examinations as at Cambridge it was not quite as neglected as Tatham's outburst might suggest. The tradition of giving lectures on experimental philosophy which David Gregory had begun was maintained throughout the century; it did not, however, receive the official sanction of a university readership until 1810 (*ibid*:660, 670). Moreover, although students for the BA were traditionally examined on 'the Sciences' laid down in the Laudian Statutes of 1636 – logic, rhetoric, Euclid and morals – some colleges also included natural philosophy in their curriculum (*ibid*.: 475–77). Such college instruction in natural philosophy could, however, be slow to take account of the achievements of the 'Scientific Revolution'. At Christ Church, the largest and most influential of the Oxford colleges, Aristotelian-based seventeenth-century scholastic texts such as those of Burgesdijk (1632 [*383]) or Caspar Bartholinus (1698 [*382]) continued to be used until the 1760s. When, finally, these Aristotelian residues were abandoned around the middle of the century the ensuing vacuum was filled at Christ Church not so much by a modernised form of natural philosophy but rather by increased attention to the classics (Bill 1988 [*43:308–10]).

One brake on the study of modern natural philosophy at Oxford was that there was little attention to the study of mathematics. At Cambridge, mathematics was actively promoted both by the examination system and because it had largely usurped the position enjoyed by logic. At Oxford, by contrast, logic remained a staple of undergraduate instruction as it had been at most European universities since the High Middle Ages. Logic had traditionally been the means of testing whether new propositions were consistent with the received learning which it was the universities' function to preserve and transmit (Howell 1971[*17:5]). As the new science challenged such a largely static view of the body of knowledge and as it emphasised other methods of arriving at knowledge than syllogistic enquiry, so, too, logic lost its traditional role as the core of an undergraduate's studies. However, at a university such as Oxford where the new science made only limited inroads, the ascendency of logic was not so severely challenged. Moreover, syllogistic logic remained a useful means of testing whether novel theological positions could be combined with the traditional teachings of the Church – an important function in a university such as Oxford where theological disputation and apologetic remained a vital part of its functioning as a pillar of the established order in Church and State. Cambridge divines also took their turn in defending the Church against its foes. However, since, at Cambridge, the latitudinarian theological tradition was stronger such apologetic more often took the form of natural theology which could be combined with the findings of the new science as an aid to illustrating the truth of the Argument from Design – a form of theology which required little recourse to either classical logic or Church tradition.

At Oxford the study of logic was reinforced by the publication in 1691 of Henry Aldrich's long-lived text [*319] – a work which was to remain a standard text at Oxford passing through at least fourteen editions before it was finally replaced in the early nineteenth century by another neo-Aristotelian text by Richard Whateley (1825 [*325]). Significantly, Aldrich was linked with a group of High Church champions of the position of the Established Church as it came under increasing challenge in the wake of the Revolution of 1688 and the Act of Toleration of 1689 (Hiscock 1960 [*37]), for classical logic was part of the standard armory of such divines. Though Aldrich took account of recent developments in logic he largely followed the Aristotelian orthodoxy – thus he cites the Port Royal *Ars Logitandi* of Arnauld but was critical of its disrespect for traditional logic. The appeal of Aldrich's work lay in its brevity and clarity in the presentation of Aristotle's ideas and in its reaffirmation of the traditional view that logic should form an essential part of an education in the liberal arts. Aldrich recognised that classical logic had no part to play in the advancement of the new science but he and other Oxford logicians throughout the eighteenth century continued to argue that it should continue to be studied as a means of training the mental faculties – this despite the mounting criticism that classical logic was based on deduction rather than the induction employed by the experimental and observational sciences (Howell 1971 [*17:52, 57]; Bill 1988 [*43:266]).

But, though Oxford chiefly devoted itself to classical Aristotelian logic, more recent developments in the art of reasoning, especially those associated with Locke, left their mark on the university's studies. As early as 1695 Locke reported that an unnamed Oxford fellow was preparing an abridgement of his work 'for the use of young scholars, in the place of an ordinary system of logick'. Such a move came as a surprise to Locke for, as he commented, 'From the acquaintance I had of the tenor of that place, I did not expect to have it get much footing there' (Locke 1708 [*11:109]). But these early attempts to intrude such alien principles were resisted by the university at large for, in 1703, the heads of colleges outlawed the study of the *Essay* – a move prompted, so James Tyrrell told Locke, by the fact that students were reading too much of the new philosophy, something which was undermining scholastic logic and the disputations (Sutherland and Mitchell [*42:570]). But this prohibition proved a dead letter and by around the mid-eighteenth century undergraduates were defending such Lockean propositions as 'Whether science [knowledge] consists in the perception of relations among ideas?' (*ibid.*:570f.).

The essential incompatibility of Locke's work, with its emphasis on deriving knowledge from experience, and the traditional syllogistic logic which attempted to deduce propositions from first principles, was naturally downplayed by Oxford tutors – even though Locke himself had dismissed scholastic logic as 'this artificial Ignorance, and *Learned Gibberish*' (*ibid.*). According to the orientalist, William Jones, when he was at University College in 1765 the tutor would, when reading Locke to his pupils, pass over every passage critical of traditional logic (Wordsworth 1877 [30:127]). A more sophisticated way of dealing with the problem of combining old and new forms of logic was adopted in the textbook by Edward Bentham, a fellow of Oriel College (1740 [*321]). In this work he conceded the validity of many of Locke's criticisms of classical logic but, nonetheless, argued that it was still worth studying syllogistic methods. In common with most modern logics, he argued, Locke's work was primarily concerned with the nature of the human mind rather than the art of distinguishing true from false propositions – thus Locke's enquiries 'make a part of the natural History of Man, rather than a part of Logick' (Sutherland and Mitchell 1986 [*42:573]).

Predictably, at Cambridge, Locke, with his close associations with the new science, had received an earlier and more enthusiastic welcome in the undergraduate curriculum. By about 1696 Cambridge undergraduates were defending Lockean propositions (Locke 1708 [*11:156f.]) probably as a result of the initiative of Charles Kidman of Corpus who appears to have been Locke's first advocate within Cambridge (Disney 1785 [*60:3]). Even Daniel Waterland, the vigorous defender of theological orthodoxy and an opponent of Locke's ultra-latitudinarian theology, wrote in his *Advice to a Young Student, with a Method of Study* (1730[*304]) – a work based on his experience as a tutor at Magdalene College, Cambridge – that 'Locke's *Human Understanding* must be read, being a Book so much (and I add so justly) valued, however faulty the Author may have been in other Writings'. By contrast, Waterland was dismissive of Burgersdijk's text on scholastic logic (1626 [*314]) including it largely for the sake of tradition with the lukewarm comment that 'The Use of it chiefly lies in explaining *Words* and *Terms of Art*, especially to young Beginners'. But he discouraged too much attention to this syllogistic text by adding 'as to the true art of *Reasoning*, it will be better learnt afterwards by other Books' (1730 [*304:23, 19]). Other comments from the early eighteenth century also testify to the fading away of traditional scholastic logic as a part of the undergraduate curriculum. One Cambridge undergraduate, for example, went so far as to write home in 1715 that 'As for Logick, 'tis almost banished the University & but rarely, except in the Studies & heads of the old fellows, to be met with' (Hughes 1952 [*71: I,372]).

However, as logic was redefined to encompass the study of Locke's 'way of ideas' rather than traditional syllogistic logic, it appears to have become more a part of the Cambridge undergraduate curriculum – even though it did not loom as large as at Oxford. In his *Questiones Philosophicae* (a list of sample thesis topics), for example, Thomas Johnson of Magdalene College, Cambridge proposed such Lockean propositions as 'Omnes ideae originem ducant a sensatione et reflectione' ['All ideas derive their origin from sensation and reflection'] with suggested reading based on both Locke and his critics (Johnson 1735 [*305:150]). An eighteenth-century Cambridge student notebook on logic largely consists of such Lockean propositions as 'Has the mind power to make, alter or destroy its ideas?' or 'What is a mode' (Littledale [*54]). A set of lectures on logic from late eighteenth-century Cambridge divided the study of logic 'into two Parts viz Speculative Logic, which examines the properties of the Human Mind so far as they pertain to Reasoning, & Practical Logic which lays down forms & precepts for Disputation'. Predictably, when dealing with speculative logic, the lecturer took the view that 'Speculative Logic was never carried to any great perfection before the Time of our celebrated Countryman Mr. Locke'. The lectures also include a survey of the forms of syllogistic reasoning under the heading of practical logic but, significantly, concluded – with a blow at the sister university – that 'Mathematics have been in general substituted for Practical Logic, because it furnishes more clear & certain Proofs & in this part of Public Education I consider our university as having by far the superiority over Oxford. Practical Logic is still however in use in our Schools for several Reasons' [*52].

V

Other Cambridge teachers also argued that the study of mathematics could serve as a substitute for logic. Daniel Waterland, for example, advised his students that 'the Study of the Mathematicks also will help more towards it [the Conduct of the Understanding] than any Rules of Logick' (Waterland 1730 [*304:20]). An earlier guide to students by Robert Greene of Clare College took the same view declaring that 'the Elements of Geometry' were 'the best practical Logick that can be for they inure and accustom, as it were, the Mind, to true and exact Reasons' (Greene 1707 [*303:5f.]). At Cambridge, then, mathematics (which largely meant geometry) had filled the void created by the decay of traditional scholastic logic thus enabling the university to argue that it was still complying with the letter of the Elizabethan Statutes which prescribed that undergraduates should devote two of their four years to the study of logic (Heywood 1840 [*64:6]). Oxford, by contrast, remained faithful to the heritage of Aristotelian syllogistic logic. This was something for which the Cambridge-educated Coleridge praised her in the early nineteenth century maintaining – with an aside directed at his *alma mater* – that it is 'a great mistake to suppose geometry any substitute for it' (Wordsworth 1877 [*30:84]).

Since syllogistic logic had been the method that tied together the traditional scholastic curriculum its decline at Cambridge naturally hastened the dissolution of the other remnants of the old academic order – a dissolution already set in motion by the undermining of traditional scholastic natural philosophy by the conclusions and methods of the new science. Traditional scholastic metaphysics at Cambridge had largely faded out by the eighteenth century to be replaced by the study of such contemporary philosophers as Samuel Clarke (1705 [*337] and 1706 [*338]). In the 1730 edition of his student guide Waterland treated the subject as being of rather peripheral importance writing that 'Metaphysicks are chiefly useful for clear and distinct Conceptions. Baronius [Baron, *Metaphysica Generalis* 1654 [*333]] will give a general View of their Design, and the parts belonging to them' (Waterland 1730 [*183:27]). He added, however, that the student of metaphysics might also turn to late seventeenth-century authors such as the neo-Cartesian Malebranche (1674[*334]) or the neo-Platonic Norris (1701–1704 [*336] – authors who, in the posthumous 1755 edition of Waterland's work, were replaced by Francis Hutcheson (1742 [*340]) with Baron being omitted altogether.

The most widely-studied metaphysician within Cambridge was the Cambridge-educated Samuel Clarke though, by the late eighteenth century, his popularity was fading and, with it, the study of metaphysics in general. As Bishop Edmund Law (an editor of Locke's work and master of Peterhouse, Cambridge from 1756 to 1768) wrote: 'certain flaws being discovered in the Doctor's [Clarke's] celebrated argument, *a priori* ... his doctrine fell into

V

The Teaching of Philosophy in the Eighteenth Century 11

disrepute, and was generally given up; but its downfall, at the same time, sunk the credit of that whole science, as to the certainty of its principle, which thereby received so great a shock as is hardly yet recovered. This threw us back into a more eager attachment than ever to its rival, the Mathematics' (Fuller 1840 [*63:244]).

Nor did metaphysics fare much better at Oxford despite its continued attachment to syllogistic logic. Burgersdijk's *Institutio Metaphysicae* (1640 [*331]) was commonly used within Oxford up until about the mid-eighteenth century when it was replaced by Locke's *Essay* (Mitchell and Sutherland 1986: [*42:599f.]). To judge from Christ Church, however, even the study of Locke faded as more and more attention was accorded to classics in the late eighteenth century. By 1801 Charles Hall, a future dean of Christ Church, could write that 'Locke's Essays, although frequently recommended and read privately, do not form any part of any public course here' (Bill 1988 [*43:300]).

As scholastic metaphysics and – at least at Cambridge – scholastic logic declined so, too, scholastic moral philosophy, traditionally an important part of an undergraduate's studies, also faded. The Aristotelian assumptions about the nature of the good which had been embodied in traditional scholastic metaphysics – and hence in the moral philosophy based upon it – less and less served as a guide when it came to determining right conduct. Understandably, Aristotelian moral philosophy survived longer at Oxford than at Cambridge since Aristotelian ethics and logic were naturally coupled together. As late as 1764 William Jones complained 'that he was required to attend dull comments on artificial ethics, and logic detailed in such a barbarous Latin, that he professed to know as little of it as he then knew of Arabic' (Sutherland and Mitchell 1986: [*42:568]). At Christ Church, undergraduates were prescribed, *Ethica, sive summa moralis disciplinae*, a standard scholastic ethical text by Eustachius a S. Paulo (1655 [*354]) until 1743 when it was replaced by the anonymous *Ethicae Compendium* – probably the Aristotelian compendium of ethics published at Oxford in that same year.

But the second half of the eighteenth century saw some widening of this traditional Aristotelian framework at Christ Church to include works by natural-law theorists such as the German Samuel Pufendorf (1673 [*358]) and the Genevan Jean-Jacques Burlamaqui (1748 [*362]) (Bill 1988 [*43:298f.]). Elsewhere at Oxford, too, there were indications that students were exposed to modern moral philosophers: the *Introduction to Moral Philosophy* (1745 [*361]) by Edward Bentham included references to works by Samuel Clarke, William Wollaston, Thomas Rutherforth and Francis Hutcheson the elder; the list of suggested ethical propositions for debate drawn up by Joseph Smith, Provost of Queen's College, 1730–56, are largely Aristotelian in character but include mention of authors such as Pufendorf, Francis Hutcheson and Samuel Clarke. By 1774 students were debating propositions such as 'Whether an innate moral sense is given' or 'Whether a division of moral philosophy into ethics, economics and politics may be justified' which indicate both the influence of Locke and a widening of the range of ethical discussion (Sutherland and Mitchell 1986 [*42:583n.,581, 579]). But, despite such concessions to modernity, the Ancients continued to overshadow Oxford's teaching of moral philosophy, principally Aristotle but also other classical moralists such as Cicero. Hence the complaint by Tatham in 1811 that 'the old Moral Philosophy of Aristotle, Cicero, or Epictetus, however admirable in their days, is at this day not worth a louse ... how preposterously absurd, is it, to send the Youth of a Christian University, in the nineteenth century, to learn their Moral Philosophy from Aristotle, that uncircumcised and unbaptised philistine of the Schools?' (*ibid.*:630).

While Oxford belatedly and partially reflected some of the moral philosophy of the age in its curriculum it did little to promote original discussions of ethical issues. Eighteenth-century Cambridge, by contrast, was to prove quite a fruitful environment for moral philosophy largely because the theological re-examination linked to the latitudinarian tradition appears to have also encouraged the development of forms of ethical reasoning which could be derived from reason and Scripture. Thus Thomas Balguy linked his lectures on moral philosophy at St John's from 1741 to 1758 with the teaching of Christian evidences. The tenor of his teaching is evident in his manuscript lectures entitled *A System of Morality* [*51] which largely base ethical discussion on principles derivable by reason – whether it be from natural law or an early form of utilitarianism. He argued, for example, that 'Pleasure is to be rejected, when productive of greater pain, and pain to be chosen when productive of greater pleasure'. The lectures on moral philosophy by Balguy's colleague, Thomas Rutherforth, adopted a similar approach for they owed much to natural law theorists, particularly Grotius. Nonetheless, Rutherforth also developed Balguy's utilitarian approach to ethics giving it a more specifically Christian slant: 'Every man's happiness is the ultimate end which reason teaches him to pursue', he asserted, and Christianity promised not only 'temporal happiness' but 'happiness in a life to come, upon easier conditions than the law of Moses had promised it' (Rutherforth 1754–6 [*365:chapter titles 8 and 13]). Such a position led to the Christian utilitarianism of eighteenth-century Cambridge's most widely read moralist, William Paley, who drew heavily on Rutherforth's work in his lectures while a tutor at Christ's from 1766 to 1775 (Le Mahieu 1976 [*76:124]) – lectures which became the basis of a standard Cambridge textbook (1785 [*366]).

Another mid-eighteenth-century Cambridge proto-utilitarian was John Gay of Sidney Sussex College who, in his 'Dissertation concerning the Fundamental Principles of Virtue and Morality' (1731), presented what Burtt calls 'the first clear statement of the combination of associationism in psychology and utilitarianism in morals which was to exercise a controlling influence in the development of the next century and a half of English thought' (Burtt 1939 [*12:767]). Gay was greatly influenced by Locke as his argument that the moral sense is not 'innate or implanted in us ... [but] acquired from our own observation or the imitation of others' suggests. The chief means by which such a moral sense was developed, argued Gay, was through the association of certain actions with pleasure since 'that which he [everyman] apprehends to be apt to produce pleasure, he calls *good*'(*ibid.*:785). This whole line of argument was to be developed by the Cambridge-educated David Hartley who acknowledged in the preface of his seminal work, *Observations on Man* (1749 [*364]), that it was Gay who suggested to him 'the Possibility of deducing all our intellectual Pleasures and Pains from Association'. Hartley was in turn commonly prescribed for Cambridge undergraduates in the late eighteenth century (Wordsworth 1877 [*30:122f.]).

Despite such signs of activity at Cambridge in the field of moral philosophy, the study of ethics had shrunk in its significance within the undergraduate curriculum since the seventeenth century when it had constituted a major portion of an undergraduate's studies (Costello 1958 [*74:64]). The decay in Aristotelianism led to a concomitant decline in the study of ethics since the philosophical systems that replaced it did not lend themselves so readily to defining a *summum bonum* and, with it, a system of ethics. In the seventeenth century ethics occupied more space in students' notebooks than any other subject, but, in Waterland's programme (1730 [*304]), ethics take up only about a sixth part of the second year studies and Greene in his *Encyclopaedia or a Method of Instructing Students* (1707 [*303]) spends even less time on the subject. The disputations which accompanied the Senate House Examination did require students to defend a 'moral' topic along with two drawn from natural philosophy and, by 1778, the reading on which these ethical debates were based could encompass a wide range of modern authors: Locke, Blackstone, Hume, Wollaston, Harris, Burke, Beattie, Butler, Bolingbroke and Montesquieu [*55]. However, moral philosophy, like all else, faded in significance when compared with mathematics – the subject which determined one's ultimate position in the examination.

Thus, though Oxford and Cambridge were similar in so many ways, they displayed in the eighteenth century an interesting diversity in the way in which they reshaped the scholastic tradition that both inherited. Cambridge gave particular emphasis to the study of natural philosophy and, as the century progressed, more and more narrowed its focus to those aspects of natural philosophy which were amenable to mathematical treatment. By doing so, it accelerated the disintegration of scholasticism since the methods and conclusions of the 'new philosophy' were so clearly at variance, not only with scholastic natural philosophy, but also with the Aristotelian framework of scholasticism more generally. Only the faintest remnants of scholasticism remained in its teaching of any branch of philosophy and even the disputation, the characteristic mode of examining students in the scholastic curriculum, increasingly gave way to other modes of examination. The Senate House Examination came to be more and more based on mathematics which largely assumed the functions traditionally accorded to logic. The little metaphysics which was taught in eighteenth-century Cambridge was based on recent authors such as Samuel Clarke and the teaching of ethics drew on either the natural law tradition or on early forms of utilitarianism that Cambridge authors helped to originate. Oxford, by contrast, never committed itself so whole-heartedly to the study of natural philosophy and so the conflict between the new science and scholasticism was not as acute. Well into the mid eighteenth century it retained scholastic texts for the teaching of ethics, moral philosophy and even natural philosophy. Most significantly of all, Oxford retained throughout the eighteenth century the study of syllogistic logic which helped sustain such scholastic teaching and the forms of disputation which had been for so long linked with them. The fact that Oxford did not follow Cambridge in instituting a modern examinations system until 1800 (Rothblatt 1975 [*40]) may in part have been due to the fact that the traditional system of examining students by means of disputations corresponded better with Oxford's curriculum and, in particular, with its emphasis on logic.

But, though Oxford retained remnants of the scholastic order into the eighteenth century, it had lost that confidence in the Aristotelian first principles which had traditionally served to bind together the difference elements of the scholastic curriculum and to give it that unity and system which had made it so suitable for pedagogical purposes. Oxford retained some elements of its scholastic past principally because it lacked a clear alternative curriculum. As Dean Gregory of Christ Church wrote in 1728: 'the old scholastic learning has been for some time despised, but not altogether exploded, because nothing else has been substituted in its place' (Bill 1988 [*43:57]). In the latter part of the eighteenth century Oxford showed some signs of confronting this problem by widening its philosophical instruction to include recent authors. However, it chiefly appears to have dealt with the vacuum caused by the disintegration of the old scholastic order by expanding its teaching of classics. The teaching of philosophy at eighteenth-century Oxford and Cambridge, then, encapsulates the wider response of universities throughout Europe to the intellectual upheaval prompted by the Scientific Revolution and the Enlightenment – Oxford attempting to maintain as far as possible its traditions while Cambridge largely reshaping its teaching to accommodate itself to the new intellectual climate. Both universities were, however, too closely wedded to the traditional order in Church and State to offer any fundamental challenge to the received wisdom of their society and both in their different ways tended more to be the mirrors rather than the motors of intellectual change.

The Dissenting Academies

The Dissenting Academies owed their formation to the exclusion from Oxford and Cambridge of those who did not conform to the established Church of England after the Restoration of the Church and Crown in 1660. Among those clergy, generally of Puritan sympathies, ejected from the Established Church were over one hundred dons from Oxford and Cambridge (Whiting 1931 [*94:455]). Some of this number, together with other ejected clergy who had also generally been educated at the two ancient English universities, established academies to provide an education for their fellow dissenters. The considerably lower fees charged by such academies, as compared with the universities, also made them attractive to many excluded from Oxford and Cambridge on financial as well as religious grounds (McLachlan 1931 [*93:24]). The name 'academy' was probably an allusion to Calvin's Genevan Academy; it was also intended to relieve the suspicion that the dissenters were establishing universities which would challenge Oxford and Cambridge's traditional duopoly.

To begin with, such academies were, necessarily, modest affairs generally consisting of little more than an individual tutor and a small band of students. During the eighteenth century, however, the dissenting academies grew in size to the point where they had several tutors often offering a remarkable range of subjects (Parker 1914 [*92:57f.]). Naturally, in their early stages the academic fare offered by the academies closely resembled that of Oxford and Cambridge where most of the tutors had had their education. For, as a dissenting historian put it: 'Our dissenting academies arose out of the universities ... persons educated in the universities afterward taught, to a company of non-conforming youths, what they had there learnt, and what some of them had there taught' (Vincent 1950 [*13:118]).

In the early dissenting academies, as in late seventeenth century Oxford and Cambridge, then, philosophy – comprising logic, metaphysics, ethics and natural philosophy – dominated the curriculum. The textbooks used to teach

V

such subjects were, moreover, much the same as those employed at Oxford and Cambridge. At Sherriffhales dissenting academy (which lasted from 1663 to 1697 and had some forty students), for example, the texts in logic included such standard international neo-scholastic works as that by the ubiquitous Dutch author, Burgersdijk (with a commentary by his compatriot Adriaan Heereboord (1657 [*187])), together with English works of the same genre by John Wallis (1687 [*187]) and Robert Sanderson (1615 [*312]). Despite the criticisms by Wallis of Ramus (Howell 1971 [*17:33]) – an author long under suspicion in the predominately Aristotelian universities – the academy showed an appropriate degree of nonconformity by prescribing Ramus's *Dialectica* (1543 [*311]).

The curriculum in metaphysics included a similar range of British and foreign authors from the neo-scholastic tradition such as the Scottish Robert Baron (1654 [*333]) and the Andreas Frommenius (1691 [*335]). In ethics, traditional neo-scholastic authorities like Eustachius a S.Paulo (1654 [*354]) and Adriaan Heereboord (1648 [*353]) were accompanied by the more recent and philosophically novel work of the Cambridge Platonist, Henry More (1668 [*356]). And, as in the universities, the teaching in natural philosophy yoked together philosophical contraries with texts including such neo-scholastic standard works as that by the German, Johann Magirus(1619 ([*381]), combined with Descartes' *Principia philosophiae* (1644 [*203]) and the work of his disciples, Jacques Rohault (1682 ([*387])) and Pierre-Sylvain Régis (1690 [*182]) (Parker 1914 [*92:70]).

The main difference between the early dissenting academies and the universities appears to have been in the range of subjects the academies offered. This meant that, although philosophy was the dominant study, it was not quite as all-pervading as at the universities. At Sheriffhales, for example, the philosophical subjects were accompanied by mathematics, rhetoric, geography, history, anatomy and Hebrew – subjects which generally received only scant attention in the university undergraduate curriculum. On the other hand, Sheriffhales, like most dissenting academies, paid little attention to classics in contrast to the universities (and especially to Oxford).

This attempt to cover the full gamut of modern knowledge remained a continuing feature of many of the Dissenting Academies helping to ensure that at least the better-known ones paid some attention to the different branches of philosophy. Indeed, such academies went further than the universities in attempting to offer a curriculum which provided an introduction to all fields of knowledge. But, given that the arts course at the academies was generally only three years (McLachlan 1931 [*93:25]) – in contrast to four at the universities – the price for such breadth was, inevitably, a sacrifice in the depth with which any particular subject was studied. Gilbert Wakefield, tutor at the famed Warrington Academy from 1779 to 1783, criticised the academies for attempting to make students digest 'the whole Encyclopedia in three years' and for neglecting the study of the classics. Another Dissenter made the same point more stridently by asserting that 'The grand error in almost every dissenting academy has been the attempt to teach and to learn too much' (*ibid.*:33, 40).

Nonetheless, the encyclopedic impulse to attempt to expose students to all major branches of modern learning helped to make the dissenting academies open to new forms of knowledge and to philosophical innovation. Needless to say, the extent of such intellectual adventurousness varied considerably from academy to academy. At some, the impulse to preserve the pure milk of Calvinist doctrine acted as a brake on academic experimentation. At the Attercliffe academy around the beginning of the eighteenth century the suspicion of new forms of science went so far as to result in a prohibition on the study of mathematics 'as tending to scepticism and infidelity' (*ibid.*32). But, generally speaking, the academies, in contrast to the universities with their long weight of tradition, were less affected by the breakdown of the long-entrenched scholastic curriculum and less hampered by ancient statutes prescribing particular forms of knowledge. The dissenting academies also were not obliged by statute to continue the traditional disputations which remained part of the academic culture of eighteenth-century Oxford and Cambridge reinforcing the dominance of logic at the former and mathematics at the latter. Though the holding of disputations was common in the early academies the practice faded over the course of the eighteenth century as Latin gave way to English as the medium of instruction (*ibid.*:42).

While at Oxford the traditional philosophy curriculum largely shrank to the study of logic and at Cambridge to the study of mathematics and natural philosophy with a little moral philosophy at both, the better dissenting academies did at least attempt to expose their students to the full range of philosophy. In the early eighteenth century scholastic logic gradually gave way to the study of the workings of the human mind as recounted by moderns like Locke though, for a time, as in the universities, the old and the new were combined. Thomas Secker, the future Archbishop of Canterbury, around 1711 attended a dissenting academy where the course in logic was 'so contrived as to comprehend all [the neo-scholastic] *Heerebord* (1657 [*315]), and the far greater part of *Mr. Locke's* Essay, and the *Art of Thinking* (Smith 1954 [*96:277]). At Kibworth academy in 1728 the prominent dissenting educator, Philip Doddridge, was exposed to a similar mixture of the old and the new learning logic from the neo-scholastic Burgersdijk and his tutor's system 'a great deal of which is taken from Mr Locke' (Parker 1914 [*92:144]). Naturally, many dissenting academies adopted the *Logick* [*320] of the dissenting Isaac Watts which first appeared in 1725 and was, thereafter, reprinted many times becoming probably the most widely-used logical textbook in the English-speaking world (Howell 1971 [*17:342]). While traditional in its views on the importance of deductive reasoning it self-consciously distinguished itself from scholasticism arguing that '*True Logic* is not that noisy Thing that deals all in Dispute and Wrangling, to which former Ages had debased and confin'd it' (*ibid*.:339).

As in the Scottish universities, some of the dissenting academies devoted considerable attention to the science of pneumatology. Originally this had meant the study of spiritual beings and overlapped considerably with the traditional terrain of metaphysics but, over the course of the eighteenth century, it became more and more equated with the study of the human mind. As the Scottish common-sense philosopher, Thomas Reid, put it in the preface to his *Intellectual Powers*: 'The branch which treats of the nature and operations of minds has by some been called pneumatology' (1785 [*324: Preface]). In its attention to the way in which the mind worked it therefore intermeshed with the eighteenth century's conception of logic. And, as in the scholastic curriculum where logic and ethics had formed a natural partnership, so, too, pneumatics was closely integrated with ethics. At the Kibworth dissenting academy around 1721, wrote Doddridge: 'Our ethics are interwoven with pneumatology, and make a very considerable part of it. They are mostly collected from Puffendorf and Grotius' (Doddridge, 1829 [*91 (I):43]).

Along with Continental natural law moral philosophers such as Grotius and Pufendorf some dissenting academies also taught the work of Francis Hutcheson (1742 [*360]) with its conception of an innate moral sense. Hutcheson was himself the product of the Killyleagh dissenting academy in Ireland and had run a private academy in Dublin before being elected to the chair of moral philosophy at Glasgow in 1729. Hutcheson's career and the interest in his work are further instances of the close associations between the Dissenting Academies and the Scottish universities – a natural alliance given the largely Calvinist background of most of the Dissenters. It is symptomatic, for example, that fifteen dissenting philosophy tutors received Scottish doctorates in divinity in the eighteenth century (eleven of these being from Aberdeen) (Sell 1992 [*101:83]).

However, some of the most notable philosophical products of the dissenting academies such as Richard Price and Joseph Priestley remained too firmly committed to their Lockean epistemological heritage to look favourably on the Scottish common-sense school of philosophy. In his *An Examination of the Scotch Philosophy* (1774) Priestley summarily dismissed Reid's work as 'an ingenious piece of sophistry' arguing that he and others of his ilk

V

The Teaching of Philosophy in the Eighteenth Century 19

were subverting the attractive simplicity of the Lockean theory of the mind as further developed by David Hartley. For, in its place, the Scots advanced 'such a number of independent, arbitrary, instinctive principles, that the very enumeration of them is really tiresome' (Passmore 1965 [*97:23]). This strong Lockean bias helps to account for the fact that in the dissenting academies, as in eighteenth century philosophical culture more generally, metaphysics assumed a position of waning significance. It was still taught at such notable mid-eighteenth-century academies as that of Doddridge at Northampton but it is symptomatic of the less than vital state of the discipline that instruction was based on fairly traditional textbooks such as that of Le Clerc (1692 [*318]) and Isaac Watts (1733 [*339]) (Smith 1954 [*96:279]).

In the more prominent dissenting academies the teaching of natural philosophy often went beyond the bounds generally set in the universities and more closely approximated to science in the modern sense. Along with the mathematical sciences, which largely constituted the eighteenth-century conception of natural philosophy, at least some late eighteenth-century dissenting academies were giving greater scope to the newer empirically-based sciences such as geology and chemistry. This helps to explain why two of late eighteenth-century England's greatest chemists, Joseph Priestley and John Dalton, were tutors at dissenting academies. On the other hand, it has to be remembered, that by no means all dissenting academies offered the full range of scientific instruction available at such prominent academies as Warrington (1757–1786) where Johann Reinhold Forster taught, _inter alia_, geology and Priestley conducted his chemical experiments.

For in some dissenting academies the whole focus of endeavour was on the teaching of theology, this being particularly true of those founded in the wake of the mid-eighteenth century Evangelical Revival. In such academies scientific instruction and philosophy more generally receded considerably in importance. The curriculum of about half the dissenting academies did not extend to science and the attention to other branches of philosophy could also be rather perfunctory. At the staunchly Calvinist Trevecka Academy, for example, founded in 1768 with the support of the Evangelical Countess of Huntingdon, little philosophical instruction was offered apart from some attention to logic (Sell 1992 [*101:85–7, 121n]).

The dissenting academies were, then, far from being universally at the van of knowledge despite the claims of such contemporary advocates as Priestley who, with his customary tact, wrote in his _Letter to Pitt_ (1787): 'While your Universities resemble pools of stagnant water secured by dams and mounds, and offensive to the neighbourhood, ours are like rivers, which, taking their natural course, fertilise a whole country' (Passmore 1965 [*97:8]). But though,

V

as we have seen, Oxford and Cambridge were far less impervious to modern thought than Priestly allowed, his remark contains a grain of truth. Lacking any endowments, the dissenting academies had to be much more sensitive to the consumers of their instruction than the universities and much more attuned to the marketplace of knowledge. The result, inevitably, was a wide range of standards and curricula reflecting the different clienteles of varying religious hues to which they catered. But, if an institution deserves to be judged by its finest products, the most notable dissenting academies such as that of Doddridge at Northampton or the celebrated Warrington Academy (Fulton 1933 [*95]; Smith 1986 [*99]; O'Brien 1989 [*100]) provided a remarkably wide-ranging and up-to-date philosophical education. That in doing so they were able to attract a considerable fee-paying clientele underlines the intellectual vitality of eighteenth-century England and, in particular, its Dissenting subculture.

Learned Societies

In the oligarchic world of eighteenth-century England, where power resided chiefly in the hands of the landowning class and the formal powers of the State were circumscribed as much as possible, the institution of the club or the society loomed large. For a club provided an opportunity for like-minded gentlemen to gather together without the constraints imposed by a formal government instrumentality or the Church. Membership of a club was also a way of demonstrating one's voluntary acceptance within elite society. It was also an institution which could lend itself to a multitude of different purposes from the promotion of a particular political faction to the advancement of social reforms.

Among the different ends to which the multifarious world of eighteenth-century English clubs lent itself was the promotion of knowledge. Philosophy as a formal academic enterprise did not naturally mesh with the convivial world of a club but philosophy, in the sense of a reflection on the nature of human life and society, did form part of the realm of discussion traversed in the clubs. Joseph Addison, a former fellow of Magdalen College, Oxford, saw it as part of his mission as a writer and citizen to transplant philosophy from its academic setting to the world of the clubs. As he wrote in the tenth issue of that organ of gentlemanly culture, *The Spectator:* 'I shall be ambitious to have it said of me that I have brought philosophy out of closets and libraries, schools and colleges, to dwell in clubs and assemblies, at tea-tables and in coffee houses' (Hans 1951 [*4:167]). The tone of Locke's philosophy, with its appeal to the common sense of the educated layman and its dismissal of scholastic philosophy, conveyed a similar admiration for the cultural conventions of gentlemanly society. His

Some Thoughts Concerning Education, with its programme of education intended to train a young gentleman to take his place in society, was even more explicit in its view of philosophy as a natural part of the ambience of polite society. When discussing works on natural philosophy, for example, he wrote that 'It is necessary for a Gentleman in this learned Age to look into some of them, to fit himself for Conversation' (Axtell 1968 [*16:304]). Because the philosophy of the coffee-shops and clubs was, by its nature, largely unsystematic it has left little record in the history of philosophy. The style of the Deists or of some eighteenth-century English moralists such as Shaftesbury and Mandeville, with their appeal to the values of the man of the world, conveys something of the atmosphere of this world of genteel discussion and debate. However, systematic philosophy was generally too solitary and systematic an activity to flourish in the world of the clubs.

One branch of philosophy that was, however, better suited to the sociable and collaborative character of the clubs was natural philosophy. In the absence of any effective state funding the promotion of scientific enquiry in England largely fell to the private initiative of those with the leisure and means to take an interest in science. Such gentlemen banded together to form a number of societies for the promotion of natural philosophy, the chief of which was the famous Royal Society (founded 1660). Throughout much of the eighteenth century, however, the scientific scope of the society was limited by both the funding and the interests of its gentlemanly members (Lyons 1944 [*113]). The early Baconian ideal of the Royal Society as maintaining a virtual research institute – an ideal which the early Royal Society went some way towards achieving – waned over the course of the eighteenth century (Miller 1989 [*76]).

The Royal Society's gentlemen members were chiefly interested in forms of natural philosophy which were of obvious utility (especially in the improvement of their estates) or which were sufficiently curious or unusual to provide a stimulus for polite discussion. Thus in the mid-eighteenth century the Royal Society became a forum for the study of historical antiquities as well as of natural phenomena (Berman 1978 [*68:xxi]). Tension within the society between those who saw it as chiefly concerned to promote the mathematical heritage of Newton and those gentlemen-amateurs with an interest in practical knowledge and polite avocations such as the study of antiquities came to a head in 1783–84 with a series of disputes directed at Joseph Banks who had been elected the Society's president in 1778 with the active support of the gentlemen-amateurs (Miller 1981 [*122]). Significantly, Banks remained as president until his death in 1820 and, as the longest-serving of all presidents of the Royal Society, made it a forum for the promotion and promulgation of useful knowledge and a vehicle for linking science with the concerns of government.

Under his rule the Society emphasised utility and cultural breadth rather than the scientific specialisation which was increasingly a feature of Continental (and especially French) society. Hence Banks viewed the foundation of specialist scientific societies like the Geological Society of London (f.1807) and the Astronomical Society of London (f.1820) with considerable disfavour. He did, however, take a more indulgent view of the foundation of the Linnean Society of London in 1788 and the Royal Institution in 1799 since the goals of these societies were in harmony with the gentlemanly scientific culture promoted by the Royal Society. Both these organisations were naturally attuned to the values of a landed class since the Linnean Society promoted natural history and the Royal Institution had as its object 'the Application of Science to the Common Purposes of Life' (Gascoigne 1994 ([*131:216]).

The example of the Royal Society helped to promote an association between an interest in science and elite culture beyond London in the provincial centres which were to grow in size and significance throughout the eighteenth century. The spread of such philosophical societies, as they were generally known, throughout eighteenth-century Britain came, then, to reflect the cultural centrality of science in the value-system of the governing elite (Porter 1980 [*120]). The Royal Society had established that science could be a loyal servant of the established order and a multitude of provincial centres emulated its example by establishing societies where scientific interests could be pursued in the ambience of a gentlemen's club (Winter 1950 [*114], Sturgess 1979 [*119] and Kitteringham 1982 [*73]). One such society, the Gentlemen's Society of Peterborough (founded 1730), for example, refused to allow tradesmen to become members (Jacob 1988 [*125:158f.]).

By the late eighteenth century the newer provincial centres and their attendant scientific societies – notably the Lunar Society of Birmingham (founded in the mid-1760s) and the Manchester Literary and Philosophical Society (founded 1781) (Thackray 1974 [*116]) – provided something of a challenge to this association between science and established values by linking science more closely with religious Dissenters and the more politically suspect (Golinski 1992 [*128:57f.]). However, such associations largely faded with the onset of the French Revolution – symptomatically, the Lunar Society of Birmingham was dissolved in the mid 1790s (Schofield 1963 [*115]).

The ambience of eighteenth-century philosophical societies was best suited to the empirical sciences and particularly to natural history which was naturally of interest to members of a landed class. The sense of local attachment expressed itself not only in the collecting of specimens of the local flora and fauna but also of antiquities for, like the Royal Society, such local philosophical societies were often a focus for gentlemanly 'virtuoso' culture which embraced the study of a region's past as well as its natural history. The world of gentlemanly societies – including the Royal Society – was, however, less well suited to the promotion of the mathematical sciences which required greater resources and a more professionalised approach to enquiry. This helps to explain why, then, it was France with its state-supported academies which was, in the eighteenth century, to exploit more fully England's rich Newtonian heritage.

Linked with such clubs and societies was the increasingly more prevalent diffusion of science, philosophy and culture more generally through the fast-growing periodical press and the growth of circulating libraries. The activities of the itinerant lecturers with their array of experiments and apparatus provided another means by which science became a part of the dominant culture of the eighteenth century. As with so many aspects of the period the vogue for public scientific lectures began in London in the early eighteenth century but then spread to the provinces (Inkster 1981 [*121]; Morton 1990 [*127]; Stewart 1992 [*129:120]). With an eye to increasing the size of their audiences such visiting lecturers naturally focused on the more spectacular areas of experimental philosophy – electricity particularly lent itself to the theatrical (Schaffer 1983 [*124], 1993 [*130]). Such visiting lecturers helped to stimulate an interest in natural philosophy but their itinerant and transient character meant that they could do little to sustain and institutionalise it (Jacob 1988 [*125:152]) – a consideration which again underlines the significance of the clubs and societies in consolidating the place of science within the world of elite culture.

Scotland

The Universities

At the end of the seventeenth century the Scottish universities continued to maintain a curriculum which was heavily philosophical in character. The traditional veneration for philosophical enquiry which had been inculcated into the medieval universities since virtually their foundation still shaped the character of the curriculum offered at the Scottish universities as at most of their European counterparts. What is striking about the Scottish universities, however, is that this philosophical preponderance was transmitted into the very different intellectual environment of the eighteenth century – indeed, it continued to be a distinctive feature of Scottish higher education in the nineteenth century (Davie [*97:13]). Moreover, the philosophical instruction offered in the Scottish universities moved well away from its traditional roots in the scholastic tradition to embrace new forms of philosophical enquiry linked with the major currents of the Enlightenment. By doing so, the Scottish universities themselves became major centres for the dissemination of the Enlightenment to an extent that far surpassed their English counterparts.

The fact, too, that the Scottish universities equaled, if not surpassed, other European countries in the proportion of their population that they educated further magnified their significance. For, in contrast to other countries like England, the percentage of the population attending university in Scotland actually increased over the course of the eighteenth century rising from around two per cent in 1700 to around three by the early nineteenth century (Houston 1989 [*151:49]). Over the course of the century, too, at Glasgow (and probably also at the other Scottish universities) the social origins of students considerably broadened – the percentage of students of humble birth, for example, increased from 1.94 to 47.9 from the 1740s to the 1790s (Mathew 1966 [*244]).

The flourishing state of the eighteenth-century Scottish universities and, in particular, their philosophical instruction, is the more remarkable since, for most of the seventeenth century, the Scottish universities showed few signs of being centres of philosophical innovation. The overall organisation of philosophy was based on centuries of scholastic tradition. The long-established assumption that all branches of philosophy were linked together since they derived from a common set of largely Aristotelian premises was strengthened by the mode of teaching known as the regenting system. As the etymological link between 'regenting' and 'reading' suggests it was, in the scholastic manner, a system closely tied to instruction based on the exposition of a set of standard works (Veitch 1877 [*164:82]). A regent was an instructor, usually recently graduated, who generally taught for a few years before moving on to a position in the Church or the professions – thus removing him from university life before he was in a position to produce much in the way of published work (Cant 1970 [*170]). Typically, a single regent would teach a given cohort of students throughout their entire degree. Like the tutorial system at Oxford and Cambridge, then, it assumed that one individual could master the full spectrum of university studies. This was a not unreasonable assumption in the long centuries of scholastic instruction when it was thought that all branches of philosophy belonged to an overall unity. However, it became harder and harder to sustain with the passage of the eighteenth century and the growing diversity of philosophical styles and assumptions – a consideration which helps to explain the demise of the regenting system at the Scottish universities. For the practice of regenting was abolished at Edinburgh University in 1708 – an example followed by Glasgow in 1727 and St Andrews in 1747; at the universities of Aberdeen this innovation was resisted longer with Marischal effecting the change in 1753 and King's in 1799.

The nature of philosophical instruction at the Scottish universities at the end of the seventeenth century was conveniently summarised by the astronomer and physicist, David Gregory, in a letter to Isaac Newton of 1691 (the year in which Gregory moved from his chair in mathematics at Edinburgh to that of astronomy at Oxford). Appropriately, Gregory combined his exposition of the character of the curriculum with a description of the regenting system. 'Ther is then in each Coledge besides the head ...', he wrote, 'four professors called regents each of which circulats round the whole course. the students who study under any regent continue with him four years. in the first year the greek tongue is taught in the second logick, in the third Metaphysique and Moral philosophie, in the fourth the Naturall philosophie'. The whole curriculum was, then, dominated by philosophy for even the study of Greek was intended as an aid to philosophy (Shepherd 1987 [*196:146]). Gregory did note, however, that there was some attention devoted to classical studies which, interestingly enough, fell outside the normal pattern of the regenting system – an indication of the extent to which this system of instruction was linked to the view that the unity of philosophy was best imparted by a single individual. For, continued Gregory, 'Besides these four regents there is in some coledges a fifth added who doth not circulat as the rest doe but constantly teaches Roman Authors, Rhethorick, and the beginnings of Geography and Chronology' (Turnbull 1961 [*214:157]).

But the fact that Gregory, one of the earliest of Newton's disciples and popularisers, held the chair of mathematics at Edinburgh from 1683 to 1691 was an indication of the changing intellectual and pedagogical climate of the Scottish universities. So, too, was the appointment of Gregory's mathematically even more distinguished uncle, James Gregory, as inaugural professor of mathematics at Edinburgh from 1674 to his death in 1675 following his long tenure of the same chair at St Andrew's (again as inaugural professor) from 1668 to 1674 (Cant 1946 [*255:45]). Mathematics was a subject that did not fit neatly into the categories of the scholastic philosophical curriculum and so, in the late seventeenth-century Scottish universities, it was relegated to a specialist teacher outside the normal bounds of the regenting system. The fact that it was considered sufficiently important to warrant the appointment of such specialist professorships is, too, an indication of the changing nature of the curriculum in natural philosophy as the Scottish universities, which had long been on the periphery of European intellectual life, began to devote increasing attention to the developments linked with the Scientific Revolution. By 1670, Edinburgh student dissertations for the fourth and final year of studies (the year devoted to natural philosophy) indicated that at least some students were being encouraged to abandon the familiar framework of Aristotelian physics – though the result was a confusing *mélange* of cosmological theories as some students variously adopted the Ptolemaic, Copernican, Tychonic or Cartesian cosmologies. By the 1680s, however, students turned more consistently to Descartes (Shepherd 1982 [*176:67]), a development that may well have been connected with David Gregory's appointment as professor of mathematics in 1683 (Russell 1974 [*172:128]).

The activities of the Gregorys, together with other early Scottish admirers of Newton such as David Gregory's students, John Craige and John Keill, or the physician, Archibald Pitcairne (Eagles 1977 [*223]; Guerrini 1986 [*226]), help to explain why the University of Edinburgh became an early centre for the study of Newtonian natural philosophy – even before it had been fully institutionalised in Newton's own university of Cambridge (Gascoigne 1989 [*80:145f.]). As early as 1704 a student dissertation was entirely Newtonian in character. Of the other Scottish universities, St Andrew's was most conspicuously at the van of change in natural philosophy, probably thanks to the influence of James Gregory. By around 1680, however, the Scottish universities generally seem to have largely abandoned scholastic natural philosophy in favour of Cartesianism. This in turn had given way to Newtonian natural philosophy by about 1710 (Russell 1974 [*172:129,148]).

Within the Scottish universities, as in Cambridge, Newtonianism was adopted partly because it occasioned fewer of the theological problems posed by Descartes's radical dualism. The more conservative Scottish academics also quite accurately saw Cartesianism as a danger not only to Christian orthodoxy but also to the structure of traditional academic life which had, for so long, been linked to the defence of the Church. For, in a highly integrated system such as the traditional scholastic curriculum, change in one area – such as natural philosophy – was inevitably destined to undermine the whole philosophical edifice on which teaching had been based. Such considerations help to explain why, despite the steady advance of Cartesianism in the late seventeenth century, there continued to be a rearguard action against such dangerous novelties. Such opposition surfaced most forcefully in 1690 when, in the aftermath of the political and religious upheavals caused by the Revolution of 1688, a commission was established to examine the universities (Shepherd 1992 [*182:29f.]; Hannay 1915–1916 [*253]). Thus the commissioners, as part of their drive to establish Calvinist orthodoxy in the universities, reported in 1695 that as 'For Cartesius, Rohault and others of his gang beside what may be said against their doctrine they all labour under this inconveniency that they give not any sufficient account of the other hypotheses and of the old philosophy

V

which must not be ejected, and weer never designed to be taught to students' (Phillipson 1974 [*221:426]).

Quite accurately, the commissioners took the view that philosophical innovation in an area which loomed so large in the curriculum as natural philosophy had encouraged novelties in other areas – including the highly theologically sensitive area of ethics. This was the more likely since, as part of the regenting system, the same individual would take students for all branches of philosophy and an interest in philosophical innovation was likely to colour the character of his whole teaching (Emerson 1990 [*181:17]). Along with their condemnation of Cartesian textbooks of natural philosophy, then, the commissioners also castigated those using the ethical works of the Cambridge Platonist, Henry More (an early admirer of Descartes), as being 'grossly Arminian, particularly in his opinion *de libero arbitrio*' (Veitch 1877 [*164:90]). In general, they concluded, many of the existing philosophical textbooks were dangerous – particularly since they were largely written by 'popish professors' who 'Cunningly insinuate ye hereticall tenets mixing them with their philosophy' (Phillipson 1974 [*221:426]). Faced by such a dearth of reliably Calvinist philosophical primers they set out to produce their own, commissioning each of the Scottish universities to produce texts on the different branches of the philosophical curriculum. But the project came to naught – an indication that the study of the philosophical moderns was too firmly entrenched in the Scottish universities for the commissioners to be able to reverse the philosophical tide.

By the early eighteenth century, then, the structure of the traditional scholastic curriculum had been largely subverted by innovations in natural philosophy which inevitably called into question some of the basic Aristotelian premises on which the whole pedagogical *ancien régime* had been based. This waning of the old philosophical order, together with the political and religious upheavals generated by the Revolution of 1688, helped to provide the environment for institutional innovation at Edinburgh – the most overt manifestation of which was the abolition of regenting in 1708 (Kenrick 1957 [*212]). This measure was part of a wider movement for reform of the university prompted by the Rev. William Carstares (Principal, 1703 to 1715) who persuaded the town council, which controlled the university, to follow his lead. Like so many Scots, Carstares had followed the path of international Calvinism by studying in the Netherlands (at Utrecht) and he set out to remodel Edinburgh along Dutch lines (Morrell 1976 [*222:48]). Hence the minute of the town council abolishing regenting in June 1708 speaks of the need to follow the example of 'the most famous Universities abroad' (Rendall 1978 [*148:49]) – a reference to the Dutch universities.

V

The motives behind these reforms were various: Carstares hoped to increase the number and quality of the Presbyterian clergy while the city council hoped to improve the attractiveness of its university to both Scots and foreigners. In particular, in the aftermath of the Act of Union of 1707, which merged England and Scotland's political and economic life, Edinburgh and other Scottish universities hoped to attract greater numbers of English students excluded from Oxford and Cambridge either on religious financial grounds (for Edinburgh and Glasgow cost about one-fifth of the English universities in the early eighteenth century (Houston 1989 [*151:50])). The mercantilist motive of increasing the university's market share also helps to account for the city council's role in the establishment of a medical faculty in 1726 (Morrell 1983 [*179]; Emerson 1993 [*228]) – an innovation which again owed much to Dutch models (notably to Leiden).

At Glasgow, too, the Dutch example was influential in the early eighteenth century (Phillipson 1987 [*270:228]) as the university had to deal with the dissolution of the traditional scholastic curriculum and a rapidly changing political and religious climate. The introduction to Scotland of a Dutch style of natural jurisprudence based particularly on the work of the Dutch Hugo Grotius and the German Protestant, Samuel von Pufendorf, was largely the work of Gershom Carmichael. Drawing on these Continental models Carmichael did much to help establish the tradition of original ethical enquiry which was to be one of the most outstanding features of the eighteenth-century Scottish universities and of the Scottish Enlightenment generally. In many senses Carmichael was a transitional figure. His academic career at Glasgow spanned the transition from the regenting system to the professorial system which was to provide the institutional basis for that specialisation and original enquiry which made the universities a centre of the Scottish Enlightenment. In 1694 he was appointed a philosophical regent and, two years before his death in 1729, he became Glasgow's inaugural professor of moral philosophy.

As regent, he showed a close interest in the developments in natural philosophy which played a major part in undermining the old scholastic order – something which probably helps to explain why, in the early 1690s, Carmichael was criticised before the Commission established to examine the universities (Emerson 1990 [*181:19]; Moore and Silverthorne 1984 [*248:2]). Early in his career, Carmichael's teaching in natural philosophy was influenced by Descartes but, by the early 1700s, he had adopted Newtonianism. When Newton's work was still a novelty in the English universities he introduced his students to its major conclusions through Samuel Clarke's progressively more Newtonian three editions of Rohault's textbook of Cartesian natural philosophy [*387] – this philosophically schizoid work being the first available student text on Newtonian natural philosophy. Carmichael's sympathy for the character of contemporary natural philosophy is also evident in his praise of 'ye two great Hinges of Natural Philosophy, or rather ye two constituent parts of it ... Mathematical Demonstration and Experimental' (Donovan 1971 [*171:13]).

Innovation in natural philosophy was accompanied by innovation in moral philosophy as Carmichael established within Glasgow and the Scottish universities generally the Continental traditions of natural jurisprudence – in many ways a natural intellectual transplantation since Scottish law, like that of most Continental countries (but unlike that of England), was based on Roman law. Thanks to Carmichael, Pufendorf's *De Officio Hominis et Civis* became the set text in moral philosophy at Glasgow producing his own edition with notes (1724 [*359]) which Francis Hutcheson (Carmichael's successor in the chair of moral philosophy at Glasgow) considered of greater value than the original. Appropriately, in the preface to this work, he paralleled the changes in natural philosophy over the previous century with those in moral philosophy – another reminder of the way in which the highly integrated nature of the traditional scholastic curriculum meant that change in one area of philosophy was likely to be accompanied by change in another. The chief engine of change in moral philosophy Carmichael saw as being the work of Grotius which had been explicated in a more systematic manner by Pufendorf. Thanks to both these authors it had become evident that moral philosophy should be regarded as part of natural jurisprudence and that ethics should be derived from a demonstration of an individual's civic duties and social circumstances. This was not to say that ethics should be secular in character for ultimately the moral philosopher should be guided by the will of God as imprinted in the natural order or as revealed in Scripture (though Carmichael conceded that the latter often only offered limited guidance on specific issues (Moore and Silverthorne 1983 [*247])).

Carmichael's insistence on the importance of reason in the study of moral philosophy provided the appropriate intellectual climate for Francis Hutcheson's seminal teaching in moral philosophy when he succeeded Carmichael at Glasgow. Like Carmichael, too, Hutcheson drew from his study of natural philosophy the moral that ethics should be characterised by the empiricism that had produced such notable advances in the study of nature. In contrast to the traditional *a priori* approach to ethics (still evident in English theorists like Samuel Clarke) Hutcheson argued that the best course for a moral philosopher was not to attempt to deduce a set of moral laws from a set of given principles but rather to proceed in Baconian fashion to attempt to derive such principles *a posteriori* from a close study of human nature (Veitch 1858 [*210:xix]; Rendall 1978 [*148:75]). Hutcheson's enquiries led him to conclude that there were

V

a number of universal characteristics which could serve as a guide to proper behaviour, the chief of which was an innate human moral sense. The chief manifestation of this moral sense was an awareness of the common good to which he gave the name, benevolence. This, in turn, provided Hutcheson with the ethical foundation for the view, as he expressed it in the famous phrase which became the hall-mark of utilitarianism, that the function of the moralist was to establish what was most conducive to 'the greatest happiness of the greatest number'.

Such a natural inclination to the Good, Hutcheson saw, in unCalvinist fashion, as an indication of the hand of the Creator in the fashioning of human nature. 'Virtue itself', he wrote, 'or good dispositions of the mind, are not directly taught, or produced by instruction, but they are the effect of the great Author of all things, who forms our nature for them' (Teichgraeber 1978 [*149:88]). Since benevolence is such an intrinsic feature of human nature it follows that human beings are most likely to achieve happiness when they act in a manner which will advance universal benevolence; that, as Hutcheson, put it, virtue is 'the surest happiness of the agent' (Campbell 1982 [*246:169]). Indeed, for Hutcheson the workings of the 'Practical Disposition to Virtue Implanted in our Nature' provided a parallel with the 'Principle of Gravitation, which perhaps extends to all Bodys in the Universe' (Norton 1966 [*168:104]) – another instance of the way in which the momentous developments in natural philosophy provided a catalyst for the reexamination of other branches of philosophy.

Generally speaking, the goal of attaining universal benevolence was best achieved by allowing individuals to follow their own interests since the interests of individuals and the larger society were likely to overlap – a point that lay at the base of Adam Smith's conception of the way in which social and economic life ought to be allowed to proceed. However, Hutcheson had not totally abandoned the bleak realism of Calvinist anthropology and he considered that, since human nature was in some ways still corrupt, the restraint provided by the state was necessary for a well-ordered society. Hence, he wrote that 'The necessity of civil power must arise either from the imperfection or depravity of men or both' (Teichgraeber 1978 [*149:123]). Since Hutcheson's conception of ethics was so closely attuned to his understanding of the way in which individuals related to their larger society his work helped to promote that tradition of social enquiry which was to be such a notable feature of the Scottish Enlightenment (Bryson 1968 [*144]). Thus Hutcheson was very conscious that the forms that human benevolence took were socially conditioned so that the student of ethics might widen the province of the subject to embrace the study of different societies (Donovan 1971 [*171:54]). Such a generous view of the role of the moral philosopher helps to explain how Adam Smith laid the foundations for his work in social and economic theory while holding Hutcheson's former post as professor of moral philosophy from 1752 to 1764.

Hutcheson's successors also followed his example of lecturing in English rather than Latin (Knox 1953 [*143:12]). It was a transformation that underlined Hutcheson's break with academic tradition more generally and, in particular, his abandonment of the complex array of subtle distinctions that was embedded in the scholastic Latin which had been the traditional medium of instruction (Veitch 1877 [*164:208]). Such innovations were reinforced by the popularity of his textbook on moral philosophy (Hutcheson 1747 [*360]) based on a posthumous edition of his lectures (Coutts 1909 [*241:219]). Thanks to Hutcheson's influence moral philosophy came to assume a pre-eminent role in the undergraduate curriculum at Glasgow; at Edinburgh, by contrast, it remained relatively neglected until Adam Ferguson was appointed professor of moral philosophy in 1764 (Sher 1990 [*227:121]; Phillipson 1983 [*224:91]).

The fact that Hutcheson could exercise such influence within the University of Glasgow with a human-centred system of ethics based ultimately on an optimistic assessment of human nature and its capacity to know what was good is an indication of the waning strength of orthodox Calvinism and its pessimistic anthropology. A major milestone in this transformation of the religious and intellectual climate within the University of Glasgow was the failure of the attempt in 1715 to censure John Simson, the Professor of Divinity, for heresy. Simson answered his critics by conceding that he had advanced 'some things wholly new' but argued that this, in itself, was not culpable unless he could be shown to be in error by 'Scripture and reason'. Nonetheless, the Assembly of the Church of Scotland reported in 1717 that Simson's views were not soundly based on Scripture and attributed 'too much to natural reason and the power of corrupt nature'. By 1720, however, Simson had enough support largely to silence his opponents within the Assembly (Cameron 1982 [*245:123f.]) – thus setting the scene for the rise of the Scottish 'Moderate' divines who, like their latitudinarian counterparts within the Church of England, espoused a form of theology which gave ample place to the role of reason and tended to downplay the doctrine of original sin.

The dominance of the 'Moderate' party within the Church and the universities (Sher 1985 [*225]) provided an intellectual environment in which moral philosophy and natural philosophy assumed particular importance in the curriculum – thus strengthening the dominance of philosophy more generally within the Scottish universities. The study of both moral philosophy and natural philosophy was based on the assumption that the will of God could be discerned through the study of nature and human society as well as by means of the more traditional route of the study of the Scriptures. Such a belief in the efficacy of human reason to arrive at a knowledge of the Divine Will, as manifest in Creation, was at the heart of the Moderates' theological position. It is not surprising, then, that the study of natural theology and moral philosophy were closely linked. 'The lectures in the *Moral Philosophy* class consist of three principal divisions', wrote Thomas Reid of late eighteenth-century Glasgow, the second and third of which were ethics and natural jurisprudence but, tellingly, 'The first comprehends natural theology: or the knowledge, confirmed by human reason, concerning the being, perfections, and operations of God' (Reid 1799 [*240:734]). Instruction in ethics ultimately rested, then, on the assumption that there was a Divine Purpose which human beings could, at least to some extent, discern.

However, the arena in which such ethical enquiry took place had a more secular character focusing, as it did, on two chief areas: the nature of the human mind and the character of social obligation. Both of these strands could be found in Hutcheson and, before him, in his mentor, Locke. However, as the eighteenth century advanced, the focus of teaching and writing in the Scottish universities tended more to the former with the rise of the so-called 'Common Sense' philosophers and their detailed consideration of the nature of the human mind (Sher 1985 [*225:314]). In response to sceptics like Hume, Common Sense philosophers like Thomas Reid, James Beattie (Phillipson 1983 [*224]) and Dugald Stewart set out to demonstrate that the human mind was so equipped that it could arrive at a knowledge of basic religious and moral truths (Grave 1960 [*166:4]). The ultimate warrant for confidence in the efficacy of such human powers was the premise that these were God-given (Norton 1966 [*168:279]). Such a quest to unveil the nature and capabilities of the human mind involved an ever more technical appraisal of the nature of psychology and epistemology that drew the focus of ethical discussion away from the more social and civic concerns that had characterised the earlier Scottish Enlightenment (Sher 1985 [*225:314]).

However, this civic strand was by no means extinguished. Dugald Stewart's lectures on moral philosophy, for example, as Adam Smith's successor as Professor of Moral Philosophy at Glasgow began with a detailed consideration of human intellectual powers but also proceeded to analyze the way that the human will could exercise moral choice in relation to social duties. As part of such a survey, Stewart included, in the manner of Hutcheson, a discussion of the nature of politics (Collini, Winch, Burrow 1983 [*177:26]; Phillipson 1983 [*224]). This tradition continued well into the nineteenth century. In 1831 the professor of moral philosophy at Glasgow regarded it as being 'expressly prescribed that Political Philosophy should be his subject as much as Ethical

Philosophy'. The situation was similar at Edinburgh where the curriculum in moral philosophy followed a parallel course beginning with the 'Nature of the Human Being' and then proceeding to 'Relations in which that being is placed'. With this social foundation laid, the course then considered the social and political dimensions of moral obligation with 'Duties deduced from that nature and those relations' and 'Means by which individuals and nations may promote and guard their virtue and their happiness' (Report 1831 [*162:248,130]).

Moral philosophy, then, loomed large in the undergraduate curriculum due to the increasing specialisation which the move from the regenting to the professorial system brought with it. More fundamentally, the increasing significance of moral philosophy reflected a theological climate which was sympathetic to attempts to derive a system of ethics through the application of human reason. This same theological climate also helps to explain the attention that natural philosophy enjoyed in the Scottish universities since a sympathy for natural theology provided a warrant for the belief that the study of Nature helped to bring one to a better understanding of Nature's God. Moral philosophy and natural philosophy also had the institutional advantage of being linked to the two key professional schools: law and medicine. The strong slant towards natural jurisprudence that Carmichael had helped to establish in Scottish moral philosophy strengthened the links between moral philosophy and law while natural philosophy was seen as a natural complement to the study of a form of medicine that was becoming ever more self-consciously 'scientific'.

As we have seen, too, moral philosophers quite consciously modeled themselves on the methods and, they hoped, the successes of the natural philosophers. Both fields were seen as examples of the ways in which the application of human reason and patient observation could yield Enlightenment. This view that natural and moral philosophy were natural allies in the spread of enlightened thought was a common topos in the philosophy schools of the Scottish universities. In his lectures at Marischal College, Aberdeen, in the 1740s, for example, David Fordyce stressed the essential unity of the two branches of philosophy – a unity made particularly evident at Aberdeen which still retained the regenting system with its insistence that one individual teach all branches of philosophy (Wood 1990 [*197:140]). '*Moral Philosophy*', proclaimed Fordyce, 'has this in common with *Natural Philosophy*, that it appeals to *Nature* or *Fact*; depends on Observation; and builds its Reasonings on plain uncontroverted Experiments' (*ibid.*:137).

The scope of natural philosophy was still considered as synonymous with the traditional Aristotelian conception of 'physics' – that is the science of bodies in motion. Thus Reid defined the scope of the lectures of the professor of natural philosophy in late eighteenth-century Glasgow as 'comprehend[ing] a general system of *physics*. However, as Reid was at pains to add, natural philosophy, as taught at Glasgow, took account of the rapidly changing character of the discipline 'keep[ing] pace with those leading improvements and discoveries, in that branch of science, by which the present age is so much distinguished'. This vibrant and fertile character of natural philosophy owed much to the application of the experimental method, something which received recognition at Glasgow by dividing the curriculum into two separate courses, one devoted to theory and the other to experiments. 'The apparatus for conducting the latter', added Reid with not unjustifiable pride in a university that had produced the great experimentalist, Joseph Black, and had provided the setting for the early work of James Watt, 'is believed not to be inferior to any in Europe' (Reid 1799 [*240:735]). Mathematics was the province of a separate professor but, given the mathematical character of eighteenth-century natural philosophy, there was a considerable amount of overlap between the concerns of the professor of natural philosophy and that of mathematics. Thus among the responsibilities of the professor of mathematics was to teach 'the general principles of geometry and astronomy'.

By the late eighteenth century, scientific instruction was not complete without an introduction to chemistry (Morrell 1969 [*218]; Kent 1950 [*242]), a subject which had outgrown its always tenuous connections with natural philosophy. Natural history – once the province of the gentleman amateur but increasingly scientifically respectable since the advent of the Linnaean system of classification – also began to assume a more significant place in a well-rounded education. Like chemistry, natural history had been traditionally associated with the production of drugs and so assumed particular importance at the universities of Glasgow and Edinburgh because of their important medical schools. As Reid wrote: 'The progress of botany and natural history, and the wonderful discoveries in chemistry, have now extended the sphere of these useful branches beyond the mere purposes of the physician, and have rendered a competent knowledge of them highly interesting to every man of liberal education' (Reid 1799 [*240:736]).

Despite the huge changes in the teaching of natural and moral philosophy that had occurred in the Scottish universities over the course of the eighteenth century, there still remained at the end of the century at least the outlines of the order of studies which had for some long prevailed in the traditional scholastic order. As in the seventeenth century, the study of moral philosophy preceded the study of natural philosophy which in turn was preceded by the study of logic which, even in the early nineteenth century, remained the foundation of all further philosophical studies – as it had been in the medieval universities (Report 1831 [*162:118]). As Reid wrote of late eighteenth-century Glasgow:

'*Logic* has, in general, preceded the other two [branches of philosophy] in the order of teaching, and has been considered as a necessary preparation for them' (Reid 1799 [*240:734]). An exception was Aberdeen where at both King's and Marischal the introductory study of logic had been largely replaced by the study of civil and natural history (Wood 1990 [*197:138]). This difference probably derived from the fact that, whereas at the other Scottish universities reform had been instituted early in the century, at Aberdeen wholesale reform of the curriculum did not come until 1753 when the hold of academic tradition had considerably weakened.

However, the actual content of the discipline of logic changed considerably over the course of the century. At its beginning what Reid called 'the syllogistic art, taken from the Analytics of Aristotle' still largely prevailed but this gradually faded as the work of Locke gained increasing prominence in the Scottish universities from the 1720s (Howell 1971 [*17:273]). Furthermore, the work of Hutcheson and, later, the Common Sense philosophers more and more turned the discipline of logic into an examination of the capabilities of the human mind. At late eighteenth-century Glasgow the course in logic began by offering 'a short analysis of the powers of the understanding, and an explanation of the terms necessary to comprehend the subjects of his course'; rather dismissively, the professor then added 'a historical view of the rise and progress of the art of reasoning, and particularly of the syllogistic method, which is rendered a matter of curiosity by the universal influence which for a long time it obtained over the learned world'. After this brief historical excursus the professor returned to more congenial territory with 'an illustration of the various mental operations' (Reid 1799 [*240:735]).

Nonetheless, under whatever guise, logic remained an integral part of the Scottish undergraduate system (leaving aside Aberdeen). The longevity and significance of the subject owed much to the fact that, when the old regenting system was abolished, philosophy was originally divided among two professors holding chairs of logic and moral philosophy (the latter of which originally included the teaching of natural philosophy) (Rendall 1978 [*148:49])). The former professorship, as Veitch writes, 'was devoted to the topics of Intellectual Philosophy – embracing generally Logic, Psychology, and Metaphysics, and, in the cases of St Andrew and Glasgow, Rhetoric as well' (Veitch 1877 [*164:207]). This continuing allegiance to logic (however defined) helped to distinguish the Scottish academic world from the mathematical emphasis cultivated at Cambridge which, in the nineteenth century, became more and more the dominant intellectual model of how a British university should function (Davie 1961 [*167]). Hence when, in 1831, a Royal Commission attempted to persuade the Scottish universities to go down the Cambridge path by downplaying logic

V

and increasing the emphasis on mathematics the response was a declaration of the pedagogical merits of logic combined with an assertion of Scottish identity. Logic, it was asserted in a minority report, was calculated 'to call forth in a remarkable degree the faculties of the mind, thus creating a zeal for Science and Philosophy which mere Classical and Mathematical study may fail to excite; and we are therefore decidedly of opinion that it ought to continue to hold the place which, in almost all our universities, has been assigned to it' (Report 1831 [*162:91]).

While at least vestiges of the traditional scholastic ordering of philosophical studies could be discerned in the progression from logic to moral philosophy and thence to natural philosophy, metaphysics, once an integral part of the undergraduate curriculum, had in Scotland, as in England and Europe more generally, largely faded in importance. Significantly, when the faculty of Marischal College, Aberdeen instituted a reform of the curriculum in 1753 – reform which included the belated abolition of the regenting system – it singled out metaphysics, along with logic, as subjects which, since the 'reformation in philosophy', took up time which could be better spent on 'some very useful parts of knowledge' (Withrington 1970 [*145:186]). Within the Scottish universities metaphysics survived chiefly as a part of the study of 'pneumatics', a subject generally combined with moral philosophy. The scope of pneumatics was defined in 1734 by the Senatus of the University of Edinburgh as embracing 'the being and perfections of the one true God, the nature of Angels and the soul of man, and the duties of natural religion' (Rendall 1978 [*148:49]). However, as the century progressed, pneumatics tended to lose these vestiges of metaphysics and became more akin to the study of the human faculties which the Common Sense philosophers had done so much to promote. Hence James Beattie, a disciple of Reid and professor of moral philosophy and logic at Marischal College, Aberdeen from 1760 to 1803, wrote in the introduction to his *Elements of Moral Sciences* that 'The Speculative part of the Philosophy of the mind has been called Pneumatology'(1790 [*367:I,13]).

The waning of metaphysics was one of the most striking instances of the dissolution of the traditional scholastic curriculum in the course of the eighteenth century. And, though the other branches of the old philosophical order – moral philosophy, natural philosophy and logic – survived in name, their content was almost totally changed. For it was the great achievement of the eighteenth-century Scottish universities to transform the philosophical heritage it had received from the previous centuries into new forms – forms which intermeshed with the mind of the Enlightenment and which helped to make Scotland a major centre of intellectual innovation in analyzing the changes in society and politics wrought by the eighteenth century. That

Scotland's universities played a major part in such an intellectual odyssey at a time when universities elsewhere in Great Britain and, indeed elsewhere in Europe (with rare exceptions such as Göttingen), tended more to echo than to originate the major currents of thought of the age underlines the innovative character of Scottish society and the flexibility of that much-decried but long-lived institution – the university.

Learned Societies

Part of the reason for the pre-eminence of the universities in eighteenth-century Scottish life was that their manifest differences from Oxford and Cambridge helped to distinguish Scotland from an England with which it had been merged politically under the terms of the Act of Union of 1707. The universities, like the Church and the Scottish system of Roman-based law, helped to make Scotland more than an outer province of Great Britain and kept alive a sense of Scottish national identity. For Edinburgh, the transition after 1707 from a city that had been a national capital to one which, in some senses, was just another provincial town was a particularly painful one – but one which was alleviated by Edinburgh's success in creating for itself a new identity. For, over the course of the first half of the century, Edinburgh became the focus of a whole set of learned societies which, together with the university, helped to secure for Edinburgh the dignity of the capital of a Scottish Republic of Letters (Jones 1983 [*178:90]).

The clientele for such societies was drawn from the professional and aristocratic classes of Scotland – men who, in earlier times, would have concentrated their energies on the political chase, a pursuit no longer so readily engaged in with the abolition of the Edinburgh-based Scottish parliament. The desire to establish one's position in society through the pursuit of the enlightened culture promoted by learned societies was strengthened by a strong sense of the need to redress Scotland's economic backwardness and to close both the cultural and economic gap with its larger and more assertive Southern neighbour (Phillipson, 1974 [*221]). Such a goal necessitated the application to Scotland of the techniques of agricultural and commercial improvement which were bearing such diverse and bountiful fruits in England. Hence one of the first and most influential of Edinburgh's societies was known as The Honourable the Society of Improvers in the Knowledge of Agriculture (founded 1723). It was to be the first of a series of such bodies whose improving goals were apparent in their titles: Edinburgh Society for Encouraging Arts, Sciences, Manufactures and Agriculture (founded 1755), the Edinburgh Society for the Importation of Foreign Seeds and Plants (founded 1763), the Edinburgh

V

Society for the Investigation of Natural History (founded 1782), the Highland Society (founded 1784) and the Agricultural Society of Edinburgh (founded 1790).

In the minds of the Edinburgh literati the goals of improvement and the promotion of national prestige required philosophical underpinnings. Hence such practical societies were closely associated with societies devoted to the promotion and dissemination of philosophical enquiry. The first such society (and the virtual forerunner of other Scottish learned societies) was the Rankeian Club which was founded in 1716 and lasted until 1774 and which, significantly, took as its goal 'mutual improvement by liberal conversation and rational inquiry' (McElroy 1969 [*263:22]). The original impetus for its foundation reflected the impact of the Act of Union and the consequent increased exposure to things English – for it was intended to provide a forum for discussing such recent English philosophers as Samuel Clarke, Joseph Butler, the English Deists and that product of the Anglo-Irish Ascendancy, George Berkeley (Emerson 1979 [*268:156]).

Significantly, the Rankeian Club was established within the University of Edinburgh for, in contrast to England, the Scottish learned societies were generally closely linked with the universities. For it was the universities that provided them with the bulk of their members and which helped to strengthen that philosophical impulse that was so notable a feature of both the Scottish universities and the learned societies. The close association between the universities and the learned societies became more pronounced as the Universities of Edinburgh and Glasgow were reformed and revitalised during the 1720s and 1730s (Jones 1983 [*178:99]). This transformation also helped to give greater prominence to new forms of philosophical enquiry in both the universities and the societies which drew much of their intellectual nutriment – as well as their membership – from the universities.

The Select Society, which was founded in 1754 by some of the more recent products of the Scottish universities led by Adam Smith and David Hume (Phillipson 1974 [*221:444]), placed philosophical enquiry at the centre of its goals – goals which were summarised as 'the pursuit of philosophical inquiry, and the improvement of the members in the art of speaking' (McElroy 1969 [*263:48]). The chief aim of such philosophical discussion was to reflect on the nature of citizenship and the ways in which Scotland could best undertake that process of modernisation and improvement which was essential if it was not to be overwhelmed both economically and intellectually by its more populous Southern neighbour (Phillipson 1974 [*221:445]; Emerson 1973 [*264]). Such philosophical enquiry led in turn to practical action to advance the cause of improvement: in 1755 the Select Society founded the Edinburgh Society for

Encouraging Arts, Sciences, Manufactures and Agriculture with the goal of offering prizes for useful innovations and in 1763 it was instrumental in the establishment of the Edinburgh Society for the Importation of Foreign Seeds and Plants (Shapin 1974 [*266]).

The symbiosis between learned societies and the universities was strengthened by the fact that the activities of the learned societies were generally seen as complementing rather than competing with the formal instruction offered by the universities. The Society for the Improvement of Medical Knowledge, for example, was founded in 1731 by a group of Edinburgh medical professors and physicians who wished to have a forum in which to discuss recent advances in medicine. Their example was, in turn, emulated by their students who, in 1734, formed the Medical Society (Macarthur 1993 [*273]). Both these societies were encouraged by the university to the point where they became an integral part of Edinburgh intellectual life: the former evolving into the Philosophical Society of Edinburgh (founded 1737) while the latter was incorporated in 1778 as the Royal Medical Society of Edinburgh. Student initiative also prompted the formation of the Speculative Society in 1764 which became one of the major forums for philosophical debate within the University of Edinburgh.

The need for such philosophical debating societies perhaps in part derived from the changes in university life which led to the abandonment of the traditional practice of the formalised disputation, an integral part of the traditional scholastic curriculum (Veitch 1877 [*164:213]). Eighteenth-century philosophical instruction was generally lecture-based leaving little room for the active discussion which such societies could generate. Moreover, such societies could act as a forum not only for students but also for staff wishing to explore new modes of philosophical enquiry. The Philosophical Society of Aberdeen (1758–1773), for example, played a major role in the development of the Common Sense school of philosophy. At Glasgow, the university seemed to be less sympathetic to such societies than at Edinburgh or Aberdeen, but there, too, there were a number of short-lived student philosophical discussion societies. In 1776, for example, there were three such societies: the Eclectic, the Dialectic and the Academic (McElroy 1969 [*263:122]).

The relative dearth of learned societies at Glasgow reflects not only the attitude of the university but the fact that Edinburgh could draw on a larger pool of educated professionals who had the education, means and leisure to join such societies (Shapin 1974 [*267:5]). Edinburgh was the centre of the Scottish legal system (and therefore had the further advantage of the huge Advocates' Library) and it was also where the Assembly of the Church of Scotland was held. It was natural, too, that Edinburgh's ancient dignity as the traditional political capital of Scotland should be reflected in its position as the

capital of Scotland's republic of letters. Thus Scotland's principal academy –
The Royal Academy of Edinburgh (founded 1783) – was, as its name suggests,
closely identified with Edinburgh. The aim of the academy, in contrast to the
Royal Society of London, was to act as a body for promoting all forms of
polite learning not simply natural philosophy. As a 'literary and philosophical
society', then, the Academy had much in common with the learned societies of
provincial England which, like Edinburgh, were too small to support specialist
societies for both science and literary pursuits in the manner of London.

However, the ambivalent status of post-Act of Union Edinburgh as a former
capital reduced to the status of a provincial town (*ibid.*:3) was brought out in
the way in which the Academy also looked to European models. Thus William
Robertson, Principal of the University of Edinburgh (which, predictably,
took an active role in the Academy's foundation), urged its founders to model
the institution on those European academies which cultivated 'every branch
of science, erudition and taste' (McElroy (1969 [*263:79]). Such broad aims
had been a part of the pre-history of the Academy. For, in the 1730s, The
Society for the Improvement of Medical Knowledge had attracted so diverse
a range of interests that Professor Colin MacLaurin suggested the formation
of a general learned body which could cultivate fields as diverse as natural
knowledge and Scottish antiquities (Emerson 1979 [*268]). Consequently, in
1737, the Edinburgh Philosophical Society was founded which, in turn, was
transformed into The Royal Academy of Edinburgh in 1783. However, the
attempt to fuse together the literary and scientific interests of Scotland's
literati under the aegis of the Academy had limited success for the literary
section disappeared from the Academy's transactions after 1798 (McElroy
1969 [*263:81]). The increasingly aristocratic tone of the Academy, which was
accentuated by the conferral of its royal charter in 1783, also prompted the
formation of the short-lived Academy of Physics of Edinburgh (1797–1800).
This was dedicated to Newtonian physics and contemptuous of what its virtual
founder, Henry Brougham, called the 'abominable politics, trifling pursuits and
vile aristocracy' which characterised both the Royal Society of London and the
Royal Academy of Edinburgh (Cantor 1974 [*265:116]).

The Scottish learned societies, then, helped to ensure that philosophical
enquiry (including science) became part of the civic culture of Enlightenment
Scotland. Such knowledge was, in turn, pressed into service in the great
mission of the Scottish literati: the modernisation of Scotland through
the techniques of improvement in order to defend national honour and to
ensure that Scotland did not become a poor outer province of England. Such
improvement necessitated not only such practical measures as the application
of better agricultural methods or the promotion of new forms of commerce

but, more fundamentally, required a modernisation of Scotland's political and social culture. This meant that the philosophical innovations promoted by the Scottish universities needed to be related to the practical problems of the age – a task in which Scotland's learned societies played a critical role by bringing to bear the philosophical principles of criticism and experience on the institutions which sustained the larger society. That, in doing so, they not only successfully nurtured Scotland's economic modernisation but also maintained a sense of Scottish separate identity in an age when the danger of being swamped by English culture was strong, indicates the vitality and pervasive influence of Scotland's learned societies.

Ireland

Trinity College, Dublin

As a resolutely Protestant institution Trinity College, Dublin was at its foundation in 1591 greatly indebted to the Puritan-tinctured Cambridge of the late sixteenth century. It was, in the words of the seventeenth-century historian, Thomas Fuller, a '*colonia deducta*' of Cambridge largely deriving its statutes from that university and drawing its first five provosts from thence (Mullinger 1888 [*67:135]). In the different conditions of the eighteenth century, too, Trinity College retained a strong political and intellectual affinity with Cambridge, both universities siding ideologically with the Hanoverian regime and its defence of the Protestant succession. For Trinity, the university of the Anglo-Irish Protestant Ascendancy, the Hanoverian regime represented a bulwark against the claims of the Catholic Stuarts and their Irish Catholic supporters. Like Cambridge, Trinity was, then, impregnated by the Whiggish principles linked with the defence of the Hanoverian order – as a pamphleteer of 1783 put it, since the accession of the House of Hanover 'the College of Dublin has ever peculiarly merited the name of a Whig university' (McDowell and Webb 1982 [*292:72]). Whiggish political principles were linked at Trinity, as at Cambridge, with the institutionalisation of the intellectual systems of Newton and Locke the rationality of which was seen as consonant with the claims of the orderly Hanoverian state to be based on reason as against the political mystifications of the Stuarts.

Over the course of the eighteenth century, then, the world-view of Newton and Locke gradually became more and more deeply entrenched at Trinity particularly since, as at Cambridge, the work of these two great luminaries of the English Enlightenment was seen as supportive of Anglican

V

Christianity rightly understood. By 1759 a pamphleteer could write that 'The Newtonian Philosophy; the excellent Boyle's experimental Philosophy, and Mr Locke's Metaphysicks prevail much in the College of Dublin' (Maxwell 1946 [*290:149]). At the beginning of the century this connection between Lockean epistemology, Newtonian natural philosophy and Anglican apologetics had helped to shape the outlook of Trinity's most famous philosophical offspring, Bishop George Berkeley, an undergraduate there from 1700 to 1704 and a fellow from 1707 to 1724.

Though advanced spirits like Berkeley might assimilate the work of Locke and Newton early in the century the diffusion of their work to the rank and file of Trinity students was a much more gradual process. For, as at Oxford, so, too, at Trinity, the traditional scholastic order had been effectively shored up in the early seventeenth century by Archbishop Laud who became Chancellor in 1633 and instituted a new set of statutes expunging Ramist influences and reasserting an orthodoxly Aristotelian curriculum (McDowell 1947 [*291:14]). The consequence of this, together with natural academic conservatism reinforced by Ireland's relative isolation, was that in the early eighteenth century the general outlines of the traditional scholastic curriculum could be discerned. And, given the dominance of philosophy in this traditional curriculum, the consequence was that Trinity undergraduates were given a reasonably comprehensive introduction to most of the major branches of philosophy. In changed form this wide philosophical coverage was to continue throughout the century and, indeed, eighteenth-century Trinity provided a rather broader philosophical education than either Oxford or Cambridge. It also combined this with a wider scope for classical studies than prevailed across the Irish Sea (at least at Cambridge) (Clarke 1959 [*16:160]; Stanford 1941 [*288]).

At Trinity, then, the hold of scholastic tradition remained strong for much of the century. Thus if one examines the undergraduate curriculum for around 1736 much of the framework of the traditional scholastic order still survived. The primacy of logic was maintained with this discipline being prescribed for the first two years while natural philosophy was set for the third year and ethics and metaphysics for the fourth. Trinity also still retained an attachment to the world of international scholastic scholarship which had transcended the bitter confessional divisions of the seventeenth century. The texts in logic ranged from that of Dutch Protestants like Franco Burgesdijk (1626 [*314]) and Jean Le Clerc (1692 [*318]) to the Polish Jesuit, Martin Smiglecius (1618 [*313]). In ethics and metaphysics the French Cistercian, Eustachius a S.Paulo (1654 [*354]), was combined with the staunchly Protestant Scottish Robert Baron (1654 [*333]) and the English Bishop Robert Sanderson (1647 [*352]).

The main concession to the outlook of the Enlightenment was the work of Pufendorf (1673 [*358]).

Appropriately, in natural philosophy the hand of the moderns was more evident. For, in most universities throughout Europe, it was the developments of the Scientific Revolution which were to undermine the traditional curriculum in natural philosophy and, with it, much of the scholastic philosophical curriculum more generally. Like many universities of the late seventeenth and eighteenth centuries, Trinity attempted to marry the old and the new in the realm of natural philosophy the result being, as one undergraduate critic remarked in 1703, a farrago of conflicting hypotheses from Aristotle, Descartes, Colbert (1678 [*386]), Epicurus, Gassendi, Malebranche and Locke (McDowell and Webb 1982 [*172:32]). Even by 1736 the set texts in natural philosophy were largely pre-Newtonian in character though they took account of the scientific developments of the sixteenth and seventeenth centuries and, in particular, the work of Descartes. Thus the set texts included works by seventeenth-century Dutch authors such as Le Clerc (1696 [*388]) and Bernadus Varenius (1650 [*384])(though the latter work had been scantily revised by Newton in 1681 and, more fully, by his disciple, James Jurin, in 1712). The most recent work was a rather pedestrian textbook on astronomy by the undistinguished Edward Wells (1712 [*393]) (Mc Dowell l947 [*291:17–23]; McDowell and Webb 1982 [*292:46–49]).

Mathematical studies were slow to take root at Trinity – a situation which acted as a brake on the range and modernity of its curriculum in natural philosophy. Geometry was not introduced until around 1760 and algebra not until about 1808 (McDowell and Webb 1982 [*292:19]). This dilatory approach to mathematical instruction may have owed something to the strength of logical studies at Trinity for mathematics (and particularly geometry) were frequently regarded as an alternative to logic. Thus when Euclid was introduced into the undergraduate curriculum at Trinity it was combined with the study of logic in the first year (Anon.1833 [*285:215]). The study of logic was consolidated by the fact that over the course of the late seventeenth and eighteenth centuries two of the college's provosts – Narcissus Marsh and Richard Murray – produced logical textbooks (1679 [*316] and 1782 [*323])) which became a staple part of the curriculum. These works were largely Aristotelian in content but made some concessions to the age in their format and clarity of presentation (Furlong 1942 [*289:40f., 45]). By the later eighteenth century such traditional logic was combined with the study of Locke (Anon. 1833 [*285:215]) despite the incongruity of exposing undergraduates to both traditional logic and an author so critical of such a discipline.

By this period, too, Locke's political writings were prescribed in the fourth year as part of the moral philosophy section of the curriculum along with the work of the mid-eighteenth-century Genevan natural law theorist, Jean-Jacques Burlamaqui (1748 [*362]). With the coming of the French Revolution, however, Locke's contractarian views of government came to seem rather more suspect and we find in 1798 Thomas Elrington (fellow and later provost) including in his preface to a Dublin edition of Locke's *Essay on Government* the claim that 'the venerable advocate of political freedom has sometimes unguardedly expressed himself' and that the unwary could interpret him as favouring democracy (McDowell 1947 [*291:30]).

By the late eighteenth century Locke's intellectual partner, Newton, was well-entrenched in the curriculum with the work of his loyal disciple, John Keill, being set as the textbook in natural philosophy together with works based on lectures by two Newtonian-inclined fellows of Trinity, Richard Helsham (1739 [*399]) and Hugh Hamilton (1767 [*401]). The increasing attention paid to science over the course of the eighteenth century can also be traced in the foundation of such chairs as the Professor of Natural and Experimental Philosophy (1724), Professor of Mathematics (1762), Professor of Chemistry (1785) and Professor of Botany (1785).

Overall, then, eighteenth-century Trinity offered an undergraduate curriculum which provided an introduction to the chief branches of philosophy (though metaphysics there, as elsewhere, diminished in importance over the course of the century). Its curriculum tended to lag behind the major intellectual developments of the day especially those occurring on the Continent. Nonetheless, it played its part in inculcating the heritage of Locke and Newton and, with it, a respect for reason and a belief that true religion and science could work in partnership.

With the conspicuous exception of Bishop Berkeley, Trinity did not, however, produce any philosophical writings of any note. Part of the explanation for this may lie in the fact that there, as at Oxford and Cambridge, all tutors taught the full range of subjects and thus did not develop the specialised philosophical expertise that existed, for example, in the Scottish universities. As an early nineteenth-century critic put it: at Trinity 'the great evil of the system is the want of division of labour: every tutor lectures on the same things' (Anon. 1833 [*285:232]). But perhaps what Trinity lost in depth of philosophical enquiry it to some extent compensated for in the breadth of a curriculum. For it introduced its undergraduates both to the learning of the ancients through its comprehensive classics curriculum and to at least those moderns most in tune with the ideological outlook of a Protestant Ascendancy – an Ascendancy

V

which, however inappropriately, regarded itself as part of a political and religious order based on the principles of reason and order.

Learned Societies

Since Ireland was virtually a colonial society with a thin layer of Protestant rulers over a subject Catholic population the likely clientele for philosophical clubs and societies was a limited one. Nonetheless, the Protestant Ascendancy did support a few societies which helped to disseminate philosophical concerns beyond the confines of the university. With their close links with England some members of the Ascendancy attempted to transplant into the Irish context some of the Enlightenment culture which in England was given wide currency by philosophical societies or societies devoted to the improvement of agriculture or economic life more generally.

The earliest such society was the Dublin Philosophical Society which lasted from 1683 to 1708 and which drew on the example of the Royal Society. Just as the early membership of the Royal Society drew heavily on the alumni of Oxford and Cambridge, the bulk of the Dublin Philosophical Society's membership was drawn from alumni of Trinity College Dublin – eighteen of the society's foundation members being graduates of that bastion of the Ascendancy (Hoppen 1970 [*297:53]). The society, like many of its English philosophical counterparts, showed little interest in the direct application of science to such economically central areas as agriculture or textiles (*ibid.*:152). Its concern was more with the intellectually ennobling character of science as a conduit for Enlightenment values.

However, such practical concerns for the improvement of the economy and society more generally were taken up by the Dublin Society for Improving Husbandry, Manufactures and other Useful Arts founded in 1731. Again the links with Trinity College Dublin were strong: the first meeting of the society was held in the college's 'Philosophical Rooms' and virtually all its members were graduates. By the terms of the constitution each member of the society was required 'to choose some particular subject either in Natural History or in Husbandry, Agriculture or Gardening, or some species of manufacture or other branch of improvement' and to report his reading or his experiments thereon' (Royal Dublin Society 1932 [*296:xxx, 11]).The Society actively sought to promote innovation by publishing a regular account of its proceedings and by establishing prizes for useful inventions. Its role in fostering much-needed improvement within Irish agriculture and industry was recognized in 1746 by an annual government subsidy of five hundred pounds (Lecky 1913–19 [*295 (I): 298f.]).

V

Appropriately, when, in 1766, the society eventually established its permanent premises it located itself alongside the Royal Irish Society which in 1786 was incorporated as the Royal Irish Academy. The Academy provided a permanent forum for the activities once pursued by the Dublin Philosophical Society as well as for other aspects of gentlemanly culture such as the study of antiquities (Lecky 1913–19 [*295 (II):505]). Like the other eighteenth-century Irish learned societies, it was an institution that proclaimed the Irish Protestant Ascendancy's links with the Enlightenment culture of a Whiggish political establishment on which its power rested.

Original version: John Gascoigne: 'Philosophie an den britischen Universitäten und in den gelehrten Gesellschaften'. In: *Grundiss der Geschichte der Philosophie. Begründet von Friedrich Ueberweg. Die Philosophie des 18. Jahrhunderts*, Vol. 1/1: Grossbritannien und Nordamerika. Niederlande. Ed. by Helmut Holzhey and Vilem Mudroch. Basel: Schwabe Verlag, 2004. pp. 3–45.

BIBLIOGRAPHY

The bibliographical format follows the original German edition (which is a slightly abridged translation of this English original). I am grateful to the editors, Helmut Holzhey and Vilem Mudroch, along with Francis Cheneval and Wolfgang Rother, for undertaking the arduous task of checking and standardising these references.

Background [*1–*3a].

England [*11.–*132].
 General [*11.–*18].
 Oxford [*30–*43].
 Cambridge [*51–*80].
 Dissenting Academies [*91–*101].
 Learned Societies [*112–*132].

Scotland [*143–*273].
 General [*143.–*152].
 Universities in General [*162–*182].
 Aberdeen [*193–*198].
 Edinburgh [*209–*229].
 Glasgow [*240–*248].
 St Andrews [*253–*256].
 Learned Societies [*263–*273].

Ireland [*285–*297].
 Trinity College, Dublin [*285–*292].
 Learned Societies [*295–*297].

Textbooks [*301–*401].
 General works [*301.–*305].
 Logic [*311.–*325].
 Metaphysics [*331.–*340].
 Moral philosophy [*351.–*367].
 Natural philosophy [*381.–*401].

Background

1. Charles Schmitt, 'Towards a reassessment of Renaissance Aristotelianism', *History of Science*, 11 (1973), pp. 159–193.

V

2. Lawrence Brockliss, 'Philosophy teaching in France 1600–1740', *History of Universities*, 1 (1981), pp. 131–168.

3. John Gascoigne, 'A reappraisal of the role of the universities', *Reappraisals of the Scientific Revolution*, ed. D. Lindberg and R.S. Westman (Cambridge, 1990) pp. 207–260.

3a. Jean-Pierre Schobinger (ed.), *Grundriss der Geschichte der Philosophie. Die Philosophie des 17 Jahrhunderts*, Vol. 3 *England* (Basel, 1988) xxxiv + 874pp.

England

General

11. John Locke, *Some familiar letters between Mr Locke, and several of his friends* (London, 1708) 540pp.

12. Edward A. Burtt (ed.), *The English philosophers from Bacon to Mill* (New York, 1939) xxiv + 1041pp.

13. Wilfred Alfred Leslie Vincent, *The state and school education 1640–1660, in England and Wales: a survey based on printed sources* (London, 1950) 156pp.

14. Nicholas Hans, *New trends in education in the eighteenth century* (London, 1951) 261pp.

15. Martin L. Clarke, *Classical education in England* (Cambridge, 1959) 233pp.

16. James L. Axtell (ed.), *The educational writings of John Locke. A critical edition with introduction and notes* (Cambridge, 1968) 442pp.

17. W.S. Howell, *Eighteenth century British logic and rhetoric* (Princeton, 1971) xii + 742pp.

18. Geoffrey N. Cantor, *Optics after Newton. Theories of light in Britain and Ireland, 1740–1840* (Manchester, 1983) ix + 257pp.

Oxford

30. Christopher Wordsworth, *Scholae academicae. Some account of studies at the English universities in the eighteenth century* (Cambridge, 1877) xii + 435pp.

31. Falconer Madan, *Oxford books: a bibliography of works relating to the University and the City of Oxford*, 3 vols. (Oxford, 1895–1931).

32. John Richard Green and George Thomas Roberson, *Studies in Oxford history chiefly in the eighteenth century*, ed. by Charles L. Stainer (Oxford, 1901) xx + xxiii + 382pp.

33. J.A. Venn, 'Matriculations at Oxford and Cambridge, 1544–1906', *Oxford and Cambridge Review* (1908), pp. 48–66.

V

34. Charles E. Mallett, *A history of the University of Oxford*, 3 Vols. (London, 1924–27) 448; 502; 530pp.
35. Strickland Gibson (ed.), *Statuta antiqua Universitatis Oxoniensis* (Oxford, 1931) cxxii + 668pp.
36. William R. Ward, *Georgian Oxford. University politics in the eighteenth century* (Oxford, 1958) x + 296pp.
37. Walter G. Hiscock, *Henry Aldrich of Christ Church 1648–1710* (Oxford, 1960) 75pp.
38. Edward Harold Cordeaux and Denis Harry Merry, *A bibliography of printed works referring to the University of Oxford* (Oxford, 1968) xxvii + 809pp.
39. William Norman Hargreaves-Mawdsley, *Oxford in the age of John Locke* (Norman, 1972) xi + 132pp.
40. Sheldon Rothblatt, 'The student sub-culture and the examination system in early 19th century Oxbridge', *The university in society*, Vol. 1, ed. Lawrence Stone (Princeton N.J., 1975), pp. 247–303.
41. Lawrence Stone, 'The size and composition of the Oxford student body 1580–1909', *The university in society*, Vol. 1, ed. L. Stone (Princeton, N.J., 1975), pp. 3–110.
42. Lucy Sutherland and Leslie George Mitchell (eds.), *The history of the University of Oxford*, Vol. 5, *The eighteenth century* (Oxford, 1986) xix + 949pp.
43. Edward G.W. Bill, *Education at Christ Church Oxford 1660–1800* (Oxford, 1988) 367pp.

Cambridge

(Manuscript Sources)

51. Thomas Balguy, 'A system of morality' (St John's College, Cambridge, MS 545).
52. William Bennet, 'Lectures on logic', 2 Vols. (Emmanuel College, Cambridge, MSS 196–7).
53. William Derham, 'Recollections of Isaac Newton' (King's College, Cambridge, Keynes MS 130).
54. Joseph Littledale, Logic notebook (St John's College, Cambridge, MS not folioed).
55. Moderator's book for 1778 (Trinity College, Cambridge, MS R. 2.81, (45).

(Printed Sources)

59. Benjamin Hoadly (ed.), The *works of Samuel Clarke*, 4 vols. (London, 1738–42).

V

60.   John Disney, *Memoirs of the life and writings of Arthur Ashley Sykes* (London, 1785) xxiv + 367pp.

61.   George Dyer, A *history of the University of Cambridge*, 2 vols. (London, 1814) 268; 452pp.

62.   George Dyer, *The privileges of the University of Cambridge; together with additional observations on its history, antiquities, literature and biography*, 3 vols in 2. (London, 1824) 630; 260; 200pp.

63.   Thomas Fuller, *The history of the University of Cambridge and of Waltham Abbey*, ed. James Nichols (London, 1840) xxiv + 668pp.

64.   James Heywood (ed.), *Collection of statutes for the University of Cambridge*, 2 vols. (London, 1840–1845) 359, 43; 246, 238pp.

65.   Charles H. Cooper, *Annals of Cambridge*, 5 vols. (Cambridge, 1842–1908) 452; 619; 614; 712; 656pp. – Vol. 1: 695–1546, Vol. 2: 1546–1602, Vol. 3: 1603–1688, Vol. 4: 1688–1849, Vol. 5: 1850–1856, Index.

66.   Christopher Wordsworth, *Social life at the English universities in the eighteenth century* (Cambridge, 1874) 727pp.

67.   James B. Mullinger, *A history of the University of Cambridge* (Cambridge, 1888) xvi + 232pp.

68.   Denys A. Winstanley, *The University of Cambridge in the eighteenth century* (Cambridge, 1922) vi + 349pp.

69.   Denys A. Winstanley, *Unreformed Cambridge: a study of certain aspects of the University in the eighteenth century* (Cambridge, 1935) xi + 411pp.

70.   Hester Jenkins and D.Caradog Jones, 'Social class of Cambridge University alumni of the 18th and 19th centuries', *The British Journal of Sociology*, 1 (1950), pp. 93–116.

71.   Edward Hughes, *North country life in the eighteenth century: the north-east, 1700–1750* (London, 1952) 435 pp.

72.   Ben R. Schneider, *Wordsworth's Cambridge education* (Cambridge, 1957) 298pp.

73.   Edward W. Strong, 'Newtonian explications of natural philosophy', *Journal of the History of Ideas*, 18 (1957), pp. 49–83.

74.   William T. Costello, *The scholastic curriculum at early seventeenth-century Cambridge* (Cambridge, Mass. 1958) 221pp.

75.   M.A. Hoskin, '"Mining all within": Clarke's notes to Rohault's, *Traité de Physique*', The *Thomist* 24 (1961), pp. 353–363.

76.   D.L. Le Mahieu, *The mind of William Paley. A philosopher and his age* (Lincoln, Nebraska, 1976) xi + 215pp.

77.   Henry J. Steffens, *The development of Newtonian optics in England* (New York, 1977) vii + 190pp.

78.   John Gascoigne, 'Mathematics and meritocracy: the emergence of the Cambridge Mathematical Tripos', *Social Studies of Science*, 14 (1984), pp. 547–584.

79. John Gascoigne, 'Politics, patronage and Newtonianism: the Cambridge example', *Historical Journal*, 27 (1984), pp. 1–24.

80. John Gascoigne, *Cambridge in the age of the Enlightenment. Science, religion and politics from the Restoration to the French Revolution* (Cambridge, 1989) xi + 358pp.

Dissenting Academies

91. Philip Doddridge, *Correspondence and diary of Philip Doddridge*, 5 vols., ed. John Doddridge Humphreys (London, 1829) 488; 520; 560; 576; 552pp.

92. Irene Parker, *Dissenting Academies in England. Their rise and progress and their place among the educational systems of the country* (Cambridge, 1914) xii + 168pp.

93. H. McLachlan, *English education under the Test Acts* (Manchester, 1931) xvi + 344pp.

94. C.E. Whiting, *Studies in English Puritanism from the Restoration to the Revolution, 1660–1688* (London, 1931) xvi + 584pp.

95. J.F. Fulton, 'The Warrington Academy (1757–1786) and its influence upon medicine and science', *Bulletin for the Institute of the History of Medicine*, 1 (1933), pp. 50–80.

96. Joe William Ashley Smith, The *birth of modern education: the contribution of the dissenting academies 1660–1800* (London, 1954) xii + 329pp.

97. John A. Passmore (ed.), *Priestley's writings on philosophy, science and politics* (New York, 1965) 352pp.

98. David C. Humphrey, 'Colonial colleges (USA) and English Dissenting Academies: a study in transatlantic culture', *History of Education Quarterly*, 12 (1972), pp. 184–197.

99. Barbara B. Smith, *Truth, liberty, religion. Essays celebrating two hundred years of Manchester College* (Oxford, 1986) xxiv, 325pp.

100. Padraig O'Brien, *Warrington Academy 1757–1786* (Wigan, 1989) 164pp.

101. Alan P.F. Sell, 'Philosophy in the eighteenth-century dissenting academies of England and Wales', *History of Universities*, 11 (1992), pp. 75–122.

Learned Societies

112. Charles R. Weld, A *history of the Royal Society*, 2 Vols. (London, 1848) 527; 611pp.

113. Henry Lyons, *The Royal Society 1660–1940. A history of its administration under its charters* (Cambridge, 1944) x + 354pp.

114. H.J.J. Winter, 'Scientific associations of the Spalding Gentlemen's Society during the period 1710–50', *Archives Internationales d'Histoire des Sciences*, 3 (1950), pp. 77–88.

115. Robert E. Schofield, *The Lunar Society of Birmingham. A social history of provincial science and industry in eighteenth-century England* (Oxford, 1963) x + 491pp.

V

116. Arnold Thackray, 'Natural knowledge in cultural context: the Manchester model', *American Historical Review* 79 (1974), pp. 672–709.

117. Ian Inkster, 'Science and society in the metropolis: a preliminary examination of the social and institutional context of the Askesian Society of London, 1796–1807', *Annals of Science*, 34 (1977), pp. 1–32.

118. Morris Berman, *Social change and scientific organization: The Royal Institution, 1799–1844* (London, 1978) 224pp.

119. R.P. Sturgess, 'The membership of the Derby Philosophical Society 1783–1802', *Midland History* 4 (1979), pp. 212–229.

120. Roy Porter, 'Science, provincial culture and public opinion in Enlightenment England', *British Journal for Eighteenth Century Studies*, 3 (1980), pp. 20–46.

121. Ian Inkster, 'The public lecture as an instrument of science education for adults', *Paedagogica Historica*, 20 (1981), pp. 80–112.

122. David P. Miller, 'Sir Joseph Banks: an historiographical perspective', *History of Science*, 19 (1981), pp. 284–292.

123. Guy Kitteringham, 'Science in provincial society: the case of Liverpool in the early nineteenth century', *Annals of Science*, 39 (1982), pp. 329–348.

124. Simon Schaffer, 'Natural philosophy and public spectacle in the eighteenth century', *History of Science*, 21 (1983), pp. 1–43.

125. Margaret C. Jacob, *The cultural meaning of the Scientific Revolution* (New York, 1988), xii + 274pp.

126. David P. Miller, '"Into the valley of darkness": Reflections on the Royal Society in the eighteenth century', *History of Science*, 27 (1989), pp. 155–166.

127. A.Q. Morton, 'Lectures on natural philosophy in London, 1750–65: S.C.T. Demainbray (1710–82) and the 'inattention' of his countrymen', *British Journal for the History of Science*, 23 (1990), pp. 411–434.

128. Jan Golinski, *Science as public culture. Chemistry and Enlightenment in Britain, 1760–1820* (Cambridge, 1992) 342pp.

129. Larry Stewart, *The rise of public science. Rhetoric, technology, and natural philosophy in Newtonian Britain, 1660–1750* (Cambridge, 1992) 453pp.

130. Simon Schaffer: 'The consuming flame: electrical showmen and Tory mystics in the world of goods', in *Consumption and the world of goods*, ed. John Brewer and Roy Porter (London, 1993), pp. 489–526.

131. John Gascoigne, *Joseph Banks and the English Enlightenment. Useful knowledge and polite culture* (Cambridge, 1994) 324pp.

132. Larry Stewart and P. Weindling, 'Philosophical threads: natural philosophy and public experiment among the weavers of Spitalfields', *British Journal for the History of Science*, 28 (1995), pp. 37–62.

Scotland

General

143. Henry M. Knox, *Two hundred and fifty years of Scottish education 1696–1946* (Edinburgh, 1953) 253 pp.
144. Gladys Bryson, *Man and society: the Scottish inquiry of the eighteenth century* (New York, 1968) ix + 287pp.
145. Donald J. Withrington, 'Education and society in the eighteenth century' in *Scotland in the age of improvement*, ed. Nicholas T. Phillipson and Roslalind Mitchison (Edinburgh 1970), pp. 169–199.
146. Nicholas Phillipson, 'Towards a definition of the Scottish Enlightenment', in *City and society in the eighteenth century*, ed. Paul Fritz, David Williams (Toronto, 1973), pp. 126–147.
147. Anand C. Chitnis, *The Scottish Enlightenment: a social history* (London, 1976) 279pp.
148. Jane Rendall, *The origins of the Scottish Enlightenment* (London, 1978) vii + 257 pp.
149. Richard F. Teichgraeber, *Politics and morals in the Scottish Enlightenment* (Ph.D thesis, Brandeis University, 1978) 302pp.
150. Charles Camic, *Experience and enlightenment. Socialization for cultural change in eighteenth-century Scotland* (Chicago, 1983) x + 301pp.
151. R.A. Houston, 'Scottish education and literacy, 1600–1800: an international perspective' in *Improvement and enlightenment*, ed. T.M. Devine (Edinburgh, 1989), pp. 43–61.
152. Daniel Brühlmeier, Helmut Holzhey, Vilem Mudroch (eds.), *Scottische Aufklärung. 'A hotbed of Genius'* (Berlin, 1996) 157pp.

Universities, General

162. *Report Made to His Majesty by a Royal Commission of Inquiry into the State of the Universities of Scotland. Ordered, by the House of Commons, to be printed 7 October 1831* ([London], 1831) 436pp.
163. James McCosh, *The Scottish philosophy: biographical, expository, critical from Hutcheson to Hamilton* (London, 1875), 481pp.
164. John Veitch, 'Philosophy in the Scottish universities', *Mind*, 2 (1877), pp. 74–91, 207–234.
165. George S. Pryde, *Scottish universities and the colleges of colonial America* (Glasgow, 1957) 55pp.
166. Selwyn A. Grave, *The Scottish philosophy of common sense* (Oxford, 1960) 262pp.

V

167. George E. Davie, *The democratic intellect: Scotland and her universities in the nineteenth century* (Edinburgh, 1961) xx + 352pp.

168. David Fate Norton, *From moral sense to common sense: An essay on the development of Scottish common sense philosophy, 1700–1765* (Ph.D. thesis, University of California, San Diego, 1966) xiv + 303pp.

169. Ronald G. Cant, 'The Scottish universities and Scottish society in the eighteenth century', *Studies on Voltaire and the Eighteenth Century*, 58 (1967), pp. 1953–66.

170. Ronald G. Cant, 'The Scottish universities in the seventeenth century', *The Aberdeen University Review*, 43 (1970), pp. 223–233.

171. Arthur L. Donovan, *Philosophical chemistry in the Scottish Enlightenment. The doctrines and discoveries of William Cullen and Joseph Black* (Edinburgh, 1971) x + 343pp.

172. John L. Russell, 'Cosmological teaching in the 17th century Scottish universities', *Journal for the History of Astronomy*, 5 (1974), pp. 122–32, 145–154.

173. Richard Olsen, *Scottish philosophy and British physics 1750–1880: a study in the foundations of the Victorian scientific style* (Princeton, 1975) vii + 349pp.

174. Roger L. Emerson, 'Scottish universities in the eighteenth century', *Studies on Voltaire and the Eighteenth Century*, 167 (1977), pp. 453–474.

175. Ronald G. Cant, 'Origins of the Enlightenment in Scotland: the universities' in *The origins and nature of the Scottish Enlightenment*, ed. Roy H. Campbell and Andrew S. Skinner (Edinburgh, 1982), pp. 42–64.

176. Christine M. Shepherd, 'Newtonianism in Scottish universities in the seventeenth century' in *The origins and nature of the Scottish Enlightenment* ed. Roy H. Campbell and Andrew S. Skinner (Edinburgh, 1982), pp. 65–85.

177. Stefan Collini, Donald Winch and John Burrow, *The noble science of politics: a study in nineteenth-century intellectual history* (Cambridge, 1983) x + 385pp.

178. Peter Jones, 'The Scottish professoriate and the polite academy, 1720–416' in *Wealth and virtue. The shaping of political economy in the Scottish Enlightenment* ed. I. Hont and M. Ignatieff (Cambridge, 1983), 89–117pp.

179. Jack B. Morrell, 'Medicine and science in the eighteenth century' in *Four centuries. Edinburgh university life 1583–1983* ed. Gordon Donaldson (Edinburgh, 1983), pp. 38–52.

180. Roger L. Emerson, 'Natural philosophy and the problem of the Scottish Enlightenment', *Studies on Voltaire and the Eighteenth Century*, 242 (1986), pp. 243–292.

181. Roger L. Emerson, 'Science and moral philosophy in the Scottish Enlightenment' in *Studies in the philosophy of the Scottish Enlightenment* ed. M.A. Stewart (Oxford, 1990), pp. 11–36.

182. C. Shepherd, 'A national system of university education in seventeenth-century Scotland' in *Scottish universities: distinctiveness and diversity* ed. Jennifer J. Carter and Donald J. Withrington (Edinburgh, 1992), pp. 26–33.

Aberdeen

193. John M. Bullock, *A history of the University of Aberdeen 1495–1895* (London, 1895) viii + 220pp.
194. Robert Rait, *The universities of Aberdeen. A history* (Aberdeen, 1895) xii + 382pp.
195. Nicholas Phillipson, 'James Beattie and the defence of common sense' in *Festschrift für Reiner Gruenter* ed. Bernard Fabian (Heidelberg, 1983), pp. 145–154.
196. Christine Shepherd, 'The arts curriculum at Aberdeen at the beginning of the eighteenth century' in *Aberdeen and the Enlightenment* ed. Jennifer J. Carter and Joan H. Pittock (Aberdeen, 1987), pp. 146–54.
197. Paul B. Wood, 'Science and the pursuit of virtue in the Aberdeen Enlightenment' in *Studies in the philosophy of the Scottish Enlightenment* ed. M.A. Stewart (Oxford, 1990), pp. 127–50.
198. Roger L. Emerson, *Professors, patronage and politics: the Aberdeen universities in the eighteenth century* (Aberdeen, 1992) xv + 181pp.

Edinburgh

209. Robert Henderson, 'Short account of the University of Edinburgh', *Scots Magazine*, 3 (1741), pp. 371–374.
210. John Veitch, *A memoir of Dugald Stewart* in *The collected works of Dugald Stewart* ed. William Hamilton, Vol. XI (Edinburgh, 1858).
211. Alexander Grant, *The story of the University of Edinburgh during its first 300 years*, 2 vols. (London, 1884) 384; 510pp.
212. Isabel Kenrick, *The University of Edinburgh 1660–1715. A study in the transformation of teaching methods and curriculum* (Ph.D. Thesis, Bryn Mawr, 1957) 157pp.
213. D. McKie, 'Some notes on the students' scientific society in eighteenth-century Edinburgh', *Scientific Progress*, 49 (1961), pp. 218–241.
214. Herbert W. Turnbull (ed.), *The correspondence of Isaac Newton*, Vol. 3, 1688–1694 (Cambridge, 1961) xviii + 445pp.
215. Douglas Nobbs, 'The political ideas of William Cleghorn, Hume's academic rival', *Journal of the History of Ideas*, 26 (1965), pp. 575–586.
216. David B. Horn, *A short history of the university of Edinburgh, 1556–1889* (Edinburgh, 1967) x + 228pp.

V

217. R.D. Thornton, 'The University of Edinburgh and the Scottish Enlightenment', *Texas Studies in Literature and Language*, 10 (1968), pp. 415–422.

218. Jack B. Morrell, 'Practical chemistry in the University of Edinburgh 1799–1843', *Ambix*, 16 (1969), pp. 66–80.

219. Jennifer J. Carter, 'The making of Principal Robertson in 1762. Politics and the University of Edinburgh in the second half of the eighteenth century', *Scottish Historical Review*, 49 (1970), pp. 60–84.

220. Jack B. Morrell, 'The University of Edinburgh in the late eighteenth century: Its scientific eminence and academic structure', *Isis*, 62 (1971), pp. 158–171.

221. Nicholas T. Phillipson, 'Culture and society in the eighteenth-century province: the case of Edinburgh and the Scottish Enlightenment' in *The University and Society* ed. L. Stone, Vol. 2 (Princeton, N.J., 1974), pp. 407–448.

222. Jack B. Morrell, 'The Edinburgh Town Council and its university 1717–1766' in *The early years of the Edinburgh Medical School* ed. R.G.W. Anderson and A.D.C. Simpson (Edinburgh 1976), pp. 46–65.

223. Christine M. Eagles, 'David Gregory and Newtonian Science', *British Journal for the History of Science*, 10 (1977), pp. 216–225.

224. Nicholas T. Phillipson, 'The pursuit of virtue in Scottish university education: Dugald Stewart and Scottish moral philosophy in the Enlightenment' in *Universities, society and the future* ed. N. Phillipson (Edinburgh, 1983), pp. 82–101.

225. Richard B. Sher, *Church and university in the Scottish Enlightenment. The moderate literati of Edinburgh* (Edinburgh, 1985) xix + 390pp.

226. Anita Guerrini, 'The Tory Newtonians: Gregory, Pitcairne and their circle', *Journal of British Studies*, 25 (1986), pp. 288–311.

227. Richard B. Sher, 'Professors of virtue: the social history of the Edinburgh moral philosophy chair in the eighteenth century' in *Studies in the philosophy of the Scottish Enlightenment* ed. M.A. Stewart (Oxford 1990), pp. 87–126.

228. Roger L.Emerson, 'Medical men, politicians and the medical schools at Glasgow and Edinburgh 1685–1803' in *William Cullen and the eighteenth century medical world* ed. Andrew Doig, J.P.S. Ferguson, I.A. Milne and R. Passmore (Edinburgh, 1993), pp. 186–215.

229. Roger L. Emerson, 'The "affair" at Edinburgh and the "project" at Glasgow: the politics of Hume's attempt to become a professor' in *Hume and Hume's connexions* ed. M.A. Stewart and John P. Wright (Edinburgh, 1994), pp. 1–22.

Glasgow

240. Thomas Reid, 'A statistical account of the University of Glasgow' in T. Reid, *Philosophical works* (Edinburgh, 1799), Vol. II, pp. 721–739.

241. James Coutts, A *history of the University of Glasgow from its foundation in 1451 to 1909* (Glasgow, 1909) xii + 615pp.

242. Andrew Kent (ed.), *An eighteenth century lectureship in chemistry* (Glasgow, 1950) xiv + 233pp.

243. John D. Mackie, *The University of Glasgow, 1451–1951. A short history* (Glasgow, 1954) xii + 341pp.

244. W.M. Mathew, 'The origins and occupations of Glasgow students', *Past and Present*, 33 (1966), pp. 74–94.

245. J.K. Cameron, 'Theological controversy: a factor in the origins of the Scottish Enlightenment' in *The origins and nature of the Scottish Enlightenment* ed. Roy H. Campbell and Andrew S. Skinner (Edinburgh, 1982), pp. 116–130.

246. T.D. Campbell, 'Francis Hutcheson: 'Father' of the Scottish Enlightenment' in *The origins and nature of the Scottish Enlightenment* ed. Roy H. Campbell and Andrew S. Skinner (Edinburgh, 1982), pp. 167–185.

247. James Moore and Michael Silverthorne, 'Gershom Carmichael and the natural jurisprudence tradition in eighteenth-century Scotland' in *Wealth and virtue. The shaping of political economy in the Scottish Enlightenment* ed. Istavan Hont and Michael Ignatieff (Cambridge, 1983), pp. 73–87.

248. James Moore and Michael Silverthorne, 'Natural sociability and natural rights in the moral philosophy of Gerschom Carmichael' in *Philosophers of the Scottish Enlightenment* ed. Vincent Hope (Edinburgh, 1984), pp. 1–12.

St Andrew's

253. Robert Hannay, 'The visitation of St. Andrews University in 1690', *Scottish Historical Review*, 13 (1915–6), pp. 1–15.

254. Ronald G. Cant, 'The St. Andrew's university theses, 1579–1747. A bibliographical introduction', *Transactions of the Edinburgh Bibliographical Society*, 2 (1941), pp. 105–150; 263–272.

255. Ronald G. Cant, *The university of St. Andrews. A short history* (Edinburgh, 1946) xii + 156pp.

256. William C. Dickinson (ed.), *Two students at St. Andrew's 1711–1716* (Edinburgh, 1952) 76 + 94pp.

Learned Societies

263. Davis Dunbar McElroy, *Scotland's age of improvement: A survey of eighteenth-century literary clubs and societies* (Pullman, Washington, 1969) 175pp.

264. Roger L. Emerson, 'The social composition of enlightened Scotland: the Select Society of Edinburgh, 1754–64', *Studies on Voltaire and the Eighteenth Century*, 114 (1973), pp. 291–329.

V

265. G. Cantor, 'The Academy of Physics at Edinburgh', *Social Studies of Science*, 5 (1974), pp. 109–134.

266. Steven Shapin, 'The audience for science in eighteenth-century Edinburgh', *History of Science*, 12 (1974), pp. 95–121.

267. Steven Shapin, 'Property, patronage and politics of science: the founding of the Royal Society of Edinburgh', *British Journal for the History of Science*, 7 (1974), pp. 1–41.

268. Roger L. Emerson, 'The Philosophical Society of Edinburgh 1737–1747', *British Journal for the History of Science*, 12 (1979), pp. 154–191.

269. Roger L. Emerson, 'The Philosophical Society of Edinburgh 1748–68', *British Journal for the History of Science*, 14 (1981), pp. 133–176.

270. Nicholas T. Phillipson, 'Politics, politeness and the anglicisation of early eighteenth-century Scottish culture' in *Scotland and England, 1286–1815* ed. R.A. Mason (Edinburgh, 1987), pp. 226–246.

271. Roger T. Emerson, 'The Scottish Enlightenment and the end of the Philosophical Society of Edinburgh', *British Journal for the History of Science*, 21 (1988), pp. 33–66.

272. Roger L. Emerson, 'Sir Robert Sibbald, Kt, the Royal Society of Scotland and the origins of the Scottish Enlightenment', *Annals of Science*, 45 (1988), pp. 41–72.

273. Donald C. Macarthur, 'The first forty years of the Royal Medical Society and the part William Cullen played in it' in *William Cullen and the eighteenth century medical world* ed. Andrew Doig, J.P.S. Ferguson, I.A. Milne and R. Passmore (Edinburgh, 1993), pp. 247–251.

Ireland

Trinity College Dublin

285. Anon, 'University of Dublin', *The Quarterly Journal of Education*, 6 (1833), pp. 201–237.

286. William B.S. Taylor, *A history of the University of Dublin* (London, 1845) xix + 540pp.

287. John W. Stubbs, *The history of the University of Dublin from its foundation to the end of the eighteenth century, with an appendix of original documents...* (Dublin, 1889) xii + 429pp.

288. William B. Stanford, 'Classical studies in Trinity College, Dublin, since the foundation', *Hermathena*, 57 (1941), pp. 3–24.

289. Edmund J. Furlong, 'The study of logic in Trinity College, Dublin', *Hermathena*, 60 (1942), pp. 38–53.

290. Constantia Maxwell, *A history of Trinity College, Dublin, 1591–1892* (Dublin, 1946) xiii + 299pp.
291. Robert B. McDowell, 'Courses and teaching in Trinity College, Dublin, during the first two hundred years', *Hermathena*, 69 (1947), pp. 9–30.
292. Robert B. McDowell and D.A. Webb, *Trinity College Dublin 1592–1952. An academic history* (Cambridge, 1982) xxiii + 580pp.

Learned Societies

173. W.E.H. Lecky, *A history of Ireland in the eighteenth century*, 5 vols. (London, 1913–19), 471; 517; 548; 473; 560pp.
174. Royal Dublin Society, *Bicentenary souvenir, 1731–1931* (Dublin, 1932), 80, lxxiipp.
175. Karl T. Hoppen, *The common scientist in the seventeenth century. A study of the Dublin Philosophical Society, 1683–1708* (London, 1970) 297pp.

Textbooks

General Texts

301. Franco Burgersdijk, *Idea philosophiae tum naturalis tum moralis, sive epitome compendiosa utriusque; et Aristotele excerpta & methodice disposita* (Leiden, 1622–3; Oxford, 1631, 1637, 1641, 1654, 1667).
302. Pierre-Sylvain Régis, *Système de philosophie, contenant la logique, la métaphysique, la physique et la morale*, 3 vols. (Paris, 1690) 480, 648, 544pp.
303. Robert Greene, *Encyclopaedia, or a method of instructing pupils* (Cambridge, 1707) 8pp.
304. Daniel Waterland, *Advice to a young student, with a method of study* (London, 1730), 32pp. (also London, 1761; Cambridge 1760; Oxford 1755, 1760, 1761).
305. Thomas Johnson, *Questiones philosophicae in justi systematis ordinem dispositae ... ad calcem subjicitur appendix de legibus disputandi* (Cambridge, 1735) viii + 204pp. (also Cambridge, 1741).

Logic

311. Petrus Ramus, *Dialecticae institutiones ...* (Paris, 1543, 1560; London, 1574, 1576, 1669, 1689; Cambridge, 1584, 1640, 1669, 1672; Edinburgh, 1637) – under the title: *The logicke* (London, 1574, 1581, 1626 1632).

V

312.   Robert Sanderson, *Logicae artis compendium* (Oxford, 1615)lv + 382pp. (also Oxford, 1618, 1631, 1640, 1657, 1664, 1668?, 1672, 1680, 1700, 1705, 1707, 1741, 1841).

313.   Martin Smiglecius (Marcin Śmiglecki), *Logica ... selectis disputationibus et quaestionibus illustrata ... in qua quicquid in Aristotelico Organo vel cognitu necessarium, vel obscuritate perplexum ... pertractatur,* 2 Vols. (Ingolstadt, 1618; Oxford, 1634, 1638, 1658).

314.   Franco Burgersdijk, *Institutionum logicarum libri duo* (Leiden, 1626, 1632; Cambridge, 1637, 1644, 1647, 1666, 1668, 1680; Oxford, 1644, 1660; London, 1651).

315.   Adriaan Heerebord, *Epitome logica; seu explictio, tum per notas, tum per exempla, synopsis, logicae Burgersdicinae* (Leiden, 1657) 334 pp (also London, 1658, 1676; Cambridge, 1663, 1680).

316.   Narcissus Marsh, *Institutiones logicae in usum juventutis Academicae Dubliensis* (Dublin, 1679) 268pp. (also Dublin, 1681, 1697).

317.   John Wallis, *Institutio logicae, ad communes usus accommodata* (Oxford, 1687) 262 pp. (also Oxford, 1699, 1702, 1715, 1729).

318.   Jean Le Clerc, *Logica, sive ars ratiocinandi* (London 1692) – Under the title: *Logica, ontologia, et pneumatologia* (Cambridge, 1704).

319.   Henry Aldrich, *Artis logicae compendium* (Oxford, 1691) 99pp (also Oxford, 1692, 1696, 1764, 1750, 1771, 1793, 1804, 1823, 1828, 1841, 1848, 1850, 1852, 1856, 1862).

320.   Isaac Watts, *Logick: or the right use of reason in the enquiry after truth* (London, 1725) vi + 534pp. (also London, 1726, 1733, 1736, 1740, 1745, 1793, 1797, 1801, 1809, 1911; Edinburgh, 1807; Boston, 1812).

321.   Edward Bentham, *Reflexions upon logick* (Oxford, 1740, 1755).

322.   Isaac Watts, *The improvement of the mind: or, a supplement to the art of logick* (London, 1741) xv + 362pp (also London, 1743, 1782, 1787, 1795, 1804, 1809, 1814, 1817, 1819, 1821, 1822, 1825, 1826, 1856, 1868; Boston, 1826).

323.   Richard Murray, *Artis logicae compendium in usum iuventutis collegii Dubliensis* (Dublin, 1782) 81pp. (in English: Dublin, 1812, 1847).

324.   Thomas Reid, *Essays on the intellectual powers of man* (Edinburgh, 1785) xii + 766pp. (also Edinburgh, 1803, 1853, 1865, 1941).

325.   Richard Whateley, *Elements of logic, comprising the substance of the article in the* <u>*Encyclopaedia Metropolitana*</u> (London, 1826) xxxii + 340 (also London, 1827, 1832, 1836, 1840, 1844, 1848, 1853; Oxford, 1827; Cambridge, 1874).

Metaphysics

331.   Franco Burgersdijk, *Institutionum metaphysicarum libri duo* (Leiden, 1640; Hague, 1657 [ed. A Heereboard]; Oxford, 1675).

V

332. René Descartes, *Principia philosophiae* (Amsterdam, 1644) 22 + 310 pp. (also Amsterdam, 1650, 1656, 1664, 1672, 1678, 1685).
333. Robert Baron, *Metaphysica generalis* (Leiden, 1654) 392pp. (also Oxford, 1658; 1669; Cambridge, 1685).
334. Nicholas Malebranche, *De la recherche de la verité* (Paris, 1674–1675) – Latin translation: Geneva, 1685, 1753 – English translation: London and Oxford, 1694, 2 vols.).
335. Andreas Frommenius, *Synopsis metaphysica* (Oxford, 1691) 389pp. (also Oxford, 1704).
336. John Norris, *An essay towards the theory of the ideal or intelligible world*, 2 parts (London, 1701–4), 452, 564pp.
337. Samuel Clarke, *A demonstration of the being and attributes of God: more particularly in answer to Mr Hobbs, Spinoza, and their followers. Wherein the notion of liberty is stated, and the possibility and certainty of it proved, in opposition to necessity and fate. Being the substance of eight sermons preach'd ... in the year 1704 at the lecture founded by the Honourable Robert Boyle, Esq.* (London, 1705) 264pp (also London, 1706, 1716, 1719, 1725, 1732, 1739).
338. Samuel Clarke, *A discourse concerning the unchangeable obligations of natural religion, and the truth and certainty of the Christian Revelation. Being eight sermans preach'd ... in the year 1705, at the Lecture founded by the Honourable Robert Boyle, Esq.* (London, 1706) 405pp. (also London, 1785).
339. Isaac Watts, *Philosophical essays on various subjects, viz. space, substance, body, spirit ... with some remarks on Mr Locke's Essay on the human understanding. To which is subjoined a brief scheme of ontology, on the science of being in general with its affections* (London, 1733), xii + 403pp. (also London, 1742; 1793; Edinburgh, 1823).
340. Francis Hutcheson, *Metaphysicae synopsis* (Glasgow, 1742) 89pp. (also Edinburgh, 1744).

Moral Philosophy

351. Hugo Grotius, *De jure belli ac pacis libri tres* (Paris, 1625)739pp, (also Amsterdam, 1631, 1650, 1651, 1702, 1712; Utrecht, 1696–1703; Cambridge, 1703, 1853, 1854; London [in translation], 1682, 1715, 1738).
352. Robert Sanderson, *De obligatione conscientiae praelectiones decem; Oxonii in Schola Theologicae habitae* (London, 1647) 246pp. (also London, 1661, 1670, 1683, 1686, 1710, 1719) – In English translation (London, 1665, 1716).
353. Adrianus Heerebord, *Collegium ethicum, in quo tota philosphia moralis aliquat disputationibus* (Leiden, 1648; London, 1658) viii + 150pp. (also London, 1666, 1671, 1677, 1693).
354. Asseline Eustachius a S. Paulo, *Ethica: sive summa moralis disciplinae, in tres partes divisa* (Cambridge, 1658), viii + 150pp. (also London, 1666, 1671, 1677, 1693).

V

355.   Robert Sanderson, *De obligation conscientiae prelectiones decem: Oxonii in Schola Theologicae habitae* (London, 1660) 387pp. (also London, 1661, 1682, 1686, 1696, 1710, 1719) – In English translation (London 1660, 1851, 1877).

356.   Henry More, *Enchiridion ethicum, praecipua moralis philosophiae rudimenta* (London, 1668) 229pp. (also London, 1669; Amsterdam, 1679, 1695) – Under the title: *An account of virtue; or Dr Henry More's abridgement of morals* (London, 1690) 268pp.

357.   Samuel von Pufendorf, *De jure naturae et gentium libri octo* (Lund, 1672) 1227pp. [in English translation] (London, 1703,1716, 1729, 1749; Oxford 1710).

358.   Samuel von Pufendorf, *De officiio hominis et civis juxta legem naturalem libri duo* (Lund, 1673) 240 pp (also, 1701, 1737, 1748, 1758; Cambridge, 1782, 1735; Edinburgh, 1724) – Under the title: *The whole duty of man according to the law of nature* (London, 1691, 1716, 1735; Dublin 1716).

359.   Gershom Carmichael, *S. Puffendorfii de officio hominis ... juxta legem naturalim, libri duo. Supplementis et observationibus ... auxit ... G. Carmichael* (London, 1724) xi + 536pp.

360.   Francis Hutcheson, *Philosophiae moralis institutio compendiaria, ethices et jurisprudentiae naturalis elementa continens* (Glasgow, 1742) iv + 347pp. (also Glasgow, 1745, 1755; Rotterdam, 1745; Dublin, 1787) – Under the title: *A short introduction to moral philosophy* (Glasgow, 1747, 1753, 1764, 1772; London, 1755).

361.   Edward Bentham, *Introduction to moral philosophy* (Oxford, 1745) 109pp. (also, Oxford, 1756).

362.   Jean Jacques Burlamaqui, *The principles of natural law ...* (London, 1748), xvi + 312pp. – Vol.2: *The principles of politic law* (London,1752), 372pp.

363.   Thomas Rutherforth, *A system of natural philosophy, being a course of lectures in mechanics, optics, hydrostatics, and astronomy*, 2 vols. (Cambridge, 1748), 496, 497–1105pp.

364.   David Hartley, *Observations on man, his frame, his duty and his expectations*, 2 parts (London, 1749, 1791, 1801–1791; Bath, 1810).

365.   Thomas Rutherforth, *Institutes of natural law; being the substance of a course of lectures on Grotius' De jure belli et pacis* 2 vols (Cambridge, 1754–1756; Philadelphia, 1799; Baltimore, 1832).

366.   William Paley, *Principles of morals and political philosophy* (Cambridge 1785; 1816; London, 1786, 1788, 1790, 1791–4, 1799, 1806, 1809, 1822, 1824, 1826, 1852, 1859; Dublin, 1793; Boston, 1801, 1825).

367.   James Beattie, *Elements of moral sciences*, 2 vols. (Edinburgh, 1790, 1793) xv + 438; viii + 688pp.

Natural Philosophy

381. Johann Magirus, *Physiologia peripatetica, libri sex, cum commentariis* (Frankfurt, 1603; London, 1619; Cambridge, 1642).

382. Casparos Bartholin, *Ionnis Magiri physiologiae peripateticae, libri sex, cum commentariis ... accessit Caspari Bartholini* (London, 1619) 662pp. – under the title: *Specimen philosophiae* (Oxford 1698).

383. Franco Burgersdijk, *Collegium physicum in quo tota philosophia naturalis aliquot disputationibus perspice & compendiose explicatur* (Leiden, 1632).

384. Bernardus Varenius, *Geographia universalis, in qua affectiones generales telluris explicantur* (Amsterdam, 1650, 1664, 1671) – New edition with additional title: *... summa cura quam plurimis in locis emendata, et xxxiii schematibus novis ... aucta et illustrata ab Isaaco Newton* (Cambridge, 1672, 1681) 511pp. – New edition with additional title: *... adjecta est appendix, praecipua recentiorum iuventa ad geographiam spectantia continens, a Jacobo Jurin* (Cambridge, 1712) 511pp. – In English translation: (London, 1683, 1733, 1734, 1736, 1765).

385. Adrianus Heereborb, *Philosophia naturalis, cum commentariis peripateticis antehac edita: nunc ... novis commentariis, partim e Nob. D. Cartesio, Cl. Berigardo, H. Regio, aliis que ... petitis, partim & propria opione dictatis, explicata* (Leiden, 1663) 256pp. (also Oxford, 1668).

386. Jacques-Nicolas Colbert, *Philosophia vetus et nova, ad usum scholae accommodata, in regia Burgundia novissimo hac biennio pertractata*, 4 vols. (Paris, 1678).

387. Jacques Rohault, *Tractatus physicus ... cum animad -versionibus Antonii Le Grand* (London, 1682, 1692, 1696; Amsterdam, 1708) – Under the title: Physica, *latine reddit et annotationculis quibusdam illustravit S[amuel] Clarke* (London, 1697, 1702, 1710, 1718) – Under the title: *Rohault's system of natural philosophy; illustrated with Dr S. Clarke's notes, taken mostly out of Sir Isaac Newton's philosophy with additions. Done into English by John Clarke*, 2 vols. (London, 1723, 1728–9, 1735).

388. Jean Le Clerc, *Physica, sive de rebus corporeis libri quinque* (London, 1696) 12 + 492pp. (also Amsterdam, 1695; London, 1708; Cambridge, 1700, 1705, 1708).

389. David Gregory, *Astronomiae physicae et geometricae elementa* (Oxford, 1702) 494 pp. (also Geneva, 1726).

390. John Keill, *Introductio ad veram physicam, seu lectiones phyiae, habitae in schola naturalis philosophiae Academiae Oxoniensis. Quibus accedunt C. Hugenii theoremata de vi centrifuga et motu circulari demonstrate* (Oxford, 1702) 191pp. (also Oxford, 1705; London [in English translation], 1720, 1733, 1745, 1758).

391. William Whiston, *Praelectiones astronomiae Cantabrigiae in scholis publicis habitae ... quibus accedunt tabulae plurimae astronomicae Flamstedianae correctae, Halleinae, Cassinianae, et Streetianae* (Cambridge, 1707) 459pp.

V

392.   William Whiston, *Praelectiones physico – mathematicae Cantabrigiae in scholis publicis habitae quibus philosophia illustrissimi Newtoni mathematica traditur, et facilius demonstratur: cometographia etiam Halleiana commentariolo* (Cambridge, 1710) 367pp. (also London, 1726) – Under the title: *Sir Isaac Newton's mathematick philosophy more easily demonstrated* (London, 1716).

393.   Edward Wells, *The young gentleman's astronomy, chronology, and dialling, containing such elements of the said arts or science as are most useful and easy to be known* (London, 1712, 1718, 1725, 1736).

394.   John Keill, *Introductio ad veram astronomiam, seu lectiones astronomicae, habitae in schola astronomica Academiae Oxoniensis* (London, 1721, in English translation 1721, 1730, 1739, 1748).

395.   John Clarke, *A demonstration of some of the principal sections of Sir Isaac Newton's Principles of Natural Philosophy* (London, 1730) xvi + 313pp.

396.   John Rowning, *A compendious system of natural philosophy, with notes, comprising mathematical demonstrations, 4 parts* (Cambridge, 1735–42; London, 1744, 1767).

397.   Roger Cotes, *Hydrostatical and pneumatical lectures … published with notes by … Robert Smith* (London, 1738, 1775; Cambridge, 1747).

398.   Robert Smith, *A compleat system of opticks in four books, viz. a popular, a mathematical, a mechanical, and a philosophical treatise: to which are added remarks upon the whole*, 2 vols. (Cambridge, 1738, 1788).
       – See also *363 above, Rutherforth, *A system of natural philosophy* … (Cambridge, 1748).

399.   Richard Helsham, *A course of lectures in natural philosophy* (London, 1739) viii + 404pp. (also 1743, 1767, 1777; Dublin, 1818).

400.   Robert Smith, *Harmonics, or the philosophy of musical sounds* (Cambridge, 1749), xv + 292pp. (also London, 1759–62).

401.   Hugh Hamilton, *Four introductory lectures in natural philosophy* (Dublin, 1767) 108pp.

VI

Motives for European Exploration of the Pacific in the Age of the Enlightenment[1]

WHAT WAS IT THAT prompted Europeans of the late eighteenth and early nineteenth centuries to condemn themselves to long and dangerous voyages in fragile wooden vessels that, as Cook found on the Great Barrier Reef, were all too vulnerable to the vagaries of unknown lands? The perennial motives of a quest for strategic and economic advantage played a large part in this, as in most ages, but what is interesting is the extent to which such motives were combined or, at least to some extent, masked by the quest for knowledge both of the natural and human world. Consequently, exploration could be regarded as consistent with the goals of the Enlightenment and the motto that Kant attributed to it: "Aude sapere" (Dare to know). In this paper the ways in which such proclaimed goals shaped the actual practice of Pacific exploration are explored.

GOD, GOLD, AND GLORY: THE SPANISH AND THE PACIFIC

The Pacific voyages of the late eighteenth century had, of course, been preceded by the extraordinary explosion outward of the Spanish and Portuguese in the fifteenth and sixteenth centuries. Their motives were evident enough and often stated: the quest for gold, God, and glory as the crusading spirit,

which had led to the reconquest of Spain from the Muslims, then spilled out onto the larger global arena with the burning ambition to claim new souls for the Holy Catholic Church and new wealth and territory for the king of Spain.

In this period, then, idealism of another sort combined with and, to some extent, colored the quest for direct economic or national advantage as religion justified action. Such idealism was given a quasi-legal form with the argument that the pope had the power, as overlord of Christendom, to authorize such conquest provided it was done with the intention of bringing more souls within the community of the faithful. A more sophisticated version of this argument was that of the great Dominican jurist Francisco de Vitoria, who argued, in lectures delivered at Salamanca in 1539, that, though the pope's temporal power might be questioned, he did have a "regulating authority." Such an authority should conform to the canons of natural law that provided the legal basis for the relations between nations—including those between the king of Spain and the Indians of the New World. In Vitoria's exposition the extension of royal power could be justified on the grounds of Christian proselytizing, but he was plainly uneasy about the use of force and preferred an empire based on trade (which came closer to the Portuguese practice) (Parry 1990: 137).

Religious idealism, then, did not always suit national needs, as the king of Spain found when his nation's reputation was blackened by the impassioned denunciation of the behavior of the conquistadors by the missionary Bartolomé de Las Casas. In this earlier phase of exploration the religious justification for conquest as a means of extending the reach of Christendom and bringing more souls to the Christian heaven did have a two-edged character. It could act as an often thinly veiled excuse for ruthless conquest and exploitation, but, in theory at least, it also acted as something of a brake on the excesses of the conquistadors. Las Casas, for example, was bold enough to admonish the Spanish king, urging him to recognize that "the only title that Your Majesty has is this: that all, or the greater part of the Indians, wish voluntarily to be your vassals and hold it an honor to be so" (Pagden 1995: 51). Although Las Casas accepted the view that Spain had a papally sanctioned mandate to spread the Gospel, it was one subject to definite ethical restrictions. These went no further than allowing Catholic monarchs to "induce the peoples who live in such islands and lands to receive the Catholic religion, save that you never inflict upon them hardships and dangers." The fact that the Spanish had not done this provided the Indians with cause to wage a "just war" (Las Casas 1992: xvi–xvii). As the scholarship of Pagden has brought out, whatever the often brutal reality in America, on the distant Iberian peninsula academics at universities such as Salamanca and Coimbra engaged in long and arduous debate about the ethics of conquest and the religious status of the peoples of the New World (Pagden 1982: 104–143, 1995: 46–51).

The almost manic determination of the conquistadors waned by the seventeenth century, and the attempt by Mendana (1568 and 1595) and Quiros (1606) to open up new territories in the Pacific in the late sixteenth and early seventeenth centuries met with little response from the Spanish Crown (Beaglehole 1966: 106). Part of the reason for this was the fact that the Pacific islands seemed to hold few riches, but even Quiros' impassioned plea to bring these new souls to Christ was received with scant enthusiasm by a regime that was increasingly the creature of its own bureaucracy and that had more than enough work cut out to absorb the vast territories added to its domains in Central and South America.

Nonetheless, the argument that the Spanish Crown had an obligation to spread the reach of Christendom could not be easily overlooked—like many ideological systems it had a life of its own that did not always conform readily to the immediate needs of the imperial power. Quiros was quietened by making vague undertakings about missionary activity that might be sponsored by the viceroy of Peru whom Quiros accompanied on his return from Spain. Though the problem of dealing with Quiros' inconvenient appeal to the governing ideals of the Spanish empire was solved with his death on that voyage in 1615, it continued to prick the consciences of the Spanish overlords. As late as 1630–1633 there were vain Franciscan appeals to the Crown to mount a mission to the Pacific islands that Mendana and Quiros had brought under Spanish gaze (Spate 1979: 142).

So, after this frenetic wave of activity, principally in the sixteenth century, Europe's involvement in the Pacific was largely quiescent apart from the Spanish consolidation of its power in the Philippines and the growth of the Dutch rigorously commercial hold on the East Indies. The initiative of the energetic Dutch governor of the East Indies, Van Diemen, to send Abel Tasman in 1642 on a voyage of exploration to New Holland and New Zealand, as the Dutch termed these territories, only confirmed the Dutch view that little wealth was to be gained from these uncharted lands. Consequently, the Dutch, like the Spanish, devoted their energies to making money out of the territories they already possessed. Other European regimes on the whole were willing to accept Spanish claims that the Pacific was part of their sphere of influence largely out of indifference—thus the Pacific remained a Spanish lake (Schurz 1922).

SCIENCE AND COMMERCE: THE AMBIVALENT
MOTIVES OF LATE EIGHTEENTH-CENTURY
PACIFIC EXPLORATION

The Pacific was, however, to be awakened
from its slumbers and brought firmly into the
mixed crosscurrents of European imperial
expansion in the period from 1763 onward.
In that year Europe concluded one of the
major chapters in the ongoing "second hundred years' war" between France and Britain
for world dominance, with Britain left largely
secure in its dominance of North America
and, to a lesser extent, of India. Pressures
that had been building up for Britain to use
its naval might more effectively in securing
new territories were now more likely to be
realized as the burden of war was removed
and as Britain could bask more self-
confidently in its great power status. Both
eighteenth-century French advocates of Pacific exploration like de Brosses and British
like Callandar and Dalrymple had urged
the possibility of a Great Southern Land, the
mass of which would balance the vast tracts
of land in the Northern Hemisphere (Dunmore 1965–1969, 1:47, 50, 190). Having
largely secured dominance in North America, Britain was anxious to do likewise in any
new large territory that the Pacific might
harbor, and it was confident that, if necessary, it could defy Spain in achieving that
goal. The result was the voyage of John
Byron from 1764 to 1766, which achieved
little, and the joint voyages of Wallis and Carteret in 1767, which opened up Tahiti to the
gaze of Europe.

Not to be outdone, the French soon afterward dispatched the voyage of Bougainville
in 1768, which further confirmed in European minds the myth of Tahiti as a new
Garden of Eden. The defeat in the Seven
Years' War made France more determined
than ever not to allow the British to gain an
advantage in the quest for new territories and
riches. Appropriately, Bougainville had been
one of those involved in the surrender of the
French stronghold of Quebec in one of the
key battles of the Seven Years' War. He was
driven, too, by the hope of reducing British
dominance in the "new world" of the Pacific,

with the Falkland Islands armed by either the
French or their then allies, the Spanish, serving as a brake on British ambitions in the
eastern Pacific (Dunmore 1965–1969, 1:59).
Such voyages were to be the curtain-raisers,
as it were, of the great voyages of Cook and,
from the French side, of the voyages of La
Pérouse (1785–1788) and, subsequently, of
D'Entrecasteaux (1791–1793) and Baudin
(1800–1804).

So the motives for exploration were, in
part, the familiar ones of great power rivalry
and the quest for new territories as sources
of wealth. But what was striking about the
Pacific voyages of the late eighteenth century,
and what distinguished them from the earlier
voyages of the Spanish and the Portuguese,
was the extent to which they were linked to
the advancement of science and knowledge
more generally. Where the Iberian explorers
justified their exploration in religious terms,
the late eighteenth-century explorers were
more likely to invoke more secular justification consistent with the worldview of the
Enlightenment. It was a transition that testified to the impact of the worldview of the
Scientific Revolution on the European elite.
Perhaps, too, the divisive consequences of
religious warfare in the seventeenth century
had prompted a quest for more secular justifications for action in exploration as in law
and politics. Scientific exploration had the
benefit of being based on canons of inquiry
that could transcend the confessional divide
and that were closely associated with notions
of natural law on which European jurists like
the Salamanca school and the Dutch Grotius
or the German Pufendorf had attempted to
erect systems of international law. The question then presents itself: how far did such an
ideology prescribe limits on action as well as
justification for it, as the earlier Spanish
appeal to Christian idealism had?

As with the Iberian exploration of the fifteenth and sixteenth centuries, there is plenty
of evidence that exploration was linked to
those hardy perennials of human nature:
commercial and strategic advantage. But,
just as the Iberian conquerors felt the obligation at least to construct a theoretical
justification for their actions, so, too, the

VI

late eighteenth-century explorers appealed to higher motives. In 1767, just after Bougainville had set off on the first of France's major voyages of Pacific discovery, one French memorialist urged further activity to discover the great southern continent. He appealed principally to commercial advantage (and, in particular, to the advantage that would accrue to France in the trade with China), but also argued that such lands would provide a great "quarry for the sciences" (Bibliothèque Nationale, NAF 9439, 52, De Lozier Bouvet, "Mémoire touchant la Découverte des Terres Australes").

When Bougainville returned, he, too, appealed to a mixture of scientific idealism and national advantage in urging further Pacific exploration in 1773. On the one hand he acknowledged the need for further surreptitious activity against the British, but, on the other, he stressed the worth of such exploration to assist in "perfecting the knowledge of the globe" (Bibliothèque Nationale, NAF 9439, 70v, Bougainville to the Minister of the Marine, 27 February 1773). At very much the same time the great naturalist Buffon was also endeavoring to advance the cause of Pacific exploration by appealing to similarly mixed motives in relation to the proposal for a second voyage by Kerguelen—the glory that could accrue to France by the promotion of scientific enquiry, which could also rebound to its commercial advantage (Archives Nationales, B/4/317, no. 111, Buffon to M. le Duc d'Auguillon, 2 January 1773).

It is interesting that there is no reference to the sort of ideals to which the Spanish had earlier appealed even though these had loomed large in another attempt, as recently as 1735, to institute a search for the fabled Gonneville Land—the great Southern Continent that Sieur de Gonneville had improbably claimed to have discovered in 1504. "The Glory of God," wrote the French India Company captain Jean-Baptiste-Charles Bouvet de Lozier in his *mémoire* to the Crown, "and the interests of religion require us to carry out this undertaking; very likely these various countries are inhabited by numerous peoples who are groaning in the shadows of death. One cannot hasten too

much to bring them the torch of the Gospel." But, he added with an aside about the Dutch and English Protestants that underlined the intertwining of the appeal to idealism with that to advantage: "one must apprehend that our neighbours who are separated from the church may forestall us in order to increase their trade" (Dunmore 1965–1969, 1 : 197).

But such attempts to mobilize the French state on religious grounds seem largely to have faded in the second half of the century and increasingly to have been replaced by appeals to the uses of Pacific exploration as a means of promoting knowledge. This reflected the extent to which the attitudes associated with the Enlightenment gained greater currency and acceptance within the French elite from around 1751, when the first volume of that great summation of the Enlightenment worldview, the *Encyclopédie*, appeared. Indeed, in France appeals to the scientific benefits of exploration seem to have been more insistent and pervasive than in England. This perhaps was a consequence of the French absolutist monarchy's close involvement with the advance of science through its support for the Académie des Sciences. By contrast, in England the monarchy's support for the Royal Society took little more than a symbolic form, and, for much of the eighteenth century, science was not as closely intertwined with the workings of the state.

Such more secular justifications for Pacific exploration rose to a crescendo with the planning of La Pérouse's great voyage. This was intended to put France's stamp on the Pacific in a manner comparable to the way in which the British presence in the Pacific had been advanced by the voyages of Cook. Indeed La Pérouse was to take further France's civilizing mission by extending the range of scientific enquiry made possible by exposure to the Pacific. Baron Gonneville, descendant of the early sixteenth-century explorer who had left the French with a lingering sense of ownership of the Great Southern Land, rose to rhetorical heights in linking the proposed voyage with the promotion of French prestige through the advancement of learning. It was a voyage that he saw as promoting the glory of the French crown because it had "no other end than that of the truth, no other motive than that of the general good" (Bibliothèque Nationale, NAF 9439, 20, Gonneville to the Minister of the Marine, 19 May 1783).

La Pérouse himself was more candid about the motives of a voyage that had grown out of a commercial venture and that had clear strategic goals. Hence such reports as that on Manila about which he commented that given a moderately sized naval and military force its conquest seemed "easy and so certain." About Chile he remarked that, if the alliance between France and Spain were abandoned, it would be easy "to advance the ruin of Spanish interests" by forming an alliance with the native peoples. Such aggressive reflections he combined with musings, in Enlightenment fashion, on the nature of natural man and admiration for Rousseau, whose bust and works he possessed (Dunmore and de Brossard 1985, 1 : 5, 220–221).

La Pérouse's voyage was one of the last major projects of the French old regime, but the Baudin expedition of 1800–1804 represented one of the many exercises of Napoleonic patronage of the sciences. But, like its prerevolutionary counterparts, it was characterized by a familiar mixture of Enlightenment idealism and national advantage. As early as 1798 Antoine Jussieu, professor of botany at the Muséum d'Histoire Naturelle, alluded to the way in which scientific and national goals might be combined in a voyage commanded by Baudin. For, he suggested, Baudin could advance the interests of natural history and "render further service to his country" by "combin[ing] geographical researches with those which interest us more particularly" (Horner 1987 : 36).

The character of some of the researches that were particularly to interest the Napoleonic state was made evident in a letter by the expedition's chief scientist, François Péron, to the commander of the Ile-de-France (Mauritius). For, he wrote, Napoleon's real object in deciding upon the expedition—the need to gather military and strategic intelligence— "was such that it was indispensable to conceal it from the governments of Europe, and

especially from the Cabinet of St. James [Britain] ... all our natural history researches, extolled with so much ostentation by the Governor, were merely a pretext for its enterprise." Flinders' voyage he characterized as having objectives similar to his own (Horner 1987 : 320). Perhaps Péron was exaggerating the military significance of the voyage for his own ends, but his letter helps to explain why the governor of the Ile-de-France kept Flinders prisoner, dismissing his protestations about the worth of scientific and geographical enquiry. Nor, in other contexts, was Péron himself reluctant to invoke the Enlightenment ideology of the advancement of science as being a pursuit that transcended the petty quarrels between nations— even if national competition provided a healthy spur to such activity. Thus he described "discoveries in the sciences" as "amongst the chief records of the glory and prosperity of nations," particularly as they were "of general utility to all" (Mackay 1989 : 116).

Given space one could multiply such examples of the admixture of Enlightenment ideology about the virtues of scientific exploration in the Pacific with the pursuit of national commercial or strategic gain from the British side. Cook's Enlightenment-inspired quest to observe the transit of Venus in Tahiti on his *Endeavour* expedition of 1768–1771 was, of course, combined with secret instructions to search for the Great South Land to advance "the honour of this nation as a Maritime Power" and with a view to "the advancement of the Trade and Navigation thereof" (Beaglehole 1955–1974, 1 : cclxxxii). Given such mixed motives on this and other Pacific voyages one can better understand the position of the viceroy of Brazil, who so angered Joseph Banks by refusing to allow him to collect botanical specimens on the grounds that this was a cover for spying.

Dalrymple, the great theorist of Pacific exploration—whose speculations urged Cook on in the vain search for the missing Great Southern Land—could, if the need arose, narrow his focus to become a good East India Company man concerned to use Pacific exploration to advance the interests of the company and, with it, Britain more generally. Accounts of voyages to New Guinea by Forrest in 1771 and Rees in 1783 prompted him to suggest ways in which this might enable the British to challenge Dutch supremacy in the East Indies. Thus he wrote to the Secret Committee of the Company that "I am very confident that wherever the English make a settlement to the Eastward, the Dutch will decline in their consequence" (National Library of Scotland, MS 1068, 89, Dalrymple to East India Company, 21 December 1795). Banks had no illusions about the way in which commercial motives entirely dominated the East India Company's actions. Hence, when the Company granted money for Flinders' proposed circumnavigation of Australia, Banks robustly informed him that "The real reason for the allowance is to Encourage the men of Science to discover such things as will be useful to the Commerce of India & you to find new passages" (Mitchell Library, A79/4, 187, Banks to Flinders, 1 May 1801).

For the Spanish the need to recast their rationale and practice in regard to Pacific exploration was particularly urgent as they faced increasing threats to their empire in the late eighteenth century. Hence the imperative to invoke the Enlightenment ideals of exploration to establish their position as a major Pacific power and to match such rhetoric with action in the form of a major, scientifically based expedition. For the Spanish had painfully learned that their claims to Pacific dominance carried little weight unless they could show that they had truly explored it in a scientific fashion. This was a view that ran counter to their traditional belief that the best method of securing control was through secrecy to ensure that no other nation could gain an advantage through their discoveries (Cook 1973:210, 528, Snow and Waine 1979:36). Mendana, for example, had encountered the Solomons in 1568, giving it that name in the belief that it might contain gold mines like those of King Solomon's mines—a claim that sent many Pacific explorers on a wild-goose chase. For, in the primitive state of cartography that then existed, Mendana had provided no effective

way of allowing others to find these territories again. As far as the eighteenth century was concerned, then, Spanish claims to the Solomons dissolved in the absence of any effective maps that translated such aspirations into reality.

Such considerations explain why, in 1788, the king of Spain accepted an offer by Malaspina, an Italian naval officer in the service of the Spanish monarchy, to mount a voyage of scientific discovery to outshine that of Cook. In his "Plan for a Scientific and Political Voyage around the World" Malaspina gently alluded to the fact that Spain had fallen behind in the race for Pacific dominance by scientific means. "For the past twenty years," he urged, "the two nations of England and France, with a noble rivalry, have undertaken voyages in which navigation, geography and the knowledge of humanity have made very rapid progress" (Engstrand 1981:45). The agenda laid down for the voyage, which lasted from 1789 to 1794, also indicates the extent to which, in the late eighteenth century, exploration was conceived in scientific terms—it was intended that it should bring "new discoveries, careful cartographic surveys, important geodesic experiments in gravity and magnetism, botanical collections, and descriptions of each region's geography, mineral resources, commercial possibilities, political status, native peoples, and customs" (Frost 1988:38). And, indeed, this great and rarely remembered voyage did achieve many of these aims though to little effect. For, on his return, Malaspina was arrested and the fruits of his voyage consigned to oblivion. His political liberalism, especially in regard to the position of the Spanish colonies in the New World, made him suspect at a court the reactionary tendencies of which had been heightened by the revulsion against the nearby revolution in France.

The Malaspina expedition, then, did little to arrest Spain's downhill slide as its claims to Pacific dominance were largely brushed aside by the British, French, and Russians. Nonetheless, it was indicative of the extent to which the view that exploration should be scientifically based had gained ground. But, alongside such Enlightenment-tinged goals, there were, needless to say, also more pragmatic ends in view. Malaspina was commissioned both to report on the state of Spain's Pacific empire and on possible threats to it— particularly from what, to Spanish eyes, was the chief interloper in the Pacific: Britain. Hence the remarks of one of Malaspina's officers on the settlement of New South Wales that clearly brings out the extent to which the Spanish could recognize the mixed motives of Pacific exploration: "The endeavours of [the] energetic [Cook], his perseverance and labours, besides enriching the sciences of geography, and Hydrography by new discoveries, have placed his nation in a position to compensate itself for the loss of North America" (Frost 1988 : 37). What was of particular concern was that the new settlement might act as a naval base for use against the Spanish (Francisco Munoz y San Clemente, *Discurso politico sobre los establicimentos Ingleses de la Nueva-Holanda* [Frost 1988 : 37]).

It is not surprising, then, that the major powers in the late eighteenth century embarked on Pacific exploration with the goal of advancing their national interests. But, just as the conquistadors had part rationalized and part justified such motives by invoking the support of the Church and the Crown, so, too, the voyagers of the late eighteenth century felt it incumbent on themselves to point to the ways in which their endeavors could be seen as advancing the cause of science and humanity. The Enlightenment provided much of the language with which to justify one's actions even though contemporaries were well aware of the range of motives of very varying highmindedness that prompted such exploration. Louis Deschamps, one of the naturalists on board D'Entrecasteaux's expedition to search for La Pérouse, summed up the ethos of the great age of late eighteenth-century Pacific voyaging as being the outcome of the Europeans' "restlessness of spirit, curiosity, ambition or greed." As a good French republican, he added, in the spirit of Enlightenment anticlericalism, that though not all governments had "the good faith to encourage philosophy, that mortal enemy of superstition & of despotism ... all have the good sense to encourage geography and natural history" (British Museum, Natural History, Botany Library, Deschamps, *Journal de voyage sur La Recherche ...,* 110).

How far can it be said that Enlightenment-based justifications for Pacific exploration required more than lip service and, to some extent at least, acted as a brake on the more obvious self-interested motives for exploration? How far, in short, was the great wave of late eighteenth-century Pacific voyaging shaped by an ideology that, like the ideology that had shaped the earlier great wave of Pacific voyaging by the Iberian peoples in the fifteenth and sixteenth centuries, required at least some element of sacrifice and restraint if it were to be believed?

The most obvious way in which the ideology of Enlightenment exploration did require some such sacrifice was financial. To mount such expeditions was enormously expensive —a cost the French and the Spanish, with their traditions of absolute government, were more willing to bear than the British, who, where possible, relied on the private initiative of naturalists like Banks or Darwin or the scientific inclinations of naval officers like Cook or Flinders.

A second sense in which it did require some restraint of national self-interest was that, to be effective, it involved at least an element of international cooperation, giving some credence to Enlightenment ideas of a cosmopolitan republic of letters. It was an ideal that Gibbon summed up in these terms: "It is the duty of the patriot to prefer and promote the exclusive interest and glory of his native country but a philosopher must be permitted to enlarge his views and to consider Europe as one great Republic whose various inhabitants have attained almost the same level of politeness and civilisation" (Schlereth 1977 : 47).

Within the context of late eighteenth-century science and exploration the most conspicuous example of such cosmopolitanism was the way in which Cook's observations of the transit of Venus in Tahiti formed part of an international network with the aim of observing the transit from as many different vantage points as possible (Woolf 1959).

For all their rivalry the European nations came to recognize the advantages that could accrue to all by sharing at least some of the knowledge and expertise that flowed from Pacific exploration. When Bougainville had to abandon his plans for a voyage to the North Pole in 1770, he was willing to accede to the Royal Society's request that he pass on his proposals, which helped to shape the eventual itinerary followed by Captain Phipps in 1773 (National Maritime Museum, MS 57/058, BOG/2). Banks, Britain's chief promoter of such voyages, could encourage Malaspina in his designs, responding enthusiastically to a plea "to co-operate towards the progress of science" (British Library, Add. MS 8097, 216, Malaspina to Banks, 20 January 1789). He also rejoiced at his successes, writing that "In this voyage botany, mineralogy, and hydrography has received much and valuable improvement." In Enlightenment fashion he also congratulated Malaspina on the bloodless character of his voyage for "Their discoveries have not cost a single tear to the human race and they have only lost three or four crew in each vessel" (Bladen 1892–1901, 3 : 289). When the herbarium of Deschamps was seized by a British corsair, it was returned to him 2 years later through the intervention of Banks (Horner 1996 : 228). With such events in mind, the French Académie des Sciences was later to acknowledge that establishing the significance of scientific expeditions based on the use of naval power had been largely initiated by Banks (Archives Nationales, BB4/998/2, Extrait de la Séance du Lundi, 4 February 1833).

Such scientific cosmopolitanism generally extended to permitting foreign access to ports —though, as the hapless Flinders discovered when he was imprisoned on Mauritius, there were limits to such indulgence during wartime. Though not averse to some spying while they were there, the Spanish and French explorers commented on the warm reception they received when they called into Sydney—a reputation confirmed when, in 1827, the British Admiralty formally granted Dumont D'Urville's *Uranie* "a friendly reception in any British Port of Settlement to which she may repair for purposes of re-

freshment or refit, or from the pursuit of the prescribed astronomical and physical observations" (Archives Nationales, BB4/998, Viscount Castlereagh for the Commissioners of the Admiralty, 16 January 1827). Part of the reason that Pacific exploration lent itself to such courtesies—giving credence to the Enlightenment aspirations that justified the voyages—was that the Pacific, as Dunmore put it, "was large enough for prizes to be available for all, and that there was little in it so valuable that it warranted a clash" (Dunmore 1965–1969, 2 : 389).

THE DECLINE OF ENLIGHTENMENT IDEALS

However, as the pace of international rivalry in the Pacific increased in the nineteenth century, particularly between the British and the French, the scientific aspect of such voyages diminished as their strategic and imperial goals became more overt (Dunmore 1965–1969, 2 : 231). By the mid-nineteenth century even the British colonial theorist Edward Wakefield accepted it as a given, as he put it in a letter to the British Prime Minister Robert Peel, "That France has the same right as England to colonise countries the population of which consists of uncivilised tribes" (British Library, Add. MS 40550, 141, Wakefield to Peel, 19 August 1844). As the nineteenth century progressed, Enlightenment ideals of scientific cosmopolitanism faded in the face of the need for voyages to contribute to colonial expansion. Perhaps, too, the incentive for voyages of scientific discovery had diminished as the museums of Europe began to fill up with samples of the flora, fauna, and minerals of the Pacific. From another perspective, too, science itself became more national in character as its scale and importance meant that government involvement became more pronounced.

The greater pace of colonial activity in the nineteenth century also meant a waning of another aspect of the ideology of Enlightenment exploration: its determination to distinguish itself from the age of the conquistadors by acting with greater humanity toward the native peoples encountered. This formed a very important part of the self-image of the Pacific voyagers. As Andrew Kippis wrote in his contemporary life of Cook: "There is an essential difference between the voyages that have lately been undertaken, and many which have been carried on in former times. None of my readers can be ignorant of the horrid cruelties that were exercised by the conquerors of Mexico and Peru; cruelties which can never be remembered without blushing for religion and human nature. But to undertake expeditions with a design of civilizing the world, and meliorating its condition is a noble object" (Kippis 1883 : 365).

When setting out on his *Endeavour* voyage, Cook, for example, was urged "to endeavour by all proper means to cultivate a friendship with the Natives" and his secret instructions included what was to prove the ironic provision that "You are also with the Consent of the Natives to take possession of Convenient Situations in the Country in the Name of the King of Great Britain" (Beaglehole 1955–1974, 1 : cclxxx, cclxxxiii). Behind such rhetoric lay a considerable change in outlook toward non-European peoples. The attitude of superiority based on religion and culture that had been evident in Spanish behavior to the peoples of the Americas or, indeed, the British to the American Indians in the seventeenth century had lessened by the eighteenth century. Once-insular Europe had begun to appreciate that there was much to learn from other peoples and cultures, though in the nineteenth century attitudes of racial superiority based on pseudoscientific grounds began to reassert themselves as the tide of European imperialism gathered force. Relative openness to other cultures and especially those of the Pacific was further strengthened by the rather mixed attitude to the benefits of civilization that was the outgrowth of one strand of Enlightenment thinking. It was this that prompted speculation about the character of the "Noble Savage" that contact with the Pacific at first heightened. This, however, was followed by subsequent disillusionment, which provided impetus for missionary activity and, subsequently, for colonization.

So, once again, we are confronted by the mixed motives of Pacific exploration. On the one hand, they were voyages prompted by the familiar spurs of great power rivalry and the quest for commercial advantage. But such ambitions have existed in all ages and what is interesting is how a period justifies to itself what often appears in retrospect as self-interest. In the great age of Pacific exploration in the late eighteenth century we have seen that the ideological justification for pursuing such ends drew heavily on the language of Enlightenment—on the possibility of promoting human progress through the exercise of reason. Nor was this entirely rhetoric—to gain Enlightenment credentials required considerable expense in the form of a complement of scientific personnel and equipment on board ship and at least some level of international scientific cooperation. Ideally, too, one should display one's Enlightened credentials by keeping to a minimum the number of native peoples killed as a consequence of European intrusion.

The remarkable series of Pacific voyages that occurred between 1764 and 1806 and included the work of Byron, Wallis and Carteret, Bougainville, Cook, La Pérouse, Malaspina, Vancouver, D'Entrecasteaux, Baudin, and Flinders took place in a very different intellectual climate than that of the earlier great age of European oceanic exploration associated with the Spanish and Portuguese entry into the Americas and Asia. It was one in which the quest for gold was still strong though the quest for God had been weakened, if not altogether extinguished. European attitudes of superiority had been lessened by long contact with other cultures, though such attitudes could be easily reawakened. Exploration was conceived of in terms that gave greater prominence to science as the need to map the territory not only in geographical terms but also in terms of flora, fauna, and even its human population became part of any claim to have effective superiority over newly encountered lands (Miller 1996). Underlying such attitudes was a confidence in the power of reason and the possibilities of progress that drew on Enlightenment roots. European intrusion into the Pacific represented not only the power of ships and firearms but also of a body of ideas that gave purpose and direction to a new phase of European expansion that for good and ill drew a whole new sector of the globe into the larger course of human history.

LITERATURE CITED

Manuscript Sources

Mitchell Library, Sydney
A79/4, 187, Banks to Flinders, 1 May 1801
Archives Nationales
B/4/317, no. 111, Buffon to M. le Duc d'Auguillon, 2 January 1773
BB4/998, Viscount Castlereagh for the Commissioners of the Admiralty, 16 January 1827
BB4/998/2, Extrait de la Séance du Lundi [of the Académie des Sciences], 4 February 1833
Bibliothèque Nationale
NAF 9439, 20, Gonneville to the Ministère de la Marine, 19 May 1783
NAF 9439, 52, De Lozier Bouvet, "Mémoire touchant la Découverte des Terres Australes"
NAF 9439, 70v, Bougainville to the Ministère de la Marine, 27 February 1773
British Library
Add. MS 8097, 216, Malaspina to Banks, 20 January 1789
Add. MS 40550, 141, Wakefield to Peel, 19 August 1844
British Museum, Natural History, Botany Library
Deschamps, *Journal de voyage sur La Recherche . . .*
National Library of Scotland
MS 1068, 89, Dalrymple to East India Company, 21 December 1795
National Maritime Museum
MS 57/058, BOG/2, Draft (n.d.) of letter by Bougainville

Printed Sources

BEAGLEHOLE, J. 1966. The exploration of the Pacific. Stanford University Press, Stanford.

BEAGLEHOLE, J. C., ED. 1955–1974. Journals of Captain James Cook on the voyages. 3 vols in 4. Cambridge University Press, Cambridge.
BLADEN, F. 1892–1901. Historical records of New South Wales. 8 vols. Government Printers, Sydney.
COOK, W. 1973. Flood tide of empire: Spain and the Pacific North West, 1543–1819. Yale University Press, New Haven.
DUNMORE, J. 1965–1969. French explorers in the Pacific. 2 vols. Clarendon Press, Oxford.
DUNMORE, J., and M. DE BROSSARD, EDS. 1985. Le voyage de Lapérouse. 2 vols. Imprimerie Nationale, Paris.
ENGSTRAND, I. 1981. Spanish scientists in the New World: The eighteenth-century expeditions. University of Washington Press, Seattle.
FROST, A. 1988. European explorations of the Pacific Ocean. Pages 27–44 *in* R. MacLeod and P. Rehbock, eds. Nature in its greatest extent: Western science in the Pacific. University of Hawai'i Press, Honolulu.
HORNER, F. 1987. The French reconnaissance: Baudin in Australia 1801–1803. Melbourne University Press, Melbourne.
———. 1996. Looking for La Pérouse: D'Entrecasteaux in Australia and the South Pacific. Melbourne University Press, Melbourne.
KIPPIS, A. 1883. A narrative of the voyages round the world performed by Captain James Cook. Beckers and Son, London.
LAS CASAS, B. DE. 1992. A short account of the destruction of the Indies, edited and translated by Nigel Griffin with an introduction by Anthony Pagden. Penguin, London.
MACKAY, D. 1989. The great era of Pacific exploration. Pages 109–120 *in* J. Hardy and A. Frost, eds. Studies from Terra Australis to Australia. Australian Academy of Humanities, Canberra.
MILLER, D. 1996. Joseph Banks, empire, and "centres of calculation" in late Hanoverian London. Pages 21–37 *in* D. Miller and P. Reill, eds. Visions of empire: Voyages, botany, and representations of nature. Cambridge University Press, Cambridge.
PAGDEN, A. 1982. The fall of natural man:

The American Indian and the origins of comparative ethnology. Cambridge University Press, Cambridge.

———. 1995. Lords of all the world: Ideologies of empire in Spain, Britain and France c. 1500–c. 1800. Yale University Press, New Haven.

PARRY, J. H. 1990. The Spanish seaborne empire. University of California Press, Berkeley.

SCHLERETH, T. 1977. The cosmopolitan ideal in Enlightenment thought, its form and function in the ideas of Franklin, Hume,

and Voltaire, 1694–1790. Notre Dame University Press, Notre Dame.

SCHURZ, W. 1922. The Spanish lake. Hisp. Am. Hist. Rev. 5 : 181–194.

SNOW, P., and S. WAINE. 1979. The people from the horizon: An illustrated history of the Europeans among the South Sea Islanders. Phaidon, Oxford.

SPATE, O. 1979. The Spanish lake. Australian National University Press, Canberra.

WOOLF, H. 1959. The transits of Venus: A study of eighteenth-century science. Princeton University Press, Princeton.

VII

Joseph Banks, mapping and the geographies of natural knowledge

'WHAT IS ENLIGHTENMENT?'. The question that Kant set out to answer in the late eighteenth century continues to echo down the centuries as both modernists and postmodernists define themselves as either with or against the 'Enlightenment project' – a project that attempts to organise our knowledge of the world around universal canons of rationality and predictability. Yet, in the eighteenth century itself – in the very age of the Enlightenment – the establishment of such universal canons was a gradual and often contested process as local and traditional ways of understanding the world were challenged by more all-encompassing modes of thought.[1] The spread of Enlightenment values often depended more on a series of beachheads than on the advance of a continuous front. National institutions were sometimes impervious to Enlightenment values while local institutions and practices were reshaped to provide a terrain in which Enlightenment values could take root.[2]

It is this transformation of geographies of knowledge that is the theme of this chapter. What follows focuses on the case of Joseph Banks – a figure whose life and works were largely devoted to the task of promoting greater order and system, either in the natural sciences, in cartography, or in the workings of the British state. And yet, as a landed gentleman, he was rooted in a culture that gave particular status to local knowledge and to the country connections with which it was bound up. Banks's significance is linked, then, with the ways in which his promotion of the mapping of knowledge could operate on different scales and different projections: from the local to the national and thence to the imperial. As one who was very conscious of his

This essay was originally Chapter 8 in *Georgian Geographies: Essays on Space, Place and Landscape in the Eighteenth Century*, eds M. Ogborn and C.W.J. Withers (Manchester University Press, 2004). To avoid unnecessary alteration of the text, the original numbering of the figures has been retained.

dignity as a member of the landed elite, Banks represents, too, an ideal of genteel learning that was sympathetic to forms of knowledge that could be readily assimilated into that world of clubs and learned societies to which he naturally belonged. The corollary to this, however, was his opposition to the growth of more differentiated forms of knowledge based on disciplinary expertise. For Banks, the map of knowledge should be a general one without those specialised markers which would set apart regions of the world of learning as the domain of a select few.

BANKS AND ENCYCLOPAEDIC KNOWLEDGE

Banks's views on the need for knowledge to form part of the general culture of the governing classes were widely shared. Just as the eighteenth-century state with the active support of its Enlightenment-trained bureaucrats sought to reduce the role of rival jurisdictions – whether of the Church, the aristocracy, the guilds or the universities – so, too, the theorists of the Enlightenment aimed to bring all knowledge under a common banner. Claims to a separate space for knowledge derived from divine knowledge or even from traditional lore were eroded by an insistence on the interrelated character of all knowledge and its need to be exposed to public scrutiny.

One of the most characteristic features of the Enlightenment mentality was the encyclopaedia with its assumption that all forms of knowledge could be brought into a methodically organised synthesis. As Yeo has argued, the tensions between the convenience of an alphabetical organisation and the need to classify knowledge in terms of fundamental organisation were generally solved by establishing a prefatory map of knowledge.[3] This metaphor of a map of knowledge was an eighteenth-century commonplace. Francis Bacon, the great mentor of the encyclopaedists, had underlined the parallel between the conquest of geographical space and the conquest of knowledge with the frontispiece to his influential *Novum Organum* (1620) depicting the ship of learning sailing beyond the Pillars of Hercules, the traditional edge of the classical world. Maps were an embodiment of many of the central features of the Enlightenment: they were empirically based, scientifically organised, synthetic in the extent to which they drew together a diverse body of knowledge and they shed light on the dark corners of the globe: all goals shared by the encyclopaedists.

The encyclopaedists, however, keen to avoid the dogmatism of their seventeenth-century rationalising predecessors and still more that of the scholastics, were prepared to acknowledge that the form that the mapping of knowledge took could be, like the mapping of the world, a matter of convenience. What mattered was that there was some system or order, not that there was some absolute unassailable Truth. As D'Alembert wrote in the *Discours*

BANKS, MAPPING AND NATURAL KNOWLEDGE

Preliminaire to the *Encyclopédie,* 'We may therefore figure to ourselves as many different Systems of Knowledge, as there are general Maps of different Projections: and each System may have some Advantage over the rest.'[4] What mattered for the encyclopaedists was that knowledge should be made available, not sheltered in the private domains of privileged corporations. Above all, knowledge should be tested by the extent to which it yielded fruit: there was no point debating its precise boundaries so long as knowledge could be put to practical effect by being rendered accessible through clear organisation.

This was a point developed by Ephraim Chambers in the encyclopaedia which D'Alembert and Diderot took as their starting point. In a telling metaphor, knowledge for Chambers was like a part of the countryside that had become fruitful and well ordered thanks to enclosure with its challenge to traditional open-field systems of agriculture: 'In the wide field of intelligibles, appear some points which have been more cultivated than the rest . . . These spots, regularly laid out, and conveniently circumscribed, and fenced round, make what we call the *Arts and Sciences.*' But, like D'Alembert, he acknowledged that the actual layout of such fruitful fields was, in some ways, a matter of convenience: 'And yet this distribution of the land of Science, like that of the face of the earth and heavens, is wholly arbitrary; and might be altered, perhaps, not without advantage.'[5]

For Chambers and for Diderot and D'Alembert, the test of a map was its utility, in particular its utility in transforming what was traditional and unsystematic into the fruitful and the orderly. The encyclopaedists' maps of knowledge were emblematic of the fundamental Enlightenment quest to make the world more comprehensible through system and classification with the ultimate goal of utilising knowledge for the Baconian ambition of ministering to 'the relief of man's estate'. But to arrive at such geographies of knowledge meant erasing traditional types of maps or, at least, incorporating them into more wide-ranging forms of projection which merged the local with the more universal (see also Clayton's discussion of mapping, Chapter 2, pp. 39–43). The ultimate example of the clash of such knowledge systems was the dismissal of indigenous maps since they had no meaning to the scientific cartographer – a clash recounted in Turnbull's account of Cook's inability to make sense of the map of the Society Islands drawn by the Tahitian, Tupaia.[6] The compilers of the great map of Egypt commissioned by Napoleon were equally dismissive of indigenous knowledge systems.[7] The ways in which Tupaia's map was superseded by that of Cook or the manner in which the French took cartographic control of Egypt are potent illustrations of Francis Bacon's adage that knowledge is power. Representation of the world in schematic form brings with it the possibility of control and effective possession.[8] In many senses, a map captures a portion of the globe making it accessible and linking it with larger pictures and understandings of the world. As William

Blaeu put it in the letter to Louis XIV which prefaced his twelve-volume *Grand Atlas* (1663), 'maps enable us to contemplate at home and right before our eyes things that are farthest away'.[9]

Bruno Latour has underlined these points with his concept of 'centres of calculation' which pull together the fruits of a 'cycle of accumulation'.[10] In such a conception, the world is reconstructed in the rational form of a map, or some other ordered schema, through scientific institutions usually located at the centre of an imperial power. Along with all the other cargo that is the natural outcome of imperial exploration, knowledge – in the form of detailed charts and precise geographical data – is brought back to the centre to be transformed into maps which make possible yet further exploration and the discovery of still more detailed information. Yet maps are but one form of such a rational reconstruction of the world. Other types of data about the natural world and its human population are also transformed into orderly models which make possible understanding and exploitation of a world increasingly denuded of its dark and inaccessible corners. Once such models are constructed in the 'centres of calculation' they make possible the establishment of networks which can seek to reconstruct the world along the lines of the models themselves. One instance of such a reordering of the world was Banks's work in moving around species of plants and animals the better to serve the economic interests of the British empire – a process made possible by the network of botanical gardens that radiated around the world from Kew.

BANKS AND CARTOGRAPHIC KNOWLEDGE

European explorers like Banks brought back to such centres of calculation intellectual capital of all kinds: outlines of coasts, variations of the compass that enabled one to distinguish between the true and the magnetic North, information about tides, winds and currents and cases and cases of natural history specimens. Such information was then incorporated into a multiplicity of maps: charts for navigation, hydrographic maps describing the sea and its currents, diagrams of magnetic variations and, of course, vast volumes describing and classifying the natural world including its human population. Such maps, through their ability to abstract features of key interest, enabled the imperial powers to draw new sectors of the world into their networks either through trade or outright annexation. Whether at home or abroad, maps provided the readiest means of drawing knowledge together into an easily comprehensible form, a particularly important *desideratum* for the encyclopaedia-minded eighteenth century.[11] In the Georgian world, few individuals knew better than Joseph Banks the political uses of knowledge and, in particular, knowledge of the natural world. If, as William Harvey claimed, Francis Bacon wrote philosophy like a Lord Chancellor, Banks conducted science as a Privy Councillor alert to the ways in which the pursuit of knowledge could serve the state. And, if natural knowledge was to be usable, he well appreciated that it must be put in orderly forms of which the map was one obvious expression.

Banks in a certain sense served his scientific apprenticeship as part of cartographic enterprises. His first voyage abroad, to Newfoundland and Labrador in 1766, was part of an expedition which included among its aims the mapping of some of the territories over which Britain was exercising more effective control, thanks to its victories in the Seven Years War. The *Endeavour* expedition (1768–71) was under the command of James Cook, the map maker who had helped to make that victory possible through his charts of the St Lawrence and whose Pacific voyages were dominated by the quest for accurate maps. Indeed, this enterprise determined the very choice of vessel, the shallow-drafted Whitby colliers, which made it possible to come close to shore to conduct the precise observations on which accurate maps depended.

On both expeditions, the presence of Banks enabled the mapping to extend to a survey of the natural history of the areas encountered, including accounts of the human population. Such maps helped to make eventual imperial control of these areas of the globe more possible, particularly as Banks brought back not only maps of their natural history in the form of classification and taxonomic description, but also physical samples. These extended beyond pressed plants and stuffed animals to 'natural curiosities' and ethnological material. Banks even sought to bring back live human beings and, although the Tahitian Tupaia died en route, Banks was closely involved with the eventual visit of the Tahitian Omai. In all these ways, Banks added to the 'cycle of accumulation' which made it possible for his house in Soho Square – the eventual nucleus of the British Museum (Natural History) – to become a 'centre of calculation' by providing in accessible form systematic knowledge of the Pacific world.

Banks continued to use his influence to promote the more accurate charting of the globe, the most significant instance being his sponsorship of Matthew Flinders's circumnavigation of what was then New Holland with the *Investigator* expedition of 1801–3. The outcome was a map of the entire Australasian continent which made its European inhabitants more conscious of the land that they had come to inhabit and more inclined to adopt the new name, 'Australia', given to it by Flinders (see Figure 8.1). It served as an instance of a more general phenomenon: the way in which, by literally giving shape to a nation, map makers could be agents in shaping a sense of national identity.[12] More effective possession of the world through mapping was an activity that was as important at home as it was abroad. As the eighteenth century saw the consolidation of the state largely under the impulse of war, it

8.1 Matthew Flinders's Chart of Terra Australis (1804). (© British Crown Copyright 1998. Reproduced by permission of the Controller of Her Majesty's Stationery Office and the UK Hydrographic Office)

became ever more important to take effective stock of the resources that the state commanded. Following the lead of France, the state which first attempted to integrate science in the conduct of government policy, the British government began to institute a programme of mapping its own territory.[13] The connections between mapping and war were evident in the very origins of this enterprise.

MAPPING BRITAIN

The British Ordnance Survey effectively began following the suppression of the Jacobite uprising of 1745–46 when it was realised that knowledge of the Highlands by the Duke of Cumberland's forces was hampered by the lack of maps of the north and west of Scotland. This prompted the Military Survey of Scotland conducted from 1747 to 1754 under the initial leadership of Colonel David Watson and completed by William Roy.[14] As a consequence of these endeavours, William Roy, who was to be one of Banks's closest allies within the Royal Society, was appointed in 1765 to the newly-created position of Surveyor General of Coasts and Engineer of Directing Military Surveys in Great Britain under the Board of Ordnance, the wing of government whose function at its foundation was to supply military equipment. The position brought with it the responsibility 'to inspect, survey and make reports from time to time of the state of the Coasts, and Districts of the Country adjacent to the Coasts of the Kingdom, and the Islands thereunto belonging', a role that makes evident the association between mapping and defence of the kingdom.[15] It was also a post that made possible the development of mapping on a more rigorously scientific basis than had been employed in the survey of Scotland which was not based on triangulation but rather sought to put in schematic form a military reconnaissance of the area.[16] The alliance between Banks and Roy, together with others linked with the early work of the Ordnance Survey such as Charles Blagden and John Lloyd, was strengthened by their close association as founding members of the Royal Society Club (which lasted from 1775 until 1784).[17] From the project's beginnings in 1783, Banks was closely involved in Roy's great enterprise of providing a sound scientific basis for the accurate mapping of Britain through the construction of a baseline which would make possible precise trigonometrical calculations (see Figure 8.2).[18]

This baseline, which eventually ran for five miles across Hounslow Heath, served as the foundation for a series of triangles that extended to Dover. At Dover, it linked with the work of the French who, under the leadership of the Cassini scientific dynasty, had pioneered the techniques of accurate measurement of the earth's surface on which the British Ordnance Survey built.[19] Not the least of Banks's services was liaising with the brilliant but difficult instrument maker, Jesse Ramsden, who developed a form of theodolite which made Roy's work possible. The project was considered sufficiently important to the state to attract a royal grant of £3000 which was administered by Banks as President of the Royal Society.[20] Banks's deputy in Royal Society matters, Charles Blagden, was dispatched to France in 1787 following Roy's request to the Royal Society that 'a Commissioner might be appointed . . . to join the French Commissioners in their co-operation for carrying the trigonometrical measurements across the straits of Dover'.[21]

The motives for undertaking this project were made clear by Roy in the *Philosophical Transactions of the Royal Society of London* in 1785. He noted that: 'Accurate surveys of a country are works of great public utility, as affording the surest foundation for almost every kind of internal improvement in time of peace and the best means of forming judicious plans of defence against the invasions of an enemy in time of war.' Surveying was a route to

VII

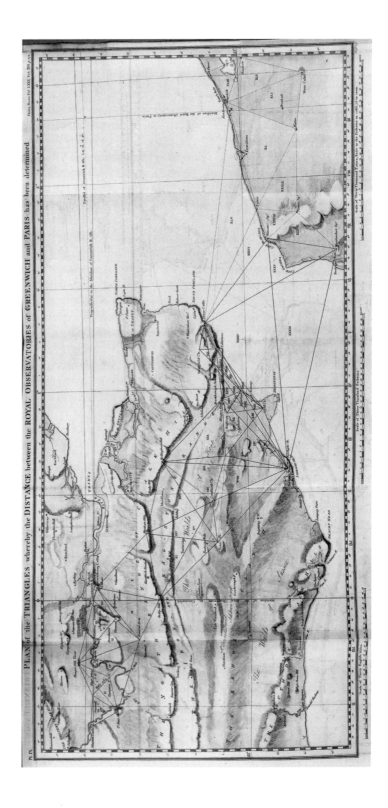

economic progress and military power. It was a means of turning knowledge into power through the full exploitation of a country's resources and of ensuring that rival powers were denied access. As Roy himself acknowledged, the initial stimulus for the project had been the 1745 Jacobite Rebellion, a fact which had prompted him to suggest that the needs of war were the most potent impulse for mapping: 'a state of warfare generally produces the first improvements in its [a country's] geography'.[22]

Such mapping had an international as well as a national character. Roy's triangulation project was one that had obvious national benefits but it had, from Banks's point of view, the further merit of fostering a scientific partnership with France since the British and French networks were connected across the Channel in 1787. Roy himself spoke of the way in which through these 'combined operations', as he termed it, 'the two most famous observatories in Europe, Greenwich and Paris, would be more accurately ascertained than they are at present'.[23] Although national considerations always came first for Banks, science retained for him a cosmopolitan character that was reinforced by the network of scientific academies of which the Royal Society formed a part.[24]

Characteristically, it was a project that was initiated by the French who had gone further than the British in linking science with the workings of government. Particularly important to the French state was the accurate mapping of its territories – an endeavour in which astronomy with its traditions of precise measurement joined with geography.[25] French geographers served the needs of a centralising French state for which accurate maps provided more effective control over its own population and the means better to defeat its enemies.[26] By linking with the British triangulation project in measuring precisely the distance between the French and the English Royal Observatories, the French scientific establishment helped both to extend the range of its own measurements and to widen the channels of scientific communication which the cosmopolitan world of the scientific academies served to foster.

MAPPING EMPIRES

With the metropolitan model of mapping and surveying established, such techniques were then gradually applied to the empire. An early transference was the work of James Rennell whose interest in applying European surveying techniques to India led to his appointment as Surveyor General of Bengal

---

facing] 8.2 'Plan of the triangles whereby the distance between the Royal Observatories of Greenwich and Paris has been determined' (based on the Hounslow baseline). From *Philosophical Transactions of the Royal Society*, 80 (1790), 272. (Courtesy of the John Rylands Library, University of Manchester)

in 1764. The fruits of his labour were embodied in his *Bengal Atlas* (1779) and a map of India that appeared in 1783. These works led to Rennell being awarded the Royal Society's Copley Medal in 1791 and elicited enthusiastic praise from Banks[27] who contrasted the accuracy of Rennell's mapping of Bengal with the relatively incomplete state of the mapping of Britain.[28] For Banks, Rennell's work was a potent illustration of the benefits that could accrue to the British empire and, more remotely, to humankind as a whole, from a scientifically-based scrutiny of the resources under the sway of British rule. Mapping required an orderly classification of the chief features of the land thus enabling its improvement and exploitation.

As Edney stresses, Rennell's map of India in many ways created the image of the subcontinent in the mind of the British imperial classes.[29] Like Flinders's map of Australia (See Figure 8.1, p. 156), Rennell's map of India gave reality to a national abstraction – the difference, of course, being that Australia largely became ethnically and demographically an extension of the British Isles. Rennell's India, on the other hand, served to foster a sense of trusteeship over an alien people lacking the scientific and organisational skills which maps such as Rennell's illustrated. Indeed, argues Edney in the manner of Latour, Rennell's work and the subsequent more accurate maps based on triangulation made it possible for the British to reassemble, as it were, a form of India in Britain. Precise cartography made it possible to construct a 'centre of calculation' in the imperial metropolis so that India could be largely ruled from afar. As Edney puts it: 'the archive could be shipped back to Britain as a symbolic appropriation of the territories and peoples in India'.[30] Godlewska makes a similar claim for the Napoleonic survey of Egypt which, as she writes, 'creates the Egypt that could be claimed and taken home and mathematically and rigorously interpreted in the silence of French libraries, laboratories and museums without the difficult complications associated with colonialism, subject peoples and the bizarreries of other cultures'.[31]

India offered the possibility of greater British self-sufficiency in key products and Rennell's maps helped to establish the route to such goals. Banks turned to Rennell, for example, for expert advice on how to reshape the ecology and economy of parts of India in order to provide Britain with the tea that it was consuming in greater and greater quantities. It was Banks's hope that this would spare the British economy the transfer of vast sums of money to purchase Chinese tea. Rennell responded enthusiastically to Banks's suggestion writing in 1788 that 'the culture of ye Tea ... would suit the Hindoos most perfectly; their patient industry, & pliable fingers would manage the whole business to great advantage'. Furthermore, continued Rennell, it might be possible not only to bypass China with regard to tea but also to transplant to British India another major product, namely silk. For Rennell, 'The culture of the Mulberry for Silk worms strikes me as being something like that of the Tea Shrub, by the historical drawings that I have seen brought from China.'[32]

Rennell's early cartographic work led in turn to the application of triangulation methods in Madras in 1802, and eventually to the Great Trigonometrical Survey of India begun in 1818. One point of continuity between the metropolitan and imperial projects was the sending to India of a Ramsden theodolite.[33] Rennell also urged the expansion of imperial cartography in another direction through his involvement with the African Association, a body founded in 1788 to promote the exploration of Africa and which largely owed its origin to Banks. The data brought back by those explorers sponsored by the Association provided the basis for maps that would make possible greater European knowledge and eventual control of the African interior. Rennell, for example, drew on the accounts of Mungo Park and others to construct in 1798 'A Map showing the progress of discovery and improvement, in the geography of North Africa' – a map that, in retrospect, reveals the still sketchy European knowledge of that continent. In the following year he contributed eleven maps to the account of Park's travels.[34] When Banks and Park collaborated on plans for a further voyage to trace where the Niger entered the sea, Rennell attempted to dissuade both from a voyage that, all too accurately, he predicted was dangerous. But Banks was not to be lightly deflected from promoting the advance of cartographic knowledge writing that 'it is by similar hazards of human life alone that we can hope to penetrate the obscurity of the internal face of Africa'.[35] Park set off on his second and last expedition to the Niger in 1805 and perished in the cause of geographical knowledge and British commercial interests.

Undeterred by such sacrifices, African exploration remained an abiding interest for Rennell. In 1813, for example, he wrote to Banks in praise of the contribution made by Johann Burckhardt on an expedition sponsored by the African Association, particularly since, as Rennell put it, 'he has given us Notices, respecting a Country that we know as little of, as of Sudan'.[36] Rennell and Banks worked together in promoting the scientific exploration of much of the globe. Along with their common interest in India and Africa, Banks consulted Rennell in 1791 over plans for the Vancouver voyage to Northwest America. In 1810 when Banks focused on the promotion of Arctic exploration, he again turned to Rennell to provide detailed advice on such matters as winds and sea currents. When Matthew Flinders was working on his account of the circumnavigation of Australia in 1811, Rennell was among the visitors when Flinders called on Banks at Soho Square.[37] The Banks–Rennell partnership was, then, a potent illustration of the extent to which the Banksian ideals of improvement at home and abroad were closely allied to the promotion of mapping as a practice of geographical enquiry that aimed to benefit Britain at home and overseas.

VII

Mapping had its local as well as its national and imperial dimensions and Banks shared the English land-owning class's ability to be evident and involved in both the affairs of the city and the country. For three or four months each summer the President of the Royal Society and adviser to government became the Lincolnshire squire with his seat at Revesby Abbey and his well-honed political connections throughout the county. Whether in the city or the country, Banks kept in touch with both worlds through diligent lieutenants whose copious correspondence he kept classified to a degree that paralleled the orderly classification of the natural world made possible by the system of Linnaeus and his successors. At Banks's family seat at Revesby, a contemporary recorded that

> Nests of drawers numbered consecutively lined the walls; and it was Sir Joseph's custom to have two catalogues, descriptive of their contents; one of which always accompanied him, the other remained at Revesby. Thus, if when in London, he required any paper contained in the drawers, he had merely to refer to his catalogue, and sending the number to his steward at Revesby, the latter was enabled, by means of the Catalogue in the office, to put his hand in a moment upon the desired document, and forward it to his master.[38]

As an improving landowner much involved in the draining of the fens, Banks was well aware of the uses to which maps could be put in consolidating estates and of effecting that move to clearer and more exclusive forms of private ownership of the land which were closely linked to the agricultural revolution.[39] One graphic illustration of the importance to Banks of such cadastral or estate maps and their iconographic role as an agent of improvement is that he is depicted as holding one such, linked to the draining of the fens, when his portrait was painted in the uniform of the Lincolnshire Volunteers (Figure 8.3). Accurate mapping enabled an improving landlord such as Banks to exploit more effectively the resources of his estate both above and below ground. Estate maps facilitated enclosure and the draining of the fens. Effective mapping brought with it the system and order embodied, for example, in the plans and surveys of his Revesby estate carried out for Banks by Thomas Stone in 1794 with their lists of tenants, field names, acreages of pasture, arable and so forth.[40] That apostle of agricultural improvement, Thomas Young, was so impressed by Banks's maps of the Revesby Estate that he incorporated them in his *General View of the Agriculture of the County of Lincolnshire* (1799, 1813). It was a work that further strengthened Banks's resolve to drain what Young called those 'horrid fens',[41] a project that was preceded by further mapping in the form of a survey by the Scottish engineer, John Rennie, with a view to establishing the best means of drainage.

8.3 Joseph Banks, by Thomas Phillips (1814). (By kind permission of Boston Borough Council)

Sub-surface geological mapping enabled Banks to take advantage of what mineral wealth was available, in particular, the deposits of lead and coal on his Overton estate in Derbyshire. Soon after he inherited this estate from his uncle, Robert Banks-Hodgkinson, Banks proceeded to have a local surveyor, George Nuttall, map the area paying particular attention to possible coal deposits. Nuttall's map was to serve as a model which helped stimulate the later surveys of John Farey which formed part of the pre-history of the Geological Survey of Britain.[42] Farey's position from 1792 as agent to the Duke of Bedford is further indication of the close links between agricultural improvement and mapping. Significantly, too, Farey's major work, the two-volume *Survey of the County of Derby including a General View of its Agriculture & Minerals* (1811–13), was published under the aegis of the Board of Agriculture, the chief body dedicated to the promotion of agricultural improvement (with the enthusiastic support of Joseph Banks). This *magnum opus*, which was commissioned by Banks, grew out of Farey's geological transect of the Lincolnshire coast and the area around the Overton estate. Farey's geological mapping was combined with Nuttall's land survey in a hand-coloured map which was exhibited in 1808–9 'as a matter of Science' at Banks's house in Soho Square.[43]

Farey's work was given scientific shape and direction by the principles of William Smith who helped to establish the ordering of English geological strata. By so doing, Smith helped promote the transformation of sub-surface geological mapping into a theoretically-based cartography of the successive ages of the earth. It brought to bear on the routine concerns of mine owners and estate owners the findings derived from the wider scientific context concerning the ways in which the landscape had changed over time. Mapping here drew attention to the role of time as well as space by establishing in pictorial form the inferences made possible by the study of outcrops. Smith's ordering of strata provided the nucleus which Farey expanded to develop a more comprehensive mapping of the world beneath the feet of improving landlords such as Banks who were his patrons.[44]

Banks's patronage of Farey was consistent with his attempts to encourage other improving landlords to use the techniques of geological surveying to promote the better exploitation of mineral resources. In 1784, for example, he had advised Lord Palmerston on the best method of undertaking 'the tryal intended to be made for Coal in your Lordships neighbourhood'. In return he requested 'an exact account of the Strata passed through with the depth of Each'.[45] Coal, that most basic of all commodities in fuelling the industrial revolution, remained an abiding preoccupation of Banks. This is apparent from his 1797 'Essay regarding the coal trade, the number of persons it employs, revenue gained from it & the importance of the coal trade to the naval strength of Great Britain'.[46] As Banks was well aware, the location of Britain's coal reserves required detailed mapping. This fact explains why, in 1794, Banks acquired a chart giving depths of a mine at Newcastle.[47] It accounts also for his enthusiastic support for William Smith's *A Delineation of the Strata of England and Wales, with Part of Scotland, exhibiting the Collieries and Mines*, which was published in 1815 and dedicated to Banks.[48]

MAPPING STATE POWER

Mapping – whether above or below ground – was one of the most potent ways in which the modernising state could take command of its resources.

BANKS, MAPPING AND NATURAL KNOWLEDGE

The late eighteenth-century state provided itself in other ways with charts of its population and its economic activities.[49] As John Brewer has stressed, the increasing cost of war made a more accurate knowledge of a nation's resources – and, ultimately, its tax-paying capacity – more and more vital. 'To be more effective government required greater knowledge', writes Brewer, 'skilled government needed more detailed and precise information'.[50]

The close conjunction between the beginnings of the Ordnance Survey in 1791 and the first British census in 1801 reflects the fact that both sprang from the same impulse: to provide a more secure framework of knowledge on which to construct state policy. Both maps and a census represent the world (natural and human) in abstracted terms which remove the data away from their local setting. Both are ways in which the state can better define itself by establishing its boundaries and the extent of population within those boundaries. Such a connection between mapping and the beginnings of national censuses is apparent in the correspondence between Banks and John Rickman, the man who conducted the first British census in 1801. As part of his more general investigations of the impact of population on agriculture, Rickman wrote to Banks on several occasions seeking, for example, to borrow General Roy's map of Scotland.[51] The common thread that drew Banks and Rickman together was that of agricultural improvement. Banks was prepared to acknowledge that agricultural change could disrupt traditional ways of life but argued that its benefits would ultimately extend to all. Banks discussed with Rickman the impact of the increase in the number of sheep in Scotland on the human population taking the sanguine view that, in the long run, the evicted population might be better off since the ultimate outcome of such Highland clearances might well be that those so displaced 'may reappear with their mouths full of mutton in our manufactories'.[52] Rickman was of like mind: 'I quite agree with you in the fitness of dislodging Highlanders for Sheep farming.'[53] Such clearances were a potent example of the way in which agricultural improvement and the techniques of surveying were closely allied as the land was put to more efficient use by its landowners. They also served as an exemplar of the way in which modernising scientific knowledge could be corrosive of traditional ways of life whether in Britain itself or in the empire more generally.[54]

Rickman's work extended to considering the effects of such a transformation on the area's human population – in effect the mapping of demographic change and resources. More accurate data on the distribution of population, like improvements in surveying, were intended to promote better planning to assist economic growth and political order. Rickman's work formed part of that same drive for order and system that informed the quest for systems of classification in the natural world. As Rickman told Banks in 1805, 'The favourite object of my life, is to distribute England into such orderly Divisions or Districts, that information may be obtained and good government enforced in the most effectual manner.'[55]

The systematisation of weights and measures in Great Britain from 1826 also forms part of the same consolidation of the state and its central authority as the definition of territory through mapping and the establishment of the extent and character of its population through the census. One instance of this connection is illustrated by the French metrical system which grew out of work to establish the exact shape of the earth. The metre was one tenmillionth of the quadrant of the terrestrial meridian which ran through France.[56] In England, William Roy's work on establishing the scientific basis on which the mapping of the Ordnance Survey was based was also linked with the legal definition of the standard yard.[57] Banks's close associate within the Royal Society, Henry Kater, was working in the same tradition when he combined measurement of longitude with advice to the Russian government on weights and measures.[58] Kater's main contribution to the reform of British practice was, however, to provide accurate data so that weights and measures could be based on the amplitude of a swinging pendulum.[59]

By the early nineteenth century, however, the need for some scientifically-based system had become urgent, particularly given the example of the recently-introduced French metric system. In 1802, Banks assured a French scientific correspondent of the Royal Society's willingness to adopt any system 'whether it is discovered in France in England or elsewhere on Condition however that the Principles on which it is Founded are simple & sufficiently correct to allow it in case of need to be reconstructed with rigorous exactitude in every part of the Globe'.[60] Hostility to things French and anti-revolutionary sentiment, however, stood in the way of the introduction of this system. From the deliberations of a parliamentary committee chaired by Banks between 1817 and 1819, the imperial system of weights and measures based on the pendulum rather than the quadrant finally emerged in 1826. Banks sought to justify this as a matter of contemporary expediency: 'lest it [the quadrant] should in Future be remeasured & proved different from what is now believd to be the amount'.[61]

Banks's exertions in improving the mint also assisted the more effective standardisation of one of the most potent symbols of state power: the issuing of coin of the realm.[62] In the tradition of Sir Isaac Newton, who had served both as Master of the Mint and as President of the Royal Society, Banks was called upon to provide scientific advice to government in relation to the national coinage. He served on the Privy Council Committee on Coinage from 1787 and took an active part in the recoinage of copper money in 1797. This recoinage was facilitated by Banks's friendship with Matthew Boulton whose steam engines provided the technological power to reform the Mint's antiquated practices. When the problem of wear of metal in coins was referred to Banks, he arranged for a series of experiments to be conducted by the Royal Society under Henry Cavendish and Charles Hatchett, the results of which, despite fears about confidentiality, were published in 1803.[63] In 1819, a year before his death, Banks was called upon to chair a committee to enquire into methods of preventing the forgery of banknotes. Such endeavours were a yet further reflection of that same mentality that prompted late eighteenth-century states to establish more clearly the boundaries of their powers.

BANKS AND THE CLASSIFICATION OF KNOWLEDGE

This quest for the better ordering of the world through maps, censuses and uniform systems of weights and measures was linked with the eighteenth-century preoccupation with classification. This had its most conspicuous scientific manifestation in the Linnean system for bringing order to the natural world. To Linnaeus, his taxonomic order was a 'map of nature'.[64] For botanists like Banks engaged in coming to terms with what, for Europe, was the new world of the Pacific, the Linnean system gave a sense of purpose and direction as thousands of natural specimens were fitted onto a map of the natural world which could become comprehensive without being overwhelmed by the information it contained. Banks's herbarium at Soho Square, for example, contained some thirty thousand specimens arranged according to Linnean principles.[65] Banks's vast collection, together with those amassed at Kew, formed a centre of calculation which made possible the reordering of the world as plants were moved around the world to promote economic growth.[66]

There were, it is true, debates about whether the Linnean system of classification, based as it was on artificial principles, was the best method of reflecting the actual realities of the natural world. In 1817, three years before his death, Banks commented to Sir James Edward Smith, the founder of the Linnean Society, that:

> I fear you will differ from me in opinion when I fancy Jussieu's natural orders to be superior to those of Linnaeus. I do not however mean to allege that he had even an equal degree of merit in having compiled them – he has taken all Linnaeus had done as his own; and having thus possessed himself of an elegant and substantial fabric, has done much towards increasing its beauty, but far less towards any improvement in its stability.[67]

As Banks's words suggest, debates as to its merits did not substantially challenge the Linnean achievement and were more an invitation to refine and improve it rather than to demolish it.

Accurate charts and botanical and zoological collecting were all part of a common venture: to know as much of the world as possible. The better the maps and the more comprehensive the charting of the animal, vegetable and mineral kingdoms the more subject such areas became to the European metropolitan powers. In the late eighteenth century, voyages of exploration were more and more justified by Enlightenment values of curiosity and the quest for knowledge. Such motives were, of course, combined with the familiar quest for strategic or economic advantage.[68] Cook's great *Endeavour* voyage was prompted by the desire to participate in the observation of the transit of Venus and had the additional scientific advantage of assisting Banks's work in natural history. But it was also intended to promote the more accurate mapping of New Zealand and the east coast of Australia, and thus to lay the foundation for the eventual incorporation of these territories into the British Empire. Explorers and naturalists like Banks and his clients expanded the realm of science and of knowledge generally. They provided maps of geography and natural history that opened up new sectors of the globe, and they also provided the intellectual resources that made effective imperialism possible. Exploration made possible the building of centres of calculation that made the world appear more rationally explicable but also recreated it in the image of an elite imbued with such Enlightenment ideals. A world thus constituted made the task of empire building both more manageable and more familiar.

Banks's involvement in the expedition of George Vancouver from 1791 to 1795 reflects this dual quality. The voyage itself was intended to consolidate a British presence on the western coast of America and Vancouver's mapping was a means to that end (see Chapter 2). To ensure the quality of the mapping and thus to secure its authority for scientific and imperial ends, Banks was directly involved in drawing up the surveying instructions along with Rennell and William Bligh.[69] When he gave instructions to Archibald Menzies, the voyage's naturalist, Banks dwelt on the extent to which the voyage would 'promote the interest of Science, & contribute to the increase of human knowledge'. Banks further instructed Menzies to assess 'whether, should it any time hereafter be deemed expedient to send out settlers from England, the Grains, Pulse and Fruits cultivated in Europe [that] are likely to thrive, and if not what kind of produce would in your opinion be the most suitable'.[70] Maps and empires were made to appear natural allies.

If imperial botanical geography was a further instance of Banks's enthusiasm for bringing order to the world, there were, nevertheless, limits to his willingness to endorse the changing map of natural knowledge. Reared in the traditions of the gentleman collector with its long pre-history of the virtuoso, Banks was unsympathetic to the growing claims for disciplinary specialisation. For him, science was a unified continent. Attempts to divide it into different realms ran counter to a gentlemanly ethos where all forms of knowledge were open for edification or practical use without the obstacles of excessive specialist expertise or organisations. Such views accounted for his strong opposition to the attempt to found scientific bodies that might challenge the Royal Society. The foundation of the Geological Society in 1807 and of the Astronomical Society in 1820 led the aged Banks to remark bitterly: 'I see plainly that all these new-fangled associations will finally dismantle the Royal Society, and not leave the old lady a rag to cover her.'[71] Banks's *cri de coeur* was in some ways that of a man of the old regime hostile to the spirit of reform and revitalisation that was beginning to reshape the political and intellectual institutions of a Britain being remoulded by the forces unleashed by the French and industrial revolutions.

Yet it was also a plea for the unity of knowledge, a plea that came naturally to a natural historian who regarded all objects, whether animal, vegetable or mineral, as being linked by their common location. Such a sense of the unity of nature was further promoted by the way in which Banks, as a member of the landed elite, had a sense of the importance of local knowledge as a way of understanding the natural and civil history of the countryside on which power and prestige were ultimately based.[72] Although, for Banks, as for other eighteenth-century naturalists,[73] such a sense of the interconnected nature of the geography of knowledge was local in its origin, it was given an imperial and even global dimension through his promotion of scientific exploration. In some ways, this was an insight that he shared with Alexander von Humboldt, another explorer whose sense of the unity of knowledge had been heightened by his dedication to the promotion of natural history in its widest sense. As Cannon writes, 'this idea of the interconnectedness of things was to be the theme that ran like a high voltage current through [Humboldt's] *Kosmos*'.[74]

For Banks, the man of the Enlightenment, the scientific specialisation and professionalisation which was becoming apparent in the early nineteenth century meant that the map of knowledge was being subdivided into separate compartments.[75] This was a development he saw as undermining that drive to overall order which sustained his own enquiries. It meant, too, a move away from the dominant goal of late eighteenth-century geography which was, as Godlewska puts it, to establish 'the reflection of the unity and coherence of the world'.[76] This was an ambition to which Humboldt remained committed. For Banks, too, a lifetime devoted to the promotion of improvement and of system was based on a firm belief in the unity of knowledge: its increasing fragmentation meant a new form of intellectual cartography which the ageing President of the Royal Society was ill-equipped to undertake.

NOTES

1 S. Harris, 'Introduction: thinking locally, acting globally', *Configurations*, 6 (1998), 131–9, quote on p. 136; L. Daston, 'The ethos of Enlightenment', in W. Clark, J. Golinski and S. Schaffer (eds), *The Sciences in Enlightened Europe* (Chicago: Chicago University Press, 1999), 495–504, quote on p. 502.

2 C. W. J. Withers and D. N. Livingstone, 'Introduction: on geography and Enlightenment', in D. N. Livingstone and C. W. J. Withers (eds), *Geography and Enlightenment*
   (Chicago: Chicago University Press, 1999), 1–32; C. W. J. Withers, 'Towards a history
   of geography in the public sphere', *History of Science*, 37 (1999), 45–78; W. Clark, J.
   Golinski and S. Schaffer, 'Introduction', in Clark, Golinski and Schaffer (eds), *Sciences
   in Enlightened Europe*, 3–31.
3 R. Yeo, *Encyclopaedic Visions: Scientific Dictionaries and Enlightenment Culture*
   (Cambridge: Cambridge University Press, 2001).
4 N. Fisher, 'The classification of the sciences', in R. C. Olby, G. N. Cantor, J. R. R.
   Christie and M. J. S. Hodge (eds), *Companion to the History of Modern Science*
   (London: Routledge, 1996), 853–85, quote on p. 862.
5 *Ibid.*, 861–2.
6 D. Turnbull, 'Cook and Tupaia, a tale of cartographic *méconnaissance*?', in M. Lincoln
   (ed.), *Science and Exploration in the Pacific: European Voyages to the Southern Oceans
   in the Eighteenth Century* (Woodbridge: Boydell & Brewer, 1999), 117–31.
7 A. M. C. Godlewska, 'The Napoleonic survey of Egypt: a masterpiece of cartographic
   compilation and early nineteenth-century fieldwork', *Cartographica*, 25 (1988), 1–
   171.
8 For an overview of these issues, see J. B. Harley, 'Maps, knowledge and power', in
   D. Cosgrove and S. J. Daniels (eds), *The Iconography of Landscape: Essays on the
   Symbolic Representation, Design and Use of Past Environments* (Cambridge: Cambridge
   University Press, 1988), 277–312.
9 D. N. Livingstone, *The Geographical Tradition: Episodes in the History of a Contested
   Enterprise* (Oxford: Blackwell, 1993), 98.
10 B. Latour, *Science in Action: How to Follow Scientists and Engineers through Society*
   (Milton Keynes: Open University Press, 1987), 215–37. On the application of Latour's
   work to Joseph Banks, see D. P. Miller, 'Joseph Banks, empire, and "centers of calculation" in late Hanoverian London', and J. Gascoigne, 'The ordering of nature and the
   ordering of empire: a commentary', in D. P. Miller and P. H. Reill, *Visions of Empire:
   Voyages, Botany, and Representations of Nature* (Cambridge: Cambridge University
   Press, 1996), 21–37 and 107–13 respectively.
11 M. H. Edney, 'Reconsidering Enlightenment geography and map making: reconnaissance, mapping, archive', in Livingstone and Withers (eds), *Geography and
   Enlightenment*, 165–98.
12 J. Black, *Maps and Politics* (Chicago: Chicago University Press, 1997).
13 C. C. Gillispie, *Science and Polity in France at the End of the Old Regime* (Princeton:
   Princeton University Press, 1980); R. Hahn, *The Anatomy of a Scientific Institution: The
   Paris Academy of Science, 1666–1803* (Berkeley: University of California Press, 1971).
   For a discussion of the continuation of one aspect of this tradition under Napoleon,
   see A. M. C. Godlewska, 'Napoleon's geographers (1797–1815): imperialists and soldiers
   of modernity', in A. M. C. Godlewska and N. J. Smith (eds), *Geography and Empire*
   (Oxford: Blackwell, 1994), 31–55.
14 S. Widmalm, 'Accuracy, rhetoric, and technology: the Paris–Greenwich triangulation,
   1784–88', in T. Frängsmyr, J. L. Heilbron and R. E. Rider (eds), *The Quantifying Spirit
   in the Eighteenth Century* (Berkeley and Oxford: Science History Publications, 1990),
   179–206; C. W. J. Withers, 'Situating practical reason: geography, geometry and mapping in the Scottish Enlightenment', in C. W. J. Withers and P. Wood (eds), *Science
   and Medicine in the Scottish Enlightenment* (East Linton: Tuckwell Press, 2002), 54–78.
15 R. A. Gardiner, 'William Roy, surveyor and antiquary', *The Geographical Journal*, 143
   (1977), 439–50.

16 C. Close, *The Early Years of the Ordnance Survey* (Newton Abbot: David & Charles, 1969), 5–29.
17 Gardiner, 'William Roy', 445.
18 On this project, see Widmalm, 'Accuracy, rhetoric, and technology', *passim*.
19 G. R. Crone, *Maps and their Makers: An Introduction to the History of Cartography* (Folkestone: Dawson, 1978), 86–8; N. J. W. Thrower, *Maps and Man* (Englewood Cliffs, NJ: Prentice Hall, 1972), 75–80.
20 H. B. Carter, *Sir Joseph Banks 1743–1820* (London: British Museum, 1988), 203–4.
21 Royal Society, CMO (Council Minutes), Vol. 7, 276, 29 June 1787.
22 W. Roy, 'An account of the measurement of a base line on Hounslow Heath', *Philosophical Transactions of the Royal Society of London*, 75 (1785), 385–478, quote on p. 385.
23 *Ibid.*, 389.
24 On Banks and the tensions between national and cosmopolitan uses of sciences, see J. Gascoigne, *Science in the Service of Empire: Joseph Banks, the British State and the Uses of Science in the Age of Revolution* (Cambridge: Cambridge University Press, 1998), 147–65.
25 A. M. C. Godlewska, *Geography Unbound: French Geographic Science from Cassini to Humboldt* (Chicago: Chicago University Press, 1999), 68–71.
26 Widmalm, 'Accuracy, rhetoric, and technology', 201.
27 Close, *The Early Years*, 37.
28 W. A. Seymour, *A History of the Ordnance Survey* (Folkestone, Kent: Dawson, 1980), 1.
29 M. H. Edney, *Mapping an Empire: The Geographical Construction of British India, 1765–1843* (Chicago: Chicago University Press, 1997), 9 and 333.
30 *Ibid.*, 337.
31 A. M. C. Godlewska, 'Map, text and image: the mentality of enlightened conquerors – a new look at the *Description de l'Egypte*', *Transactions of the Institute of British Geographers*, 20 (1995), 5–28, quote on p. 5.
32 Dawson Turner copies (DTC) of Banks's correspondence in the Botany Library, British Museum (Natural History), 23 volumes, Vol. 6, 101–2, Rennell to Banks, 22 December 1788.
33 Crone, *Maps*, 101.
34 DTC, Vol. 11, 275–6, Rennell to Banks, 22 August 1799.
35 Carter, *Sir Joseph Banks*, 425.
36 DTC, Vol. 18, 288, Rennell to Banks, November 1813.
37 Carter, *Sir Joseph Banks*, 261, 448, 507.
38 C. R. Weld, *A History of the Royal Society, with Memoirs of the Presidents*, 2 volumes (London: J. W. Parker, 1848), Vol. 2, 11–12.
39 On Banks as an improving landlord, see J. R. Farnsworth, 'A History of Revesby Abbey' (unpublished PhD thesis, Yale University, 1955); J. Gascoigne, *Joseph Banks and the English Enlightenment: Useful Knowledge and Polite Culture* (Cambridge: Cambridge University Press, 1994), 185–236; W. M. Hunt, 'The Role of Sir Joseph Banks, KB, FRS, in the Promotion and Development of Lincolnshire Canals and Navigations' (unpublished PhD thesis, Open University, 1986).
40 Lincolnshire Archive Office, RA 2/B/16, T. Stone, Survey of Revesby.
41 Carter, *Joseph Banks*, 392.
42 *Ibid.*, 344–5.
43 *Ibid.*, 398. See also Sutro Library, San Francisco (SL), Banks MSS, Geol. 1:2, Map, 9 September 1808, 'For explaining the Faults, shewn in J. Farey's small map of the great Limestone District in Derbyshire, made for Sir Joseph Banks.'

44 R. Porter, *The Making of Geology: Earth Science in Britain, 1660–1815* (Cambridge: Cambridge University Press, 1977), 179.
45 Yale University, Beinecke Library, Osborn Files, Banks to Palmerston, 28 March 1784.
46 SL, Banks MSS, Coal 1:20, 22 March 1797.
47 SL, Banks MSS, Coal 1:12, 1794.
48 Carter, *Joseph Banks*, 397.
49 J. Black, *Maps and History: Constructing Images of the Past* (New Haven: Yale University Press, 1997), 15.
50 J. Brewer, *The Sinews of Power: War, Money and the English State, 1688–1783* (London: Unwin Hyman, 1989), 221.
51 DTC, Vol. 16, 86, Rickman to Banks, 25 July 1805.
52 DTC, Vol. 16, 62, Banks to Rickman, 19 June 1805.
53 DTC, Vol. 16, 66, Rickman to Banks, 26 June 1805.
54 On Banks's connections with Sir George Mackenzie and agricultural improvement and clearance in the Highlands of Scotland, see C. D. Waterston, 'Late Enlightenment science and generalism: the case of Sir George Steuart Mackenzie of Coul, 1780–1848', in Withers and Wood (eds), *Science and Medicine in the Scottish Enlightenment*, 301–26.
55 DTC, Vol. 16, 66, Rickman to Banks, 26 June 1805.
56 R. Zupko, *Revolution in Measurement: Western European Weights and Measures since the Age of Science* (Philadelphia: The American Philosophical Society, 1990), 165–9.
57 R. D. O'Connor, *The Weights and Measures of England* (London: HMSO, 1987), 249–50, 253.
58 J. Hoppit, 'Reforming Britain's weights and measures, 1660–1824', *English Historical Review*, 108 (1993), 82–104.
59 Weld, *History of the Royal Society*, Vol. 2, 262.
60 British Library, Add. MSS 8099, f.140v, Banks to A. Lebland, 30 January 1802.
61 Yale University, Sterling Library, Historical Manuscripts Collection, Banks Correspondence, Banks to Blagden, 31 March [1817].
62 Gascoigne, *Science in the Service of Empire*, 121–3.
63 H. Cavendish and C. Hatchett, 'Experiments and observations on the various alloys, on the specific gravity, and on the comparative wear of gold', *Philosophical Transactions of the Royal Society of London*, 93 (1803), 43–194.
64 Edney, 'Reconsidering Enlightenment geography', 186.
65 H. B. Carter, *Sir Joseph Banks (1743–1820): A Guide to Biographical and Bibliographical Sources* (Winchester: St Paul's Bibliographies and the British Museum, 1987), 242.
66 On Kew Gardens as a site for promoting imperial 'improvement' from Banks's day to the twentieth century, see R. Drayton, *Nature's Government: Science, Imperial Britain, and the 'Improvement' of the World* (New Haven: Yale University Press, 2000).
67 P. Smith, *Memoir and Correspondence of the late Sir James Edward Smith*, 2 volumes (London: Longman, Rees, Orne Brown, Green and Longman, 1832), Vol. 1, 498, Banks to Smith, 25 December 1817.
68 On this, see J. Gascoigne, 'Motives for European exploration of the Pacific in the age of the Enlightenment', *Pacific Science*, 54 (2000), 227–37.
69 D. Clayton, 'On the colonial genealogy of George Vancouver's chart of the northwest coast of north America', *Ecumene*, 7 (2000), 371–401.
70 DTC, Vol. 7, 197, printed in N. Chambers (ed.), *The Letters of Sir Joseph Banks: A Selection, 1768–1820* (London: Imperial College Press, 2000), 128.
71 J. Barrow, *Sketches of the Royal Society and the Royal Society Club* (London, 1849), 10.

72 On this, see V. Jankovic, 'The place of nature and the nature of place: the chorographic challenge to the history of British provincial science', *History of Science*, 38 (2000), 79–113.

73 A. Cooper, 'From the Alps to Egypt (and back again): Dolomieu, scientific voyaging, and the construction of the field in eighteenth-century natural history', in C. Smith and J. Agar (eds), *Making Space for Science: Territorial Themes in the Shaping of Knowledge* (Basingstoke: Macmillan, 1998), 39–63.

74 S. F. Cannon, *Science in Culture: The Early Victorian Period* (New York: Dawson and Science History Publication, 1978), 105.

75 T. Broman, 'The Habermasian public sphere and "science *in* the Enlightenment"', *History of Science*, 36 (1998), 123–49.

76 Godlewska, *Geography Unbound*, 4, 239, 264.

VIII

Blumenbach, Banks, and the Beginnings of Anthropology at Göttingen

What was it that helped to make the University of Göttingen one of the seed-beds of the infant discipline of anthropology in the late eighteenth century? For it was at Göttingen in this period that the outlines of a system of classification were laid down in a manner that still shapes the way in which we attempt to comprehend the different varieties of humankind—including usage of such terms as "Caucasian". This was largely the work of one man, Johann Friedrich Blumenbach (1752-1840), professor of medicine at Göttingen from 1778 until his death in 1840. The secretary of the Parisian Academy of Sciences went so far as to write in his éloge of Blumenbach that "[i]t is to M. Blumenbach that our age owes Anthropology" (Bendyshe 1865, 49).

However, Blumenbach's innovative work in anthropology also reflected a larger academic culture within Göttingen which drew on the contributions of figures such as Christian Wilhelm Büttner (1716-1801) (who taught there) and the Forsters father and son (Johann Reinhold (1729-1798); Johann Georg Adam (1754-1794)), the former of whom contributed substantially to the ethnological collections gathered by Blumenbach and received an honorary doctorate from the University and the latter of whom was a correspondent of Blumenbach, and, indeed, related to him by marriage (Georg Forster was the son-in-law of the Göttingen professor Christian Gottlob Heyne (1729-1812), who was married to a sister of Blumenbach's wife, cf. Haase 1967, 112, 114). It was at Göttingen, too, that Christoph Meiners (1747-1810) wrote his *Grundriss der Geschichte der Menschheit* (1785).

The University of Göttingen itself had some of the advantages that arise from a relatively young institution, having been founded in 1737. There was a greater willingness to experiment with new academic forms—in particular a greater emphasis on training for the lay professions of law and medicine than was traditional in older universities where theology was more prominent in setting the academic tone of the university. Such an emphasis on lay professional training was further encouraged by the degree of competition between German universities for students, which meant that then, as now, universities sought revenue by attracting students to vocational courses. This helped to promote an academic culture which encouraged the beginnings of a system of status based on publications as well as teaching: the beginnings of the modern research university (McClelland 1980, 39-45).

Plate 4   Johann Friedrich Blumenbach (1752-1840)

Göttingen, Hanover and London

The founder of the University of Göttingen shared the dignity of being both Elector of Hanover and King of Great Britain, a reminder of the extent to which Göttingen was—in contrast to most German universities—linked to a larger, more cosmopolitan world. In the area of natural history (including anthropology) this yielded rich dividends since, as the world's leading maritime power, Britain could supply specimens of whole new areas of the globe made accessible by its naval pre-eminence. One very tangible way that Göttingen thus benefited from the Hanoverian connection was the manner in which Blumenbach was able to add to his museum in 1782 more than 350 objects from the second and third voyages of James Cook (1728-1779) thanks to a gift by George III (1738-1820) (Raabe, Schlesier, Urban 1988, 49).

And, in many cases, the British were content to collect and leave the task of theorising to others. British academic culture, with its traditions of catering to the needs of the gentleman and clergyman, often favoured the empirical rather than the theoretical. Oxford and Cambridge were not places where the professorial ideal of an individual who specialised in one particular area of knowledge flourished. Rather, teaching was the domain of college tutors who taught the full range of the curriculum before, in most cases, they moved out of the university to take up a clerical position.

In London learned activity was supported by a rich array of club-like societies of which the most important were the Royal Society and the Society of Antiquaries (Berman 1974-1975). These necessarily reflected the interests of a membership made up of gentlemen-amateurs who had no professional incentive to specialise or to advance their career through publication. The result, then, was that, though England amassed a great array of natural history specimens in the late eighteenth century, much of the task of incorporating such new data into a body of theory which would make sense of its significance was left to those outside England. Among those who took up the task were the Scots, but Germany and, in particular, Göttingen was also drawn into the larger sphere of English intellectual life.

For it was to Germany that England often looked for individuals who could help to explain the significance of the new worlds that its growing imperial power led it to encounter. This was a tribute both to the scale and sophistication of the German university world which produced both a greater number of graduates proportionally than England and a much greater range of expertise. That remarkable duo, the Forsters father and son, were, for example, the scientific observers on Cook's second great Pacific voyage from 1772 to 1775. Earlier, Göttingen had been the place where the famous expedition through Arabia led by the Göttingen graduate, Carsten Niebuhr (1733-1815), from 1761 to 1767, had been planned. Göttingen even offered special courses to meet the needs of travellers known as 'apodemics' (Ackerknecht 1955, 84).

BLUMENBACH, BANKS, AND THE BEGINNINGS OF ANTHROPOLOGY

The African Association, founded in 1788 as a means of promoting British exploration of Africa, employed two Göttingen students to venture into the unknown interior of Africa. At Blumenbach's recommendation in 1796 the Association engaged the services of Friedrich Hornemann (1772-1800) who had studied theology at Göttingen from 1791 to 1794 (Legée 1987-1988, 34). Before sending him to Africa Blumenbach ensured that he spent six months in preparation at Göttingen. There, as he assured Joseph Banks (1743-1820), the virtual founder of the African Association, Hornemann was to be instructed by him in those elements "of natural history which may make his expedition the more useful as for inst[ance] with the geognostical part of mineralogy". And, Blumenbach added, "besides this he would spend his time principally with our orientalists for the Arabian Language & c. with our Mathematicians & astronomers, & c." (Fitzwilliam Museum, Cambridge, Perceval Collection, MS 215, Blumenbach to Banks, 12 May 1796). Along with Blumenbach, Heyne, Blumenbach's brother-in-law and Arnold Hermann Ludwig Heeren (1760-1842), Heyne's son-in-law, supervised Hornemann's course.

Undeterred by Hornemann's death in 1800 (for almost all such African explorers perished) the Swiss Johann Ludwig Burckhardt (1784-1817), who studied at both Leipzig and Göttingen, also enlisted in the service of the African Association. In 1806 he arrived in London with a letter of introduction from Blumenbach to Banks, the president of the Royal Society and the virtual scientific adviser to the British government. From Blumenbach's point of view such services helped both to keep the supply of rare specimens flowing from Banks as well as helping him gain access to the data uncovered by the African explorers. As early as 1794 Blumenbach had informed Banks that he diligently searched the publications of the African Association for "some particular news concerning the corporeal singularity of the Inhabitants (in regard to the complexion, hair, characteristical feature &c)" (British Library, Add. 8098, fos. 216-7, Blumenbach to Banks, 10 March 1794).

Such links between England and Hanover were promoted by well-travelled members of Göttingen University such as Georg Christoph Lichtenberg (1742-1799) (cf. Mare and Quarell 1938) and Blumenbach who both made lengthy trips to London which laid the foundation for continued contact with English intellectual life. The connections established by the court further strengthened such ties—not least through the practical advantage of enabling specimens and correspondence to be sent securely through the diplomatic post. This was, for example, a facility much used by Banks in his copious correspondence with Blumenbach.

Banks and Blumenbach's Skull Collection

By such a route a steady stream of specimens reached Blumenbach to supplement his museum of natural history and to facilitate his researches. Appropriately, the first letter of Blumenbach to Banks which survives is one from 1783 in which he sought ethnological "curiosities" for his museum from the first of Cook's voyages (BL, Add. 8098, Blumenbach to Banks, 30 Jan. 1783).

In particular, he had access to specimens from the Pacific which, to European eyes, was a new world offering a virtual laboratory to test theories about nature in general and the development of human kind in particular. But the flow of specimens from Banks to Blumenbach encompassed all fields of natural history. Banks' contacts with Australia, for example, enabled Blumenbach to receive samples from the animal, vegetable and mineral kingdoms: a stuffed platypus—a mammal which laid eggs and which Blumenbach christened "Ornithorhynchus paradoxus" (Blumenbach 1801, 73-5)—seeds and a "rare earth" (BL, Add. 8097, fos. 362-3, Blumenbach to Banks, 9 Jan. 1791).

Blumenbach warmly thanked Banks for these specimens with an allusion to the way in which, in Germany, he was cut off from such scientific riches: "How valuable in general all such *exotic* natural curiosities as the sea-otter's tail & the clay of Sydney cove, must be for an ardent lover of natural history *in the heart of the continent*". Moreover, continued Blumenbach, such specimens had the added value of being from the new world of the Pacific or, as he put it, from "the most interesting *matieres du temp*, viz. the N-W American Fur Trade & the Sydney Cove Establishment" (BL, Add. 8097, fos. 368-70, Blumenbach to Banks, 5 June 1791).

Blumenbach was particularly anxious to test his hypotheses about the nature and extent of human variation by reference to the skulls which formed the core of his physical anthropology. At first Banks could do no more than offer to send him casts from the collection of "My Friend" John Hunter (1728-1793) (Dougherty 1984, 114: Banks to Blumenbach, 20 June 1787) but, by July 1789, he had dispatched a cranium of a chief from St. Vincent Island in the West Indies which derived from Banks' client, Alexander Anderson (d. 1811), keeper of the botanical garden there (Niedersächsische Staats- und Universitätsbibliothek, Blumenbach MSS III: 31, Banks to Blumenbach, 15 July 1789). By 1793 Banks had sent him the cranium "of a male native of New Holland who died in our settlement of Sydney Cove" together with the promise to send a Tahitian one by the "next quarterly messenger" (Niedersächsische Staats- und Universitätsbibliothek, Blumenbach MSS III: 38, Banks to Blumenbach, 16 August 1793)— a reference to the royal postal service between London and Hanover. A second skull from Sydney followed in 1799. The Pacific data received both from Banks and from the Forsters prompted Blumenbach to add a new category, that of the Malay, to take account of the peoples of what later became known as Polynesia and Melanesia (Plischke

BLUMENBACH, BANKS, AND THE BEGINNINGS OF ANTHROPOLOGY

1938, 225-31). The Australian Aborigines Blumenbach was less certain how to classify and he suggested that they might occupy a middle place between the Malay and the African though, he added, they could be classed with the Africans "if it was thought convenient" (Frost 1979, 36-7).

Banks' good offices in promoting this work led Blumenbach to include in the third (1795) edition of his *De Generis Humani Varietate Nativa* a dedicatory letter to Banks. This he followed up with a letter in which he again stressed his debt to the president of the Royal Society: "by perusing the Book a little you will, I fancy, find Your name so often mentioned, that I am on more accounts indebted to You, even in these my anthropological researches, than you perhaps did imagine or remember Yourself" (BL, Add. 8098, fos. 223, Blumenbach to Banks, 1 May 1795).

Interestingly, in this letter Blumenbach employs the word "anthropology" to describe the new discipline which he did so much to establish. Elsewhere in his correspondence with Banks Blumenbach again alludes to this new discipline. The arrival of the "the precious Otaheitian Scull" in 1794 prompted Blumenbach to write of "this new proof of the generous interest you take in so liberal a way for my anthropological study" (BL, Add. 8098, fos. 216-7, Blumenbach to Banks, 10 March 1794). Again, when thanking Banks in 1797 for copies of some of his South Sea drawings, he referred to the "pretious pictures You have so generously parted with to enrich my anthropological collection" (BL, Add. 8098, fos. 318-9, Blumenbach to Banks, 2 April 1797).

The pursuit of natural history provided the most important intellectual stimulus for the early development of anthropology but the Hanoverian connection brought Göttingen into contact with another: the English virtuoso collecting tradition that was linked to a gentry culture that was less well developed in Germany. It was a tradition that was closely associated with an interest in antiquarian pursuits natural to a class whose power and status rested on its local connections and ownership of land. It was a class, too, that had sufficient wealth to accumulate antiquities and curiosities of all types— chiefly on its famed Grand Tours to the European continent (and, principally, to Italy). However, such collections also could include "artificial curiosities", as ethnographical specimens were known.

It was the fruits of such a collecting tradition in the form of Egyptian mummies that first brought Blumenbach into contact with Banks. When Blumenbach visited London in 1791 it was Banks who enabled him to view specimens in private collections as well as the British Museum. The link between such collecting and the English gentlemanly virtuoso tradition is apparent in Blumenbach's praise to Banks of the Egyptian artefacts in the collections of "your great Antiquarians Mr. Townley & Mr Knight" (BL, Add. 8098, fos. 213-4, Blumenbach to Banks, 8 Jan. 1794). Blumenbach acknowledged his debt in the form of a published letter to Banks in the Royal Society's *Philosophical Transactions*. This made evident the extent to which his interest in Egyptian antiquities was linked to his larger anthropological concerns with the varieties of the human species. For, he argued in a paper addressed to Banks, "we must adopt at least *three* principal varieties in the natural physiognomy of the ancient Egyptians". Overall he was inclined to place the Egyptians "between the Caucasian and the Ethiopian" varieties of humankind (Blumenbach 1794,191-3).

In England, then, the massive collections that British naval incursion into the new world of the Pacific made available were, to a large degree, incorporated into the larger national culture through the tradition of the gentleman collector. True, the great institutions of the London learned world, the British Museum, the Royal Society and the Society of Antiquaries often could promote rigorous analysis and an original approach which the strongly individual character of English intellectual life encouraged. But in Germany such data were more likely to be incorporated into a university system which was being revitalised by the growing wealth and self-confidence of the late eighteenth-century German states. Moreover, the extent to which local loyalties and prestige were invested in the geographically diverse universities helped make the German university system the most widespread in Europe with the greatest number of institutions. Furthermore, over the course of the late seventeenth and eighteenth centuries new foundations such as Halle (1694), Breslau (1702), Göttingen (1737) and Erlangen (1743) had increased the degree of competition among them.

Student and Teacher

Blumenbach's chief debt to his university training was the degree to which Göttingen promoted the development of a faculty of medicine which sought to attract students from Hanover and beyond by the reputation of its faculty and the extent of its resources. This had been in large measure the achievement of Albrecht von Haller (1708-1777) who had held the inaugural chair of medicine (or, more accurately, anatomy, botany and medicine) at Göttingen and brought to the University some of the lustre of the great medical teacher, Herman Boerhaave (1668-1738), under whom he had studied at Leyden. As one of Blumenbach's obituarists put it, when Blumenbach reached Göttingen, "Haller had left the place; but his reputation was everywhere" (Bendyshe 1865, 49). Blumenbach maintained contact with Haller after his return to Bern writing a preface to his *Journal of Medical Literature.*

By taking a wide view of the nature and extent of medicine Göttingen helped to facilitate the growth of an interest in natural history. This had also been promoted by Blumenbach's earlier studies at Jena where one branch of natural history, the study of the earth, had been encouraged by the mineralogical lectures of Johann Ernst Immanuel Walch (1725-1778) (Plischke 1937, 2; *Dictionary of Scientific Biography*, s. v. "Blumenbach"). Blumenbach's continuing fascination with the wider domain of natural history was to be reflected in his *Handbuch der Naturgeschichte* (1779; 12th ed. 1830). His interests in this area were later to be shared by his colleague Lichtenberg, a keen student of volcanology.

Along with the study of the animal, vegetable and mineral kingdoms the study of natural history was also to embrace the world of humankind and thus to promote the beginnings of the systematic study of anthropology. It was this wide conception of natural history to which Johann Reinhold Forster appealed in his account of Cook's second voyage, *Observations made during a Voyage around the World* (1778). In this he proclaimed that "[t]he object pursued in this work, is an investigation of NATURE, in its greatest extent" including "the Philosophical History of the Human Species"—a subject about which he was indebted to the "works of Dr. Blumenbach and Dr. John Hunter [...] [which] have furnished me with some anatomical facts" (Forster 1996, lxxvii and 9).

As a student at Göttingen Blumenbach had attended the lectures of Christian Wilhelm Büttner which encompassed the full scope of natural history including the human world—material that was given life and immediacy through the incorporation of travellers' accounts of foreign peoples. Such a background helps to explain why Blumenbach chose as doctoral dissertation in 1775 the subject "de generis humani varietate nativa" (Bendyshe 1865, 6). It was natural, then, that Blumenbach in his capacity as curator of the natural history collection (a post he held from 1776) should seek to persuade Banks not only to part with botanical, zoological and geological specimens but also with specimens that shed light on the nature and diversity of human populations. The extent to which Blumenbach regarded anthropology as a branch of natural history is evident in the way in which he thanked Banks for sending him a second skull from the Carribean with a recognition of "the noble liberal way you promote any well minded attempt, for the advancement of natural history" (BL, Add. 8097, fos. 264-5, Blumenbach to Banks, 22 Sept. 1790). Significantly, too, in the dedication to Banks of the third edition of the *De Generis Humani Varietate Nativa* (1795) Blumenbach speaks of the importance of the data collected by Banks and "cultivators of natural history and anthropology" in the Pacific (Bendyshe 1865, 150).

Such specimens were incorporated into an academic study of natural history, which had as its chief characteristic a quest for systems of classification. The great watershed had been the publication of Carl Linnaeus' (1707-1778) *Systema Naturae* in 1735 which provided a foundation on which others built in the great Enlightenment quest to make the world comprehensible and manageable through the patient and orderly pigeon-holing of the bewildering array of nature's productions. Blumenbach inherited from Linnaeus the belief that humankind should be incorporated into this larger mission and that human beings should be classified in the same way as other living things.

There were, however, important differences between the two: Linnaeus used a mixture of physical and cultural criteria to classify the varieties of humankind while Blumenbach concentrated on physical criteria and, while Linnaeus included human beings with the primates, Blumenbach allocated human beings a genus to themselves. Though in his dedication to Banks of the third (1795) edition of the *De Generis Humani Varietate Nativa,* Blumenbach wrote of the way in which Pacific exploration had made it "very clear that the Linnaean division of mankind could no longer be adhered to" (Bendyshe 1865, 150) he nonetheless elsewhere acknowledged his debt to "the immortal Linnaeus".

By the time of Blumenbach, too, the scientific status of natural history had been elevated largely thanks to the work of Linnaeus. No longer could natural history be dismissed as the pursuit of amateur butterfly catchers as the enormous range of specimens was neatly filed away in an orderly system. Such order helped to provide the basis for greater claims on the part of natural historians to the dignity of providing not simply raw data but systems of explanation.

Throughout much of the seventeenth and the early eighteenth century the map of knowledge was divided between those who cultivated natural history and the natural philosophers in a way that emphasised the lowly status of natural history. In his *Advancement of Learning* (1605) Francis Bacon (1561-1626) described the division of intellectual labour in these terms: "natural history describeth the variety of things; physique [physics or natural philosophy], the causes" (Bacon 1965, 93).

By the late eighteenth century, however, natural historians were beginning to lay claim to a greater scientific standing. Georges Louis Leclerc Buffon (1707-1788) indicated such a position in the preface of his great *Histoire Naturelle* (1749-1750) by urging natural historians to rise above their traditional role of collectors of data and to aspire to provide generalisations about the workings of nature. For, he argued, "it is not necessary to imagine even today that, in the study of natural history, one ought to limit oneself solely to the making of exact descriptions and the ascertaining of particular facts". Thus the naturalist should seek "combination of observations, the generalisation of facts" (Lyon and Sloan 1981, 121).

But it was in Germany particularly that natural history began to be reconceptualised in such a way that it was seen as providing greater explanatory power. The remark of Karl Friedrich Heinrich Marx (1796-1877) in his memoir of Blumenbach that "Natural history, not the description of nature, was the aim he placed before him" (Bendyshe 1865, 10) conveys the way in which Blumenbach sought to elevate the conceptual significance of natural history. Such views eventually set the scene for the emergence of the discipline of biology—a term which appears to have first been used in Germany by Gottfried Reinhold Treviranus (1776-1837) and, subsequently, by Jean Baptiste

Lamarck (1744-1829) in France in 1802 and not to have been employed in Britain until 1813 (*Oxford English Dictionary*, s. v. "Biology"). Such a system of explanation came through the increasingly historical manner in which the data yielded by the endeavours of the natural historians were viewed and organised. Such a transition was captured by Immanuel Kant's (1724-1804) distinction of 1775 between the description of nature ("Naturbeschreibung") and the understanding of nature in its historical development ("Naturgeschichte") (Lyon and Sloan 1981, 2).

Ideas of development and growth provided the intellectual climate out of which the concept of evolution was to emerge. And, as has often been remarked, it was in the infant discipline of anthropology that evolutionary ideas began to gain dominance before they did so in the realm of biology. Blumenbach's anthropological theorising reflects such a pre-occupation with change over time. In his textbook of natural history, for example, he envisaged the way in which the present array of species had altered over the course of the history of the world arguing that "it becomes more than merely probable that not only one or more species, but a whole organized preadamite creation has disappeared from the face of our planet". Even more explicitly he challenged the notion of a once and for all creation with the view that "a whole creation of organized bodies has already become extinct, and has been succeeded by a new one" (Bendyshe 1865, 285, 288). Such changes were the result of what Blumenbach called "degeneration" a force he saw as being as much at work in human as in animal species. Such a view of the way in which the development of humans and animals should be studied in the same way formed part of his increasingly elevated understanding of the domain of natural history. In his view natural history was now moving beyond its earlier, less analytical origins and this was particularly evident in the way in which a new discipline of anthropology was emerging. For, he wrote, "the naturalist was so very late in finding out that man also is a natural product, and consequently ought at least as much as any other to be handled from the point of natural history" (ibid., 298).

As befitted his role as professor of medicine Blumenbach placed particular emphasis on physical specimens and, of course, above all skulls. His quest for such skeletal remains extended to a disregard for native custom that today has prompted moves for ritual reburial in Australia and elsewhere. Banks, for example, commented of a cranium that he sent Blumenbach from St. Vincent that it had been obtained even though the indigenous people there regarded "any attempt to disturb the ashes of their Ancestors [...] as the greatest of crimes" (British Museum (Natural History), Biology Library, Dawson Turner Copies, vol. 6, pp. 159-60, Anderson to Banks, 3 May 1789). Such an emphasis on physical anthropology reflected the extent to which the medical faculty helped to provide a major institutional catalyst for the development of anthropology.

Blumenbach was a member of a medical faculty that encouraged an interest in comparative anatomy. Blumenbach himself claimed that his textbook, *Handbuch der vergleichenden Anatomie* (1805) was the first "to have appeared that dealt with the entire area of *anatome comparata*" (*Dictionary of Scientific Biography*, s. v. "Blumenbach"). It was not surprising, then, that Blumenbach should build on this as the number and extent of specimens expanded— largely as a consequence of his contact with the London learned world. Appropriately, Blumenbach's work on anthropology was, soon afterwards, translated into English as *A Short System of Comparative Anatomy* (1807) by William Lawrence (1783-1867)—a major figure in promoting the study of physical anthropology in Britain.

As professor of medicine Blumenbach helped to interest others in linking comparative anatomy with the study of physical anthropology. Among those who were thus influenced was Samuel Thomas Soemmerring (1755-1830) who studied medicine at Göttingen from 1774 to 1778 where his interest in medical research was stimulated by both Blumenbach and Heinrich Wrisberg (1739-1808). No doubt as a result of the contacts made possible through Göttingen's Hanoverian connection Soemmerring followed his graduation as an MD in 1778 with a trip to England and Scotland as well as Holland. During this he met some of the pioneering figures in physical anthropology such as Pieter Camper (1722-1789) and John Hunter. On his return he took up a chair at the Collegium Carolinum in Kassel. There the training in comparative anatomy he had received at Göttingen led to undertaking a study of the varying physical features of Europeans and Africans in 1784 concluding, like Blumenbach, that, despite some differences, both groups belonged to the one species. For, wrote Blumenbach in his manual of natural history, "[t]here is but one species of the genus Man; and all people of every time and every climate with which we are acquainted, may have originated from one common stock" (Blumenbach 1825, 35).

Conclusion

It was a view that highlighted the extent to which Blumenbach's work embodied some of the impulses of the Enlightenment: a belief in a common humanity shared across all peoples and cultures and a commitment to studying human society in the same way as the natural order. Most centrally, of all, his pioneering work in anthropology was actuated by the fundamental quest of the Enlightenment: to make the world comprehensible through systems of classification which would reveal order amidst complexity and which would enable control of nature for the benefit of human kind.

Indeed, anthropology based much of its early scientific credentials on the degree to which it allowed the development of systems of classification— even if, as Blumenbach was at pains to stress, these did not nullify a more basic unity of the human species. Such an emphasis on classification developed out of the study of natural history, which was fostered at Göttingen by the predominance of the medical faculty, which, in turn, reflected how much this newly-founded university had placed particular emphasis on professional education. The medical faculty also fostered the infant discipline of anthropology through the extent to which it cultivated the study of comparative anatomy—a study which provided Blumenbach with some of the basic criteria on which his system of classification was based.

Thanks to the Hanoverian connection, too, Göttingen brought together two cultures: that of the German university and the British predilection for collecting. The latter was linked to a gentlemanly virtuoso tradition which was given greater range by the extent of British contact with the Pacific world in the late eighteenth century. Blumenbach's contacts with this world of British collecting, as epitomised in his links with Banks, indicate the extent to which the marrying of these two cultures could prove scientifically fruitful. Indeed, the early beginnings of anthropology may be regarded in part as the offspring of Britain's global reach and the developing research ethic, which Göttingen did much to promote.

Bibliography

Ackerknecht, Erwin H. 1955. "George Forster, Alexander von Humboldt, and Ethnology," *Isis*, 46 (2): p 83-95.

Bacon, Francis. 1965. *The Advancement of Learning*, ed. George William Kitchin. London (Everyman's Library).

Bendyshe, Thomas. 1865. *The Anthropological Treatises of Johann Friedrich Blumenbach*. London.

Berman, Morris. 1974-1975. "'Hegemony' and the amateur tradition in British science", *Journal of Social History*, 8: 30-43.

Blumenbach, Johann Friedrich. 1794. "Observations on some Egyptian mummies opened in London. Addressed to Sir Joseph Banks", *Philosophical Transactions*, 84: 191-3.

Blumenbach, Johann Friedrich. 1801. "Some anatomical remarks on the *Ornithorhynchus paradoxus*, from New South Wales", *Medical and Physical Journal*, 6: 73-5.

Blumenbach, Johann Friedrich. 1825. *A Manual of the Elements of Natural History*. Transl. from the 10th German ed. by Richard T. Gore. London.

Dictionary of Scientific Biography, s. v. 'Blumenbach'.

Dougherty, Frank W. P. 1984. *Commercium epistolicum J. F. Blumenbachii. Aus einem Briefwechsel des klassischen Zeitalters der Naturgeschichte*. Göttingen.

Forster, Johann Reinhold. 1996. *Observations made During a Voyage Round the World*, eds. Nicholas Thomas, Harriet Guest and Michael Dettelbach. Honolulu.

Frost, Alan. 1979. "The Pacific Ocean —The Eighteenth Century's New World" in *Captain James Cook: Image and Impact*, ed. Walter Veit, Vol. II. Melbourne, pp. 36-7.

Haase, Carl. 1967. "Göttingen und Hannover. Geistige und genealogische Beziehungen im ausgehenden 18. Jahrhundert", *Göttinger Jahrbuch*, 15: 95-124.

Legée, Georgette. 1987-1988. "Johann Friedrich Blumenbach, 1752-1840. La naissance de l'anthropologie à l'époque de la Révolution française", *Histoire et Nature*, 28-29: 34.

Lyon, John and Phillip Sloan (eds.). 1981. *From Natural History to the History of Nature: Readings form Buffon and His Critics.* Notre Dame.

Mare, Margaret L. and William H. Quarell (eds.). 1938. *Lichtenberg's visit to England.* Oxford.

McClelland, Charles E. 1980. *State, Society, and University in Germany 1700-1914.* Cambridge.

Plischke, Hans. 1938. "Die Malaysische Varietät Blumenbachs", *Zeitschrift für Rassenkunde*, 8: 225-31.

Plischke, Hans. 1938. *Johann Friedrich Blumenbachs Einfluss auf die Entdeckungsreisenden seiner Zeit.* Göttingen.

Raabe, Eva, Eberhard Schlesier and Manfred Urban (eds.). 1988. *Verzeichnis der Völkerkundlichen Sammlung des Instituts für Völkerkunde der Georg-August-Universität zu Göttingen,* Teil 1: *Abteilung Ozeanien.* Göttingen.

In: *Göttingen and the Development of the Natural Sciences.* Edited by Nicolaas Rupke. © Wallstein Verlag, Göttingen, 2002, pp. 86–98.

IX

The German Enlightenment
and the Pacific

The distant mirror of the Pacific gave varying reflections from different
European vantage points. As the "new world" of the Pacific[1] came into
closer view in the period after the end of the Seven Years' War in 1763, so
too it was drawn into enlightened discourse on the proper functioning of
human society. Where previously America had largely provided the human
laboratory for exploring the beginnings and development of social institutions (as Locke had written, "in the beginning all the World was *America*"),[2] by the late eighteenth century, the Pacific was providing a fresher
alternative.[3] For the French philosophes, the reports brought back from
the Pacific provided further fuel to attack what they considered the artifi-
ciality of their social institutions and the baleful influence of religious
dogma—such, for example, formed the essential themes of Diderot's *Supplement to the Voyage of Bougainville.*

Where the French Enlightenment challenged the existing structures in
church and state, in its English and Scottish manifestations, the Enlightenment tended to seek the path of improvement. New knowledge, such as
that from the Pacific, helped shape theories about the way in which society could change and develop without major rupture. English voyagers
in the Pacific were also not quite as quick to paint Pacific societies in
the glowing colors of French explorers such as Bougainville or Philibert
Commersen, the naturalist who accompanied him: James Cook and Joseph
Banks certainly admired much of what they saw in the Pacific—and,
above all, Tahiti—but they did not shy away from underlining such less-
palatable features of Polynesian society as infanticide, frequent warfare,
and, in places, cannibalism.

The British tendency to look more cautiously than the French at the new world of the Pacific and to be less inclined to use explorers' accounts as a stick with which to beat their own society largely suited the more cautious world of the German *Aufklärung*—a movement that helped to give identity to the bewildering array of the German-speaking lands. "Germany? But where is it?" asked Goethe and Schiller in 1797, "I don't know how to find such a country."[4] True, the German-speaking world had no clear political expression apart from the ramshackle Holy Roman Empire, but it valued its traditions and separate identity, and few sought to reconstruct anew the institutions in the manner urged by some of the French philosophes.[5] Rather, the hope of most of the *Aufklärer* was that the ancient German institutions could be reshaped by a process of growth and development[6]—very much in the manner urged by many of the leading lights of the English and Scottish Enlightenments.

This affinity between Germany and Britain helps to explain the increasing interest in British thought in the last quarter of the eighteenth century, at the time when the German Enlightenment was developing its characteristic forms. French thought—and, above all, the work of the ubiquitous Rousseau—remained important, but its dominance was increasingly challenged. This was perhaps because the later French Enlightenment developed along more radical and less evolutionary lines than those embodied in the earlier work of a figure such as Montesquieu, whose sympathy for tradition and espousal of reform rather than revolution had long endeared him to German thinkers.[7] The English intellectual influence was strengthened by the Hanoverian connection with the British monarchy, particularly as the University of Göttingen (founded in 1737) came to play an increasingly important role in German intellectual life. And, of course, growing British trade meant that more and more of the world—including the German lands—came into contact with the British, as Kant could testify at the busy port of Königsberg with its many British ships and trading houses.

For politically divided Germany, Britain provided one of its major arteries to a larger world. There was no German state to finance the expeditions that took the British—or the French or Spanish—to the Pacific, bringing back in their wake detailed journals or packing case after packing case full of specimens of natural history displaying the richness and diversity of the three kingdoms of nature in its Pacific dimension. Though the Hapsburg Empire under Joseph II did contemplate a Pacific voyage, it lacked the resources to do so.[8] If the Germans were to join in the study of the Pacific, this had to be done largely through British intermediaries—something that further strengthened the tendency in the German lands to see the Pacific through British eyes.

From the British point of view, there were considerable advantages in involving Germans in their expeditions as subjects of states that were not likely to pose a challenge to British imperial designs. Furthermore, Germany provided university-trained experts in abundance, by contrast with the much more limited university presence in Britain itself. In England there were only the two largely clerical universities of Oxford and Cambridge; Scotland, by contrast, had four universities that, like those of Germany, developed in the eighteenth century strong traditions of professorial teaching and intellectual innovation. But the Scottish universities were a Lilliputian world compared with the relatively vast network of German universities, the scale of which reflected the diversity and lack of unity of Germany itself, for each princedom or principality took pride in having its own university. German culture, to an extent unparalleled elsewhere in Europe, drew strongly on university roots. This further strengthened the cautious character of the German Enlightenment since university professors (particularly those enjoying the status and privileges of German professors) were unlikely harbingers of revolution. The German universities looked on knowledge as a tool for improvement and reform and a means of revitalizing the institutions of church and state—not as an instrument for the demolition of the network of distinctive German traditions of which they formed a part.

The growing links between Germany and Britain in the late eighteenth century meant that a number of Germans were drawn into the British penetration of the Pacific. Thanks to its universities, Germany could provide experts in the rapidly growing field of natural history, as well as those with strong philological or medical qualifications. Moreover, the fruits of Pacific investigation could be subjected to professorial scrutiny in the German universities to an extent not possible in Britain, where much of the intellectual life continued to be conducted in the club-based world of such London institutions as the Royal Society or the Society of Antiquaries.[9] In Britain the political and social dominance of a landowning gentry class was reflected in an intellectual culture where the amateur with private means pursued what was of interest to him (the masculine pronoun was almost always appropriate) and his class. In Germany the number of universities meant that knowledge had become more institutionalized and professionalized, particularly as new universities, such as Göttingen, had begun to foster a research culture in which professorial advancement was linked to publication.[10]

France, too, had developed a core of professional savants through its academy and other bodies, which helps to explain why France, in contrast to Britain, had little need of German graduates in its exploration of the Pacific and other parts of the world. In France, however, such savants were

IX

employed in institutions more directly under the sway of the state than the German universities; in France, too, the polarization between such state savants and salon-going, unsalaried philosophes was a further catalyst to push Enlightenment thinking in more radical directions.

The German response to the "new world of the Pacific" in the late eighteenth century was very closely linked to that particularly German institution, the university—and especially to one particular university, that of Göttingen. Göttingen's links with the British royal house gave it ready access to British accounts of exploration, and its emphasis on professional education meant a strong medical faculty that provided fertile ground for the study of natural history. Its relatively recent origins meant, too, that the traditions of *Schulphilosophie* (university instruction in philosophy that continued the traditions of scholasticism), with its highly systematic, pedagogically organized overview of knowledge, were much less entrenched. The study of philosophy there was more likely to respond to more contemporary concerns such as the study of ethics, which, in turn, provided a stimulus for the study of other cultures and the beginnings of anthropology.[11] Travel had long been a particular preoccupation of the Göttingen professors: the great Albrecht von Haller—who established the fame of the medical faculty there as professor of anatomy, botany, and medicine from 1736 to 1753—promoted a reading circle devoted to travel literature, members of which included the distinguished Göttingen philologist, Johann Reinhold David Michaelis.[12] Out of such preoccupations emerged Michaelis's instructions for an expedition to Arabia (1761–69), sponsored by the Danish crown but including such Göttingen graduates as Carsten Niebuhr. Michaelis's *Fragen* (*Questions*) represented the best distillation of the work on the study of other cultures then available.[13]

In the period at the end of the Seven Years' War in 1763, when European exploration of the Pacific was gathering pace, Göttingen became even more concerned with the intellectual repercussions of European contact with vastly different cultures. Such issues were discussed regularly at the Königliche Sozietät der Wissenschaften (Royal Society of Sciences) and figured in the *Göttingischen Anzeigen von gelehrten Sachen* (*Göttingen Newspaper on Scholarly Affairs*).[14] Significantly, it was at Göttingen that the terms "Ethnographie" and "Völkerkunde" ("Ethnology") first emerged, being coined by August Schlözer, who took up a chair there in 1769— Schlözer claimed paternity of the concept of *ethnograpisch* in the course of a debate with Herder in 1772–73.[15]

Before taking up his post at Göttingen, Schlözer had been in St. Petersburg—a reminder that, along with the British Empire, early German theorists of ethnology drew extensively on the exploration made possible by the resources of the Russian Empire.[16] Furthermore, the need of the Russians for foreign experts to describe in scientific form the resources of their vast territories was much more acute than in the British case, and it was principally to the graduates of the German universities that the Russians turned—generally with a much stronger expectation of a direct return to the needs of government than the British displayed in the distant Pacific. Among such German experts employed in Russia was Johann Reinhold Forster who, along with his son, George, did most to implant in Germany an interest in the late eighteenth-century European encounter with the Pacific. Bored with the life of a clergyman in a small German-speaking enclave within the largely Polish-speaking lands near Danzig, Johann Reinhold Forster readily responded to the invitation in 1765 to undertake an investigation of the German-speaking colonies on the River Volga. But he only remained until 1766: foreshadowing his later experiences with the British government, Forster fell out with his Russian employers, who objected to his findings about the poor treatment of the German colonists. Nonetheless, his experience (and that of his eleven-year-old son, George, who accompanied him) in dealing with different cultures probably played some role in preparing him for his later voyage to the Pacific.

Johann Reinhold Forster's familiarity with both the Poles and the different peoples of Russia is a reminder of the extent to which Germans were often exposed to a greater range of cultures than other Europeans—which may have helped to interest them in the infant discipline of anthropology. Lacking a single state of their own, Germans frequently lived alongside other peoples and languages, which may have inclined them to be sympathetic to cultural diversity and more skeptical than the French of universalizing models of human society. As Koerner suggests, Herder's emphasis on the uniqueness and value of each culture no doubt owed much to his upbringing on the Eastern Baltic, where he was exposed to a range of cultures and languages.[17] This anthropological interest in other cultures was characteristic of the Enlightenment in the German lands, as elsewhere. When the *Berlinische Monatsschrift* was established in 1783—the year before it published the celebrated articles by Kant and Mendelssohn on the subject "What is Enlightenment"—it set as one of its goals inclusion of articles on "Description of peoples and their customs and institutions, preferably from countries close to us."[18]

The Forsters, father and son, both combined travel in the distant Pacific with firsthand observations of societies on the periphery of the German-speaking lands: along with his experiences as a child when his father lived in largely Polish-speaking lands and his subsequent time with his father in Russia, George Forster served a reluctant term as a professor at the University of Vilna from 1784 to 1787. Ironically, he was released from (what he considered to be) this exile by the offer of serving the Russian government like his father before him. Like the British, the Russians also were expanding into the Pacific and George's experiences serving the British Empire in his forays into the Pacific made him a natural choice in 1787 to serve as a scientist on a proposed four-year imperial Russian expedition, the main goal of which was to consolidate Russia's position in the Pacific. In the event, the chosen vessels were commandeered for the war against the Turks, but George Forster was nonetheless released from his post at Vilna to take up a more congenial position as the university librarian at Mainz.

Johann Reinhold Forster's response to the failure of his plans to work in the service of the Russian Empire was to turn to the British Empire. In doing so, he called on the international freemasonry of natural historians, and it was thanks to such natural historians as Daines Barrington, Thomas Pennant, and Daniel Solander that he first became established in Britain, gaining the post of tutor in modern languages and natural history at Warrington Academy—a position he held from 1767 to 1770. Naturally, he drew on his recent experiences in Russia to help underline his credentials in the field of natural history. The first of a series of natural history papers presented to the Royal Society was devoted to the natural history of the Volga[19]—a work that made plain his admiration for the "celebrated Linnaeus" and provided the first Linnaean account of the flora and fauna of eastern Russia.[20] Forster, then, expanded the territory made accessible to European science through the use of the Linnaean classificatory system—thus continuing the Linnaean tradition of scientific exploration and the mission of bringing classificatory order to more and more of the globe. The extent to which Forster's conception of the scope of natural history included the study of human society was underlined by the considerable collection of artifacts he brought with him—the sale of which was to provide a much needed source of income, just as, subsequently, the Forsters attempted to capitalize on the vast collection of Pacific artifacts they built up on Cook's second Pacific voyage.[21]

Though Johann Reinhold Forster had graduated in theology at the University of Halle, his primary interest had been medicine[22]—the faculty that provided the main institutional foundation for the study of natural history. (Appropriately, he belatedly received a doctorate in medicine at Halle in 1782, following his appointment there as professor in 1779 with particular responsibilities for natural history and mineralogy.) He reared his promising son, George (who also subsequently received a doctorate in medicine), in the study of natural history—a field that became the central intellectual core of the two men's careers and interests. In 1789, five years before his death, George Forster was to remark that "Natural history in its broadest sense and particularly anthropology have been my pre-occupation hitherto. What I have written since my voyage is for the most part closely related to these."[23] Though both Forsters benefited considerably from the institutional support provided to natural history by the medical faculties within the German universities, neither of them had much enthusiasm (or expertise) for the clinical practice of medicine. In 1785, while at the University of Vilna, George Forster candidly admitted that "I don't very much like practical medicine, [but] like very much natural history . . . above all because it leads so immediately to the good and true philosophy of man"[24]—a remark that underlines the extent to which the study of humankind was an integral part of the Forsters' conception of natural history.

For both father and son, the dominant deities in natural history were Linnaeus, who brought order, and Buffon, who brought majesty and purpose to the study of nature—a pursuit that was still establishing its scientific credentials. When George Forster published his account of Cook's second voyage, he was at pains to disabuse those who regarded the natural history that both he and his father pursued as mere specimen hunting. His father was sent to the Pacific, insisted the young George, with loftier ambitions than "being a naturalist who was merely to bring home a collection of butterflies and dried plants"; on the contrary, he was expected to provide "a philosophical history."[25] Like their great mentor, Buffon, the Forsters saw in natural history a means of making sense of nature, of illustrating the way in which the different facets of the terraqueous globe were interrelated and interdependent. In particular, the Forsters sought to locate humankind firmly in its natural setting and to demonstrate the extent to which human beings were an integral part of their environmental setting. Hence the grandiloquent opening of Johann Reinhold Forster's scientific reflections on Cook's voyage, *Observations Made During a Voyage Round the World*: "My object was nature in its greatest extent; the Earth, the Sea, the Air, the Organic and Animated Creation, and more particularly that class of Beings to which we ourselves belong."[26]

On a more microcosmic scale, George Forster later used that most distinctive feature of Tahitian flora, the breadfruit tree (the plant that launched Bligh's *Bounty* and *Providence* expeditions) to illustrate the interconnectedness of the human world and that of the vegetable kingdom in Tahiti—an approach that foreshadows present-day understandings of ecology.[27] "The history of the products of the soil," wrote Forster, "is closely interwoven with the fate of mankind and its emotions, ideas and actions. The realm of nature borders the domain of every science, and it is impossible to review the former without examining the latter."[28]

By taking such a wide view of the scope of natural history, the Forsters were responding to Buffon's admonition to rise above their traditional role of collectors of data and to aspire to provide generalizations about the workings of nature. The naturalist, urged Buffon, should seek "combination of observations, the generalisation of facts."[29] Appropriately, on their return from their voyage with Cook, Johann Reinhold Forster made the pilgrimage to visit Buffon and gave him a set of botanical specimens from the South Seas.[30] Buffon reciprocated by acknowledging the Forsters' contribution of some of the zoological specimens discussed in the sixth supplement of the *Histoire Naturelle*.[31] Later, George Forster was to promote the diffusion of Buffon in the German-speaking lands through his translations—notably the sixth volume of the great *Histoire Naturelle*.

Such an elevated Buffonian sense of the nature and scope of natural history was proclaimed by Johann Reinhold Forster in the lectures he gave while at the Warrington Academy. Though the academy—which was intended for the education of Protestant dissenters excluded from Anglican Oxford and Cambridge—was small and poorly endowed, it offered Forster his first opportunity to focus on what had been and would increasingly become his true intellectual love, the study of natural history. "The manner in which we propose to treat *Natural History*," he told his audience in his first set of natural history lectures (chiefly devoted to mineralogy), "is a *Scientific way*." It was his intention to ensure that "all things may be arranged in such a manner, that we may see their *Order*." Like Buffon, he drew a parallel between the role of the civil historian concerned with what Buffon termed "the epochs of the revolutions of human affairs"[32] and the role of the natural historian. Both, wrote Forster with an affirmation of the empiricist rigour that was to remain a feature of his work, had a duty of "veracity," but the lot of the natural historian was more difficult since he was obliged to "give such definitions or descriptions of the Natural bodies, that no body may be mistaken."

Out of such a truly scientific approach to the subject emerged both practical advantages and a more soundly based religion, for the natural historian could "by reflecting at [*sic*] the Nature and Properties of these bodies" demonstrate "the proper use and best which can be made of them, for human Society," as well as establishing that "Nature is the great book, on which the Deity with indelible Characters, has written its immense and adorable Attributes."[33] In the lectures that followed on entomology, he was at pains to rescue natural history from the condescension of those who dismissed it as the pursuit of mere "*fly-catchers*" whose "knowledge is confined to a very minute and trifling object." Again, he insisted, the study of nature brought with it a lofty sense of the order and interconnectedness of nature and, by doing so, strengthened a sense of the grandeur of the "great Works of the Architect of the Universe" and the "providential care for each of the minute parts . . . of his Creation."[34]

However, Forster's young audience evidently did not warm to these rhapsodies about the merits of natural history: Forster complained in a letter to the naturalist Daines Barrington in 1768 about "the Follies and Tricks of Young Monkeys, who are quite licentious and under no discipline and tease me with their Tricks to death."[35] It must have been with great relief, then, that Forster took up the offer to accompany Cook as a scientific observer on his second great voyage of 1772–75, the main aim of which was to establish once and for all whether there was a great South Land to balance the land masses of the northern hemisphere. For Johann Reinhold Forster and his son, George—who continued to accompany his peripatetic father, from whom he picked up an education as best he could along the way—the new world of the Pacific offered fertile ground to establish the significance of natural history as a pursuit worthy of the Enlightenment by shining the light of science on the dark corners of the earth. The fact that the Pacific was, in European terms, largely virgin territory made it a particularly important instance of the capacity of enlightened thinking to make comprehensible a major section of the globe. In the spirit of Linnaeus, the Forsters set out with the goal of making the natural order more comprehensible by the use of classification. It was a goal that complemented the activities of Cook, who sought to make the Pacific more manageable (and exploitable) by the use of maps that could use mathematical coordinates to render a whole range of geographical features in forms familiar to the European mind.[36]

The Pacific offered whole new categories of specimens from the three kingdoms of nature, animal, vegetable, and mineral, together with the possibility of linking such findings more closely to the study of humankind and its behavior. Making the most of such opportunities, the Forsters brought back large numbers of specimens reflecting both the natural and human worlds of the Pacific together with lengthy classificatory lists. Johann Reinhold Forster proudly noted in his account of the artifacts that he had collected that in addition, "Our Herbarium amounts to 6–700 extremely rare species of which hardly any specimens are found in the collections of Europe and of which 2–300 have never been seen or described by any other botanist."[37] The extent to which Forster saw himself as self-consciously continuing a Linnaean tradition is reflected not only in his use of Linnaean nomenclature but also in that he bound a copy of Linnaeus's 1762 *Instructiones Peregrinatoris* (*Instructions for a Traveler*)[38] in with his own observations on travel.

The Forsters could bring to this encounter with a new quarter of the globe an intellectual rigor strengthened by the German university system.

Though Johann Reinhold Forster's student days at the University of Halle from 1748 to around 1751 came at a less than distinguished period in that university's history, at least some attention was given there to the study of natural history and, elsewhere in Germany (and, in particular, at Göttingen), the many German medical faculties did much to promote natural history. Natural history was one incentive to cultivate an early form of anthropology; another was the strong philological traditions of the German university system that Johann Reinhold Forster had imbibed at Halle[39] and passed on to his son.

As students of both the ancient and modern languages, the Forsters naturally brought to their Pacific explorations a fascination with the new languages they encountered and attempted to use such linguistic material to answer larger questions about patterns of migration and cultural transmission within the Pacific. As Johann Reinhold Forster wrote in his *Observations Made During a Voyage Round the World*: "I took particular care in collecting the words of every peculiar nation we met with, that I might be enabled to form an idea of the whole, and how far all the languages are related to each other."[40] The Forsters grasped every opportunity to collect information on the languages of the Pacific islands, compiling word lists on every island where Cook landed[41] and busily questioning Odiddy, a Tahitian who accompanied Cook's voyage for part of the way, about the Tahitian language.

On the basis of such linguistic studies, he concluded that there was a link between the Polynesian languages and Malay though, he suggested, this probably went back to their descent from a common, more ancient language. Hence, he concluded (in line with modern scholarship) that the pattern of migration into the Pacific had been down from Southeast Asia and then eastward.[42] In a comment that underlies the fact that Johann Reinhold Forster's scientific investigations were also expected to serve imperial ends, he suggested that his linguistic work "may one day or other become usefull, if the Europeans especially should chuse to make settlements in the Islands or at least to erect here a new branch of commerce."[43]

The Forsters' fascination with the languages of the Pacific was of a piece with their strong belief in the interconnectedness of all aspects of human society and, beyond it, of nature as a whole.[44] Long after his Pacific voyage, George Forster continued to ponder the effects on human society of its being rooted in a particular place. In the fateful year of 1789, five years before his death, he argued for a distinction between all the local factors that shape society ("*lokale Bildung*") and those characteristics that all humanity shared ("*allegemeine Bildung*")—hence the title of the address "Über Lokale und Allgemeine Bildung."[45] The origins of such a position are evident in the way in which on their Pacific voyage both father and son devoted a great deal of detail to building up a strong impression of the particular characteristics of the widely different locations they encountered. Their accounts of the voyage conveyed a sense of the way that such characteristics helped to shape human societies— one of the primary manifestations of which was language. They partially agreed with Montesquieu and other theorists about the importance of climate in creating human variations, but qualified this with an emphasis on the way the particular features of a society, its location and cultural adaptations, also played an important role in explaining differences.[46] This strong sense of the integral links between all aspects of nature and of human society led to a strong suspicion of theory—or, at least, theories that did not allow for a great deal of variation.

Hostility to theorizing was strengthened by their well-developed empiricism[47]—an empiricism natural to travelers who were continually being presented with new information, particularly in relation to the remarkable fecundity of human invention and the cultural artifacts it produced.[48] The preface of Johann Reinhold Forster's *Observations Made During a Voyage Round the World* dismissed existing attempts to arrive at "The History of Mankind" as having led to "systems formed in the closet" contrasting this with his own work in which "Facts are the basis of the whole structure."[49] His son dutifully echoed such views in his preface to the *Voyage Around the World*, describing it as a work in which human nature is "represented without any adherence to fallacious systems."[50] Such affirmations of empiricist rigor may also have owed something to their close links with the English and their Baconian scientific traditions.

Both these characteristics—an emphasis on the interconnectedness of nature and a strong empiricism—later manifested themselves in George Forster's controversy with Kant in 1786 over the extent to which the human race could be divided into well-defined races. Forster responded in 1786 to Kant's "Bestimmung des Begriffs einer Menschenrace" ("Definition of the Concept of a Race," 1785) and "Mutmasslicher Anfang der Menschengeschichte" ("The Probable Beginning of the History of Mankind," 1786) with his own "Noch etwas über die Menschenrassen" ("More About the Races of Man"). Kant proposed that humankind had separated into different races as a consequence of environmental factors (and, above all, climate) operating on innate predispositions ("*Kerne*" or "germs") that had existed in the original race. He postulated that this original race was white but that it had since given rise to the black, yellow, and red races. The races thus produced could interbreed—an important point given Kant's emphasis on the unity of humankind—but once interbreeding took place, it was not possible to produce the pure strains of races as characterized by skin color, even if the parents moved to totally different environments—an indication that climate alone was not responsible for variations in skin color. For Kant, then, what distinguished one human race from another was the ability to produce offspring who did not display the characteristics of a mixed race, the inevitable outcome if parents of different races mated.

Forster responded to Kant with considerable respect, opening his essay with an acknowledgment of the way that the spread of the *Berlinische Monatsschrift* (in which Kant's essays were published) to distant Vilna— where Forster was then reluctantly situated—was an indication of the diffusion of the "Aufklärung." However, Forster did become rather indignant in response to Kant's suggestion that it was not possible clearly to establish the color of the South Sea islanders, especially as they had not produced offspring within the different climatic conditions of Europe. Forster responded from his firsthand knowledge together with the observations of previous Pacific explorers to illustrate the extent to which such characteristics as skin color were subject to all manner of variations. "In the New Hebrides," pointed out Forster, "both Bougainville and ourselves saw black, black-brown and dark-brown men." Such variations he saw as being complicated still further by the gradual workings of climate.[51] Such subtle shades of variation he thought undermined Kant's attempt to arrive at clear categories to distinguish one branch of humanity from another. Moreover, Kant's contention that races could be distinguished on the basis of the color of their offspring could be undermined by the way that light-skinned children could result from the union of black and white (especially if the father and mother's own skin color was in the middle of the wide spectrum of shades between white and black).[52]

Forster also sought to refute Kant by drawing on the anatomical work of Thomas Soemmerring, a very close friend and sometime colleague when Forster taught at the Collegium Carolinum at Kassel from 1778 to 1784. In 1784[53] Soemmerring had shown that, when compared with other members of the animal kingdom, the differences between black and white humans were almost literally skin deep. Though Soemmerring himself thought the differences were still substantial enough clearly to differentiate black and white races, Forster was inclined to regard the division of the humankind into distinct races with some skepticism, arguing for a high degree of continuity among all the different manifestations of humankind. Again, his inclination was to stress interconnectedness rather than difference and, in doing so, to question the quest by Kant and others to arrive at clear categories to explain human variation.

Moreover, he questioned other basic methodological positions taken by Kant, in particular his influential division of natural history into two branches: one (*"Naturgeschichte"*) concerned with the description and classification of nature, and the other (*"Naturbeschreibung"*) taking as its goal the distillation of general principles and fundamental interrelationships from the ever-increasing bodies of data being accumulated by natural historians. Forster's fundamental concern with the extent of variation in nature and the close links between all its different aspects made him skeptical of such a distinction. Any principles arrived at through a form of natural history devoted to the Kantian program of *Naturbeschreibung* were, in Forster's view, likely to be overwhelmed by the extent of natural variation[54]—variation that his own strongly empirically based work as a natural historian encountering the Pacific had made manifest to him.

Though Forster might dismiss the Kantian explanation for human variation, he still did have to provide some explanation for human differences—even if he was reluctant to erect hard and fast categories, such as concepts of different races. The important work of his friend, Soemmerring, may have underlined the gulf between animals and humans, but it had also placed considerable stress on the differences between blacks and whites in a number of anatomical respects, such as variations in bone structures. George Forster did not attempt to arrive at a definitive position to deal with this problem—indeed, in his debate with Kant, he suggested that a solution was best left to time and to the further accumulation of empirical information. However, in this essay, he did suggest that humankind did not have one single origin and that human variation could be therefore explained by its diverse beginnings.[55] Though this position of polygenesis could be used by apologists for slavery to justify the subordinate position of blacks as having a separate origin from whites, this was certainly not George Forster's intention since the overall burden of his work was to emphasize the unity and equality of humankind;[56] rather, his intention was to emphasize the extent of human variation and the way that such variation undermined any firm basis for the division into separate races.

However, polygenesis was a radical position in a society where the dominant Judeo-Christian tradition emphasized monogenesis: the common origin of all humankind from one pair of ancestors, Adam and Eve, and the fact that, as a consequence, human beings were, in the biblical phrase, "of one blood." This questioning of such a basic tenet of Christianity as the common Adamite origin of humankind was of a piece with what appears to have been George Forster's dismissal of Christianity more generally— particularly after he left Kassel for Vilna in 1784. Contemporaries remarked, for example, on the fact that he did not have his children baptized.[57] Taking up this monogenesist position also marks the growing divergence between father and son, Johann Reinhold Foster having written in his *Observations* (1778) that holy writ established it as "a fundamental position, that all mankind are descended from one couple."[58]

However, George Forster's writings contain little in the way of a direct attack on Christianity in the manner of some of the French philosophes. One of the features of the German Enlightenment (as of the English-speaking Enlightenments of the British Isles, North America, and Australia) was a tendency to work within the established institutions of society rather than to attack them directly. George Forster reflected the German *Aufklärer* in his inclination to remain within a system in which Enlightenment discourse was largely carried on within universities that were closely tied to the institutions of church and state. He might have had private reservations about religion, but it was inconsistent with his role as a professor to make these a source of controversy. Ironically, his last position was to be an employee of the prince-bishop of Mainz—a sympathizer with Enlightenment principles—as his university librarian.

Nonetheless, George Forster's account of his great Pacific voyage—though published in 1777 when he was only twenty-three—indicates some of the anticlerical impulses of the Enlightenment. References to the role of Providence are conspicuously few,[59] and where indigenous religions are discussed, there is a strong animus against "priestcraft"—something that could reflect both Protestant and Enlightenment sympathies. In Deistic fashion, George Forster argued that there was a true, simple and universal religion that had been corrupted by the machinations of a priestly class. His observations of Tahiti prompted him to observe that the fundamentals of their religion corresponded to that "simple and only just conception of the Deity, [which] has been familiar to mankind in all ages and in all countries." However, as a consequence of "the excessive cunning of a few individuals, those complex systems of idolatry have been invented, which disgrace the history of almost every people."[60] A visit to Tonga prompted a similar outburst: "religion is veiled in mysteries, especially when there are priests to take advantage of the credulity of mankind." Always on the lookout for the beginnings of luxury, George Forster speculated that the undermining of the simplicity and equality of Pacific societies was likely to be carried out "under the cloak of religion, and that another nation will be added to the many dupes of voluptuous priest-craft."[61]

George Forster did appear to have drawn a distinction between priest-craft and true, enlightened Christianity, for another attack on "the influence of priestcraft, whose great aim is ever to veil religion in mystery" was followed by an affirmation that true Christianity "does not wear the mysterious cloak . . . and throws a pure and steady light around. It admits of no mystery, and its true and venerable ministers have at all times assured and convinced us, that they reserved no private knowledge for themselves."[62] Forster's conception of true Christianity bears a remarkably close resemblance to the original pure religion of all humankind that preceded the Christian dispensation, and its lack of ritual and minimal role for its clergy help to explain George Foster's subsequent straying from the faith of his father.

By contrast, his father's accounts of his travels do have a more overtly Christian content, though Forster, too, expressed a measure of scorn for "priestcraft." Like his son, he regarded true religion as characterized by what he called "a noble simplicity, which the true adoration in the spirit and truth requires." However, his observations of Tahitian religion led him to add that "the greater part of mankind, when left to themselves, in their religious principles, and modes of worship, have always more or less deviated" from such religious simplicity. Humankind, then, needed guidance not to leave the path of true religion and, for Forster senior, that guidance came from "the Christian dispensation; wherein the ideas of the Deity are pure . . . excluding all priest-craft from its true and genuine votaries."[63] Johann Reinhold Forster's emphasis on religious simplicity and opposition to "priestcraft" and ritual would have caused little controversy in Protestant Germany and may have owed something to his own theological training as a reformed clergyman (though, interestingly, his formal involvement with the church seems to have faded away after leaving his parish near Danzig for Russia in 1765).[64] In his journal, he did on occasions appeal to specifically Protestant principles, arguing that these should prompt a clear departure from the colonizing practices of the popish Spanish and their "blind Zeal for their religion, conducted by bigoted ignorant Friars." By contrast, when it came to the treatment of the South Sea islanders, "Protestants, who boast the principles of reformation, should show them by their humanity and reformed conduct."[65]

What was, however, rather more controversial was the extent to which he thought it was possible to arrive at religious truth without the guidance of Revelation. For, in his *Observations,* he conceded that the Tahitians had retained much that was good despite the fact that they were innocent of the Gospel: that their religion "is in my opinion less cruel, and not so much clogged with superstition as many others, which were or still are in use among nations who are reputed to be more civilized and more improved."[66] Moreover, in the privacy of his journal, he observed that "they exercise all the Social virtues to one another, which are usual among the civilised nations. Charity, the main spring of all morality and virtue, is no where more exercised than among these people."[67] It was just this issue of the extent to which true virtue required Revelation that had, back in

1723, led to Christian Wolff being forced out of a then strongly Pietist Halle, for he had argued that the high morals of the Confucian Chinese were based on human reason rather than on Divine Revelation. Wolff had returned to a less strictly orthodox Halle in 1740 (dying there in 1754), but the aged Wolff is unlikely to have any very direct effect on Johann Reinhold Forster while he was a student there from 1748 to about 1751. Still Wolff's presence indicated a tolerance of religious liberalism that was strengthened by the rationalizing theology of Wolff's pupil, Siegsmund Baumgarten, whose theological lectures attracted crowds of students.[68]

Whether from his Halle background or simply as a consequence of having observed such a diversity of human behavior, Johann Reinhold Forster was prepared to distinguish very markedly between morality and religion, even to the extent of accommodating a remarkable degree of moral relativism. Thus, he rebuked his fellow travelers, who decried the practice of cannibalism among the New Zealand Maoris for attempting "to punish the imaginary crime of a people whom they had no right to condemn," adding that "the action of eating human flesh, whatever our education may teach us to the contrary, is certainly neither unnatural nor criminal in itself."[69]

Johann Reinhold Forster was, then, at least in some times and places, willing to judge the Pacific societies he encountered in their own terms without the use of a Western measuring rod. Both he and his son were also ambivalent about the extent to which Enlightenment conceptions of progress and improvement corresponded to what they observed in their travels and, in particular, at Tahiti. By the standards of Western development, Tahiti had quite some distance to travel down the path of progress but, nonetheless, in the eyes of the Fosters and other Pacific explorers, had many virtues that had been lost or obscured in Europe. There, wrote Johann Reinhold Forster in his travel notes, was a society in which "the Inhabitants are happy enough to have none of the artificial Wants which Luxury, Avarice and Ambition have introduced among the Europeans."[70] In his *Observations*, he even questioned whether the price of European knowledge of these societies was too high, for such contact was likely to lead to the demand for European luxuries: "If the knowledge of a few individuals can only be acquired at such a price as the happiness of nations, it was better for the discoverers, and the discovered, that the South Sea had still remained unknown to Europe and its restless inhabitants."[71]

Here again, then, appears to be the familiar phenomenon of the European traveler projecting onto the Pacific the myth of a Golden Age and, in the manner of Bougainville and Commersen, decrying in Rousseauist fashion the passage from nature to culture and with it the corrupting hand of civilization. And, yet, the Forsters were critical of Rousseau and

what they perceived as his excessively negative portrayal of the consequences of the evolution of human society[72] along such familiar stages as
the progression from hunter-gatherer to pastoral, thence to agriculture
and, finally, to a commercially based society. In his *Voyage*, George
Forster expressed reluctance to promote the views of "Rousseau, or the
superficial philosophers who re-echo his maxims."[73] Johann Reinhold
Forster conceded that the South Sea islanders might appear "to be happily
situated, something like what is said of the golden age, they may live almost without labour" but, he continued, "certain evils in great measure
counterbalance this seeming happiness, the faculty of the mind are [*sic*]
blunted, if the body is so enervated by indolence."[74]

Nonetheless, as West had illustrated in relation to George Forster's account of Tahiti,[75] there was a profound ambivalence in the Forsters' view
of the stadial progress so beloved of the Scottish social theorists of the
eighteenth century. The underlying framework of the Forster's accounts
of the Pacific is shaped by the idea of an evolutionary growth of society
toward civilization as largely perceived in European terms. But, particularly when it came to Tahiti, the Forsters were obviously rather uncertain
whether such an upward climb fully equated with the advance of morality
and human happiness—which raised worrying issues about the stadial
model so basic to their outlook and that of many of their fellow *aufklärer*.
To some extent, Johann Reinhold Forster tried to square the circle by offering a scientific explanation of why in Tahiti particularly there was so
much of value, even though it was a culture well removed from the civilizing impulses of Europe. In the manner of "the great Mr. de Montesquieu,"
he suggested that perhaps there was a climatic explanation: that by an in-
verse law, the Tahitians in their low latitudes bore out the contention
"that the human species when unconnected with the highly civilized nations, is always found more debased in its physical, mental moral and social capacity, in proportion as it is removed from the tropical regions."
But Forster was honest enough to concede that this did not fully explain
why Tahiti of all the societies they had encountered near the Equator was
particularly blessed and he had to lamely conclude that there must be
"some other cause of this remarkable circumstance."[76]

Such occasional backslidings into the language of the Noble Savage
aside, the Forsters shared with their German (and Scottish) contemporaries a preoccupation with the question of what were the wellsprings of
prosperity and ordered civilization—an intellectual agenda that colored
their whole encounter with the Pacific. The scholarly Forsters also brought
to their subject that interest in universal history which the German Enlightenment did so much to promote and, with it, a conviction that civilization was a cumulative phenomenon. As Johann Reinhold Forster put it

with italicized enthusiasm immediately after the section on Tahiti cited above: "All the ideas, all the improvements to sciences, arts, manufactures, social life, and even morality, ought to be considered as *the sum total of the efforts of mankind ever since its existence*"—all of which made the case of Tahiti, which had had so little contact with other cultures, all the more perplexing to Forster.

But his puzzlements about Tahiti did not weaken Johann Reinhold Forster's resolve to use his Pacific travels to illustrate the great theme of the advance of civilization: as he put it at the end of the preface to his *Observations*, it was his goal to show "by what steps they [human beings] may gradually emerge from the darkness of barbarism, and uniting in social compacts, behold the dawn of civilization." Conversely, his work also would provide salutary warnings about the way in which civilization could also go into reverse and "by what accidents and misfortunes men may, for want of mutual support, degenerate to savages." The intellectual equipment that came most readily to hand in giving substance to such a view was the stadial language of the French and Scottish social theorists. Johann Reinhold Forster's general reflections on the societies of the South Seas led him to conclude, for example, that "mankind, in a pastoral state, could never attain to that degree of improvement and happiness, to which agriculture, and the cultivation of vegetables, will easily and soon lead them."[77]

In a similar vein, George Forster concluded his *Voyage* with a hymn to progress urging his readers to reflect on the way his work illustrated the "blessings which civilization and revealed religion have diffused over our part of the globe" and the extent to which his fellow Europeans should give thanks for "distinguished superiority over so many of his fellow creatures, who follow the impulse of their senses, without knowing the nature or name of virtue."[78] In 1778, the year after this book was published, George moved to Kassel to take up a professorship of natural history and there marked his admission to the Société des Antiquités de Cassel with an "Antrittsrede" ("Inaugural Speech") in which he drew on his Pacific experiences to again underline his un-Rousseauian belief in the merits of the advance of civilization. For the young Forster, the same dynamics that he had observed in the Pacific had been at work in human history as a whole, and the self-evident merits of progress would mean that all humankind would eventually share a common culture: "Civilization arrives a little nearer by the same degrees in all lands, it is only the epochs which are different." Cannibalism had existed in the ancient world but had died out as it had in Tahiti and so, Forster was confident, it would eventually do so among the Maoris of New Zealand. He also drew on the familiar scale of human ascent from the Tierra del Fuegians to Europeans—pausing to praise the society that he addressed as an example of the contemporary enlightened spirit abroad within Europe.[79] Despite such cultural condescension underlying the speech—and Forster's work as a whole—was a strong belief in human equality since all peoples were linked in a common history and would eventually arrive at a common civilization, even if such a civilization might bear a remarkable resemblance to eighteenth-century Europe.

George Forster's time at Kassel from 1778 to 1784 brought him in regular contact with colleagues at the nearby University of Göttingen (from which he received an honorary doctorate), and he was a close friend of Georg Lichtenberg, with whom he edited the *Göttingisches Magazin der Wissenschaften und Litteratur* from 1780 to 1784.[80] Such academic connections were cemented in 1785 by his marriage to Therese, the daughter of Christian Heyne, a prominent classical philologist—a union that, like so many chapters in the life of the ill-starred George, ended badly, with his becoming part of a *ménage à trois* with his wife's lover. It was a marriage that linked George Forster with some of the leading figures in Göttingen, including the great naturalist and professor of medicine Johann Blumenbach, who was a brother-in-law of Heyne. George Forster was drawn to Blumenbach by common intellectual sympathies as well as these dynastic connections. When, in 1786, Forster came to draw together his long-gestating textbook on natural history, he conceded to Soemmerring that it was difficult to improve on Blumenbach's treatment in his *Handbuch der Naturgeschichte* (*Handbook of Natural History*—first edition, 1779; twelfth edition, 1830). This was true particularly in relation to physiology (and especially in its connections with anatomy)—hence, he acknowledged that the overall treatment would remain "tolerably Blumenbachisch."[81] Both Forster and Blumenbach also agreed on the extent to which physical anthropology should form a part of natural history; predictably, one of Forster's underlying themes was the interconnectedness of all of natural history (including physical anthropology) and, indeed, of all the different sciences.[82] He was also sympathetic to one of Blumenbach's key notions, the concept of the *"nisum formatiuum,"*[83] a formative force at work in nature. Both figures also shared a common opposition to the way that Kant attempted to define the division between species on criteria based on breeding—the thrust of Blumenbach's approach, by contrast, was to attempt to define species by a range of common morphological features, the so-called Totalhabitus.[84] When, for example, he came to consider human beings, Blumenbach attempted to distinguish the different human races by skull structure along with other characteristics.

However, there were also important differences between the two, for Blumenbach did follow Kant in accepting the notion that there were distinct human races[85]—something which, as Forster's exchange with Kant makes clear, was not consistent with Forster's emphasis on the gradual variations in nature. Forster's sympathy for polygenesis was also at variance with Blumenbach's robust defence of monogenesis—a position that led one of Blumenbach's English admirers to pen the following verses, entitled "On a Collection of Skulls from Different Nations":

> Blumenbach's penetrating Thought
> Through Nature manifold has sought
> The various Races of Men has trac'd
> And on his Shelves in Order plac'd. . . .
> Yet all of ev'ry Age and Land
> The Work of an Almighty Hand,
> One Flesh and Blood unites them all
> One God, and Father, each May call.[86]

Though at Halle from 1780 to his death in 1798, Forster senior also had considerable contact with the Göttingen professors, especially Johann Michaelis (with whom he corresponded from soon after his graduation at Halle, the university attended by both Forster and Michaelis) and Blumenbach, whose anthropological work Forster had praised in the preface to his *Observations*.[87] Soon after his arrival in Halle, Johann Reinhold Forster began a long and scientifically fruitful correspondence with Blumenbach, whose popular *Handbook of Natural History* became the textbook for Forster's courses.[88] Drawing on his rich collection of Pacific specimens, Forster could reciprocate for Blumenbach's scientific courtesies by providing him with geological samples from the *Resolution* voyage.[89]

More significantly, Forster's Pacific observations led Blumenbach to propose that, in addition to the existing classification of humankind into Caucasian, Mongolian, Ethiopian, and American, there should be a fifth—the Malay. Hence, he wrote in the revised 1781 edition of his *De Generis Humani Varietate Nativa* (*About the Innate Variety of Human Kind*)—originally his Göttingen MD thesis of 1775:

Finally, the new southern world makes up the fifth . . . Those who inhabit the Pacific Archipelago are divided again by John Reinhold Forster into two tribes. One made up of the Otaheitans, the New Zealanders, and the inhabitants of the Friendly Isles, the Society, Easter Island, and the Marquesas, and c. men of elegant appearance and mild disposition; whereas the others who inhabit New Caledonia, Tanna, and the New Hebrides, and c. are blacker more curly, and in dispositions more distrustful and ferocious.[90]

Thus, this prefigured the subsequent division by the French explorer Dumont D'Urville of the Pacific peoples into Polynesians and Melanesians. The third edition of 1795 supplemented these comments on the fifth

(Malay) division of humankind with further details gained (as Blumen-bach effusively acknowledged) from Joseph Banks[91]—to whom he dedi-cated the volume as a whole.

Both Forsters continued to keep the German-speaking world informed of British activities in the Pacific through their translations. From 1790 until his death, Johann Reinhold Forster acted as editor of the *Magazin von Merkwürdigen neuen Reisebeschreibungen aus Fremden Sprachen* (*Magazine of Curious Travel Accounts in Foreign Languages*, sixteen volumes, 1790–1800). This series provided an abridged account in Ger-man of Captain Hunter's voyage to Australia, including lengthy notes by Forster on the Pacific background. A posthumous translation by Forster of Vancouver's travels also appeared in this series.[92] Earlier, in 1784, Jo-hann Reinhold Forster had provided a justification for such voyages of discovery in his *Geschichte der Entdeckungen und Schiffahrten in Nor-den* (*History of the Voyages and Discoveries Made in the North*), which was dedicated to Catherine the Great as a tribute to Russia's role as a pro-moter of exploration and in a vain bid to obtain membership of the St. Pe-tersburg Academy. Though such explorations might be prompted by commercial motives, they served enlightened ends, for they "seem to have greatly contributed to the promotion of knowledge, and to the introduc-tion of milder manners and customs into society." Indeed, Forster con-cluded with an affirmation that there was a Providential purpose at work in such voyages and the general expansion of European society that ac-companied them. Exploration revealed something of "the wisdom of a supreme being, who dispenses his benefits over the whole universe, and manifests the utmost sagacity and intelligence in the accomplishment of his purposes."[93]

Johann Reinhold Forster's view of the ways that civilization and progress were promoted by exploration could be well contained within the generally unthreatening bounds of the German Enlightenment. His view of the advance of civilization was positive but nonetheless content with gradual improvement, and his *Weltanschauung* still kept a consider-able place for the workings of Providence. By contrast, his son brought to his account of the upward march of civilization a degree of enthusiasm and a secularism that was less comfortably contained within the *status quo*. George Forster so brightly portrayed the benefits of the spread of civilization and the Pacific exploration—which formed a conspicuous ex-ample of such progress—that the drab realities of the society around him must have been even more evident.

Like his father in his edition of Hunter's journal, George Forster pre-dicted a bright future for the infant colony of New South Wales. In 1786 he published his "Neuholland und die brittishche Colonie in Botany-Bay"

("New Holland [Australia] and the British Colony in Botany-Bay [New South Wales])" in response to the plans for a penal colony there—plans that did not eventuate until 1788, an indication of how *au courant* Forster was in matters relating to the British in the Pacific, even while at Vilna. For Forster, the British were extending the reach of civilization just as they had in America, and New Holland was likely to become "the future homeland of a new civilized society which, however mean its beginning may seem to be, nevertheless promises within a short time to become very important." It was a colony that offered the opportunity of realizing Enlightenment hopes for the rehabilitation of the socially deviant through exposure to a new environment. Thus, a convict there could "cease to be an enemy of society whenever he regains his full human rights."

In his *Voyage*, George Forster's recent contact with the Pacific had tempered to some extent his affirmations of the merits of the advance of progress along the path to European conceptions of civilization. By the time he came to write this essay, however, he took a strongly teleological view of human development. Where Cook and Banks had written sympathetically of the Australian Aborigines as untroubled by European preoccupation with luxuries, Forster in his "Neuholland" dismissively argued that "Among all the races which may claim to be called human, that which inhabits New Holland is the most wretched," noting in stadial fashion that they lived "without agriculture." Their way of life, he conceded, "is also in the end a path to civilization," but he saw it as a particularly long and uncertain path. In the manner of Kant, Forster took the strong anti-Rousseauian view that it was only in the conditions of civilization that Man "begins to achieve the potential with which he has been endowed in the form of his faculties, and becomes a truly human being." It followed, then, that the spread of civilization—of which the Botany Bay venture was a part—was "in the interests of mankind, and population of the whole earth with civilized inhabitants is the great goal, which we above all see before us as worthy of our efforts." He concluded the essay with an encomium to Cook—whose exploration had prompted the new colony—as a second Columbus, and, like him, one who "defines a second similar epoch in our day."[94]

This theme of Cook as a world historical figure was developed in an essay that George Forster wrote the following year (1787), entitled "Cook der Entdecker" ("Cook the Discoverer"), which formed the preface to his German translation of the official account of Cook's third Pacific voyage. Again the spread of European influence around the globe is very much linked to the spread of the Enlightenment: he warmly commends Cook as one who "has advanced his century in knowledge and enlightenment" and with it, "a golden future of general advanced knowledge." The advance of civilization to which Cook had so signally contributed Forster connected with a more general dynamic of progress that had brought with it the recognition of "the universal rights of man"—a process that he hoped European contact with the Pacific would serve to stimulate further, particularly since "enlightenment . . . advances from experience to experience without limits."

The spread of civilization could also be hastened by the foundation of European colonies in the Pacific—Forster recommending, for example, that the British send settlers to the north island of New Zealand. Such opportunities had been made possible through Cook's initiative and courage, which now served to promote "the general enlightenment of all civilized peoples." In an almost millenarian passage, Forster held out the hope that Cook's voyages ushered in a new age for humankind with "general enlightenment, [and] the joint advance of our whole kind towards a certain goal of perfection"; this he saw as enabling the "wiser Europeans" to "finally assail the old Asiatic obstinacy as well as the invincible refractoriness directed against all progress of enlightenment." Within Europe and beyond, "the limit of progressive enlightenment lies beyond our horizon," but its fruits were becoming manifest as "Tolerance and freedom of conscience announce the victory of reason and pave the way for the freedom of the press and the free study of all conditions which are important to man in the name of truth."[95]

Such chiliastic zeal for the progress of Enlightenment helps to explain why George Forster greeted the French Revolution with enthusiasm as a way of hastening the march of progress. In his view of human history, the steady growth of civilization was ineluctable, but the possibility of rapidly overcoming some of the obstacles in its way must have been intoxicating indeed. When the French Revolutionary armies conquered Mainz and the prince-bishop fled, Forster responded enthusiastically to the prospect of being able to play a part in the establishment of a republican and democratic form of government—even though his trip to Paris (where he died) led to disillusion about actual Jacobin practice.[96] But even during the Revolution, Forster was sufficiently still in sympathy with the character of the German Enlightenment to urge German rulers to introduce reforms from above and bring about change without wholesale revolution. As he put it in 1792: "One could take advantage . . . of the events in France without having to pay too steep a price for that which is good in them; the French volcano could make Germany secure before the earthquake."[97] But, of course, such a *via media* was illusory in the face of the polarization between defenders and opponents of the old regime that the French Revolution brought in its wake.

The Anglophile professors at Göttingen were generally more cautious in their political views and had considerably more reason to be content with the *status quo* than the peripatetic and generally impecunious George Forster. Both he and his father, however, brought welcome contact with the larger British world and firsthand knowledge of the exciting new world of the Pacific. Ironically, the Forsters themselves had almost systematically destroyed their highly placed contacts within England as a consequence of the wrangling over who should receive the royalties for the official account of Cook's second voyage. Johann Reinhold Forster's vain attempts to secure this lucrative publication for himself culminated in his commissioning young George to publish the egregiously ill-judged *A Letter to Lord Sandwich*, in which he accused Lord Sandwich (the first lord of the Admiralty) of being motivated by "revenge and *private pique*," the prominent naturalist Daines Barrington (a member of a well-connected aristocratic family) of not keeping his word, and the king himself of being guilty of "neglect[ing] a duty" in not responding to the Forsters' petition.[98]

That great disposer of the scientific riches of the British Empire, Joseph Banks, was not included in this fusillade, but the Forsters perceived a growing coldness from him. It was not surprising: Banks was a close friend of Sandwich and he would also no doubt have heard some of the envious remarks that Johann Reinhold Forster made about him. Banks also grew increasingly pressing about the 250 pounds that Johann Reinhold Forster owed him—though when Forster died, he forgave his penniless and long-suffering widow the debt following the intervention of Blumenbach.[99] In her effusive thanks to Banks, Forster's daughter went so far as to acknowledge that her late father "with all the great Qualifications of his head and heart, still was but too often deficient in point of Prudence and cool Discrimination."[100] As Hoare aptly puts it, Blumenbach in effect was to replace the Forsters as the recipient of a steady stream of scientifically significant natural history specimens made possible by Banks's position as president of the Royal Society and chief organizer of Britain's voyages of scientific discovery.[101] The one prominent member of the British natural history community with whom the Forsters maintained contact was the Welsh naturalist Thomas Pennant,[102] to whom George Forster poured out his heart in 1787 from Vilna, bemoaning the unwisdom of having sought "to claim as a debt [from Lord Sandwich], what on the other side was considered as the favour of a voluntary choice" and, more generally, the doleful effects on his own career of his father's actions, including the fact that "Sir Joseph's [Banks] mind appears to have been alienated from me."[103]

The Forsters may have undermined their links with the British scientific and political establishment, but they nonetheless brought with them to Germany both firsthand knowledge of the Pacific and a rich store of

Pacific natural history specimens. These included an extensive collection of ethnological artifacts, some of which the Forsters bestowed on Göttingen.[104] The university's store of Pacific artifacts was greatly expanded in 1782 when, thanks to the initiative of Blumenbach, the university received from George III a gift of more than 350 items deriving from the second and third voyages of Cook.[105] In 1799 these were supplemented (again thanks to Blumenbach's political skills) by a further 160 objects through "the purchase of the specified collection of South Seas curiosities left by Professor Forster of Halle."[106]

Quite apart from the Forsters, the Götttingen professors had their own British contacts from whom they eagerly sought further information generally about the Pacific and Cook's voyages. George Forster's friend, the polymath Georg Lichtenberg, had come to know many of the major figures involved in Pacific voyaging, including the botanist Daniel Solander, the artist William Hodges, Banks, and the Tahitian Omai—from these he brought back from his second trip to London in 1775 accounts of Cook's second voyage.[107] Lichtenberg's interest in Pacific voyaging led him to write a biographical account of Cook[108] that drew on information from this circle together with the Forsters—a work to which George Forster approvingly referred in his essay on "Cook the Discoverer" and which he praised in a letter to Banks as doing "the most ample Justice to his [Cook's] Life and Character."[109] The Orientalist Johann Michaelis grasped every opportunity to collect information about indigenous peoples and particularly those of the Pacific; he was fascinated by the role of languages and the extent to which it reflected the opinions of a people.[110] He had been in touch with Johann Reinhold Forster as far back as 1765 before he set off for Russia, and received from him reports on Cook's first Pacific voyage including a sample of bark "tapa" cloth that Banks had brought back from Tahiti. When Forster himself set out for the Pacific, Michaelis provided him with a list of possible areas of enquiry that had been drawn up for the Danish expedition to Arabia.[111]

Another of Michaelis's Pacific informants was the prominent physician Sir John Pringle, who preceded Joseph Banks as president of the Royal Society (serving from 1772 to 1778). He visited Göttingen in 1766 and in the same year Michaelis secured for him membership of the Götttingen Königliche Sozietät der Wissenschaften. Pringle responded with warm praise of both this society and the university along with commendation on the impact in England of Michaelis's Hebrew scholarship and of the "learned questions drawn up for the use of the Danish Missionaries"—the ethnological agenda that Forster was later to take with him. The return of Samuel Wallis's expedition to the Pacific in 1768 gave Pringle an opportunity to correct earlier reports from the Byron expedition of 1764–66 about the existence of giants in Patagonia, and the return of the *Endeavour* in 1771 prompted a lyrical account of Tahiti and the Society Islands (based largely on Banks's account). These islands, Pringle wrote, "may be most truly called the Fortunate Islands"—though he did qualify this Rousseauian idyll by adding that "Their greatest unhappiness is from war." Pringle was able to give Michaelis advanced intelligence of the fact that Cook's second voyage had proved the nonexistence of a great South Land for, he wrote in 1774 following the return of Furneaux's *Adventure* (the ship that had accompanied Cook's *Resolution* but had become separated), "C. Furneaux has made a most successful circumnavigation, in a very high southern latitude . . . without seeing anything but water, air and sky." He concluded, then, "that the great outlines of our globe are already drawn." Pringle continued his reports during Cook's third voyage: after announcing Cook's arrival in Cape Hope in mid-1775, he expressed the hope that Michaelis would pass on "this account to my good friends of the University that may be curious in such matters." Among the eager recipients of news from the Pacific would have been Blumenbach, to whom Michaelis later passed on a copy of Pringle's treatise on preserving the health of mariners.[112]

In the same letter, Pringle commented sympathetically on David Hume's objection to the notion of eternal punishment as inconsistent with a loving God—though adding that rather than reject the New Testament, Hume should have rejected the teachings of Calvin and Knox. Evidently, then, along with their shared interests in natural history, Pringle and Michaelis had in common a liberal attitude on theological matters. Reacting against the Pietist orthodoxy of his father, Christian Michaelis—partly as result of the rationalizing theological teaching of Baumgarten (who may also have influenced Johann Reinhold Forster)— the younger Johann Michaelis left Halle for Göttingen. There he eventually opted for a chair in philosophy rather than theology to avoid having to subscribe to the Augsburg creed and its declaration of Lutheran orthodoxy.[113]

Together with Michaelis and Lichtenberg, the other main epistolary contact between Göttingen and the English-speaking world was Blumenbach, whose standing and, with it, the number and range of his correspondents, increased up to his death in 1840. Along with the extensive and scientifically important correspondence with Joseph Banks (which has been discussed elsewhere),[114] Blumenbach developed a network of contacts who could pass on to Göttingen some of the scientific riches made possible by the expansion of British imperial power. Blumenbach was ever-anxious for skulls from all parts of the globe to assist his investigations into comparative physical anthropology, and he built up a considerable collection that his family referred to as his "Golgotha," the biblical "place of the skull."

The unearthing of the skull of the Scottish king, Robert Bruce, in 1821, for example, led to an application to the royal librarian "to send a Cast of this Celebrated Skull to the far famed Professor of Göttingen [Blumenbach] whose Museum is the wonder of the World and who is in the habit of receiving such presents from all the crowned Heads in Europe."[115] The methods his correspondents used to obtain these skulls were not overscrupulous and effectively amounted in some cases to grave robbery of indigenous sites. Blumenbach tended to adopt the same insouciance toward the supply of crania from European as well as non-European sources, being supplied in 1820 with a skull of the allegedly distinctive Irish kind, along with one from a church in Kent whose piles of skulls were supposed to go back to Anglo-Saxon times.[116]

Thanks to Banks, Blumenbach obtained skulls from the West Indies, Tahiti, and Australia. Blumenbach also collected notes on Australian Aborigines from early accounts of the British colony of New South Wales by Collins, Tench, and Hunter—accounts that he supplemented with correspondence with early visitors to the colony such as John Bigge.[117] From New Zealand he obtained a skull of a Maori prince through the *aide de camp* of His Royal Highness, the Duke of Cambridge—the *aide* declaring to Blumenbach in 1822 that "you have *friends* in Hanover and *England* sensible of *your* worth."[118] From another part of the Pacific—Peru— Blumenbach obtained a skull in 1825 thanks to the British informal empire of trade in South America, it being sent by the chief commissioner of the Anglo-Chilean Association on an English transport.[119] Four years later another South American skull arrived, this time from Demerara (Georgetown, British Guiana), again via an English contact (whose nephew Blumenbach had helped in Göttingen).[120] British expansion into Southeast Asia led to crania being sent back from Burma, and the colonial administrator, John Crawford, insisted to the officials at the British Museum that these be forwarded on to "the only proper hands, those of Dr. Blumenbach."[121]

Blumenbach's contacts also extended to England's separated brethren in the young United States. He received natural history news from Benjamin Barton, professor of medicine and natural history at the University of Pennsylvania, and (it would appear) a skull that reached him via Michaelis.[122] In 1824 a skull of a Narragansett Indian arrived thanks to a young American who was thankful for "your intercourse and—may I say—friendship while in Göttingen,"[123] followed in 1835 by a cast of the head of the Methodist preacher George Whitefield, which an enthusiastic Massachusetts phrenologist had sent on because it was supposedly "wholly deficient in the organ of *religious sentiment*."[124]

Along with Blumenbach's contacts with the British Empire (and its former transatlantic possessions) went recourse to the nearby Russian Empire for natural history specimens from its increasing territories. Appropriately, these two empires were drawn together in the dedication to the 1795 edition of Blumenbach's magnum opus, *De Generis Humani Varietate Nativa*, both to Banks and the Baron Georges-Thomas de Asch of St. Petersburg, Blumenbach's chief conduit of Russian specimens. Like so many members of the Russian scientific elite, Von Asch was of German origin, having studied medicine at Göttingen under the great von Haller. He sent Blumenbach more than forty skulls, including that of a female Georgian—the Georgians, who were celebrated for their beauty, being the exemplar for the Caucasian racial type.[125] Another Russian source of skulls was Professor Karpinsky of St. Petersburg who, as Blumenbach reported to Banks in 1792, sent four specimens of "the *Mongol* race."[126] Much later, in 1826, Blumenbach drew on the close marriage ties between the Russian imperial house and some of the German princely families to obtain two skulls from Kamtchatka on Russia's Pacific coast as a consequence of the visit of Princess Auguste of Weimar to her mother at St. Petersburg.[127]

Increasing Russian expansion into the Pacific was reflected in some of the other anthropological material that made its way to Blumenbach. The Von Asch ethnographical collection, which was integrated into the Göttingen museum, included a pickaxe from "North America opposite the Chukot Peninsula," which was reminiscent of New Zealand artifacts because the blade was made from nephrite (greenstone).[128] For information on Pacific (and other) languages, Blumenbach could draw on the work of Adam Krusenstern, commander of the first Russian circumnavigation from 1803–1806—the belated realization of the voyage on which George Forster had planned to sail. On his voyage, Krusenstern was accompanied by the Göttingen-trained physician Georg Langsdorff.[129] Another member of this expedition, W. G. Tilenau, consolidated links with Göttingen by a visit there in 1814, leaving some notes on Polynesian bark cloth samples.[130]

Thanks to such activities of the Göttingen professors and, a fortiori, to the work of the Forsters, the Pacific became well established as part of the vibrant late eighteenth-century German debate about the nature of human development and the great Rousseauian questions posed by the transition from nature to culture. In defending his fundamental thesis that each culture is unique with its own inner dynamic that prompts change and evolution, Johann von Herder drew explicitly on the Forsters. In his *Outlines of a Philosophy of the History of Man*, he referred to Johann Reinhold Forster as the "Ulysses" of the Pacific and his *Observations* as a model of its kind—adding the wish that there was a comparable *"philosophico-physical geography* of other parts of the World, as foundations for a history of man."[131]

Herder drew on Forster's Pacific reflections and other travelers' accounts of the Pacific to substantiate his arguments about the way cultures change and evolve: "A few centuries only have elapsed since the inhabitants of Germany were Patagonians; but are so no longer." With his stress on the unique value of each culture, Herder was, however, more disposed than the Forsters to emphasize what was uniquely valuable about particular Pacific societies. When, for example, he referred to the Mallicollese [Malekulans] of the New Hebrides [Vanuatu]—described in the Forsters' accounts of Cook's second Pacific voyage—he pointed out that they "display capacities that many other nations do not possess." Herder painted the Tahitians in the customarily bright colors as leading "a tranquil happy life." In the manner of Montesquieu, he attributed this largely to their climate, whereas the "less temperate" climate of Australia required the Aborigines to "live more hardly, and with less simplicity." Nonetheless, the Aborigine, too, he saw as having developed his own integrated way of life reflecting the worth of each culture: "he has united as many ways of life as his rude convenience required, till he had rounded them as it were into a circle, in which he could live happily after his fashion"—the same, he added, could be said of the New Caledonians and New Zealanders, adding in a rather un-Herderian manner, together with "even the miserable creatures of Tierra del Fuego." These examples drawn from Cook's second voyage he supplemented with material from the third voyage that, no doubt, he read in George Forster's German translation.[132]

For Kant—a voracious reader of travel accounts—the Pacific also served as a social laboratory to test assumptions about human nature and culture. As Kant wrote in his *Anthropology from a Pragmatic Point of View*: "One of the ways of extending the range of anthropology is *traveling*, or at least reading travelogues."[133] He drew on Cook's account of Tahiti, for example, to substantiate his view that wife beating in some cultures was expected by the wives themselves as a manifestation of jealousy of other men.[134] Like George Forster, Kant took the view that humankind required the amenities of civilization to live a fully moral life; consequently, he regarded some of the almost utopian accounts of life in the South Seas with rather more skepticism than Herder.[135] In his review of Herder's work, Kant did go so far as to refer to the "happy inhabitants of Tahiti," but he called into question Herder's view that they could be quite happy in their existing form of culture. For Kant, true happiness or, at least, true human fulfilment came from participation in "the ever continuing and growing activity and culture . . .

whose highest possible expression can only be the product of a political constitution based on concepts of human right, and consequently an achievement of human beings themselves."[136]

The exchange between Kant and Herder is one instance of the way that accounts of the Pacific were drawn into larger German Enlightenment debates about the nature of human society. Not only were the accounts of the Pacific to be significant in shaping discussion about human development, but the approach adopted by the Forsters and the Göttingen professors (and particularly Blumenbach), with its emphasis on basing theoretical speculation on strong empirical foundations, continued to be influential in Germany. The great heir to the Forsters was to be Alexander von Humboldt, who met George Forster while studying at Göttingen in 1789–90 with his brother, Wilhelm (who later drew on the Forsters' accounts of the South Seas languages in his influential work on comparative philology).[137]

George Forster and Alexander von Humboldt were drawn together by their shared scientific and political views, and in 1790 the two were traveling companions on a journey through the Low Countries, France, and England (where Forster introduced him to Banks).[138] Quite self-consciously Alexander von Humboldt saw himself as continuing along the same trail that George Forster had blazed by utilizing the genre of the travel account to explore the interrelationship between all the different elements of a particular location: its natural history and the manifestations of human culture. As Humboldt put it in his great *Kosmos* (1845–62): "Through my famous teacher and friend George Forster, began a new era of scientific expeditions, the purpose of which is comparative ethnology and geography." Humboldt even drew a parallel between the way George Forster had been invited to serve as a scientific observer on a Russian expedition and the way "I was invited by the Emperor Alexander in 1812 to undertake an expedition through the interior of Asia"[139]—another reminder of the important links between German science and Russian imperial expansion.

While many of Humboldt's nineteenth-century scientific contemporaries devoted themselves to refining further their particular scientific specialties, Humboldt maintained something of the integrated worldview that had characterized the Forsters' writings on the Pacific.[140] His monumental *Kosmos* was probably the nearest the nineteenth century came to presenting a total view of nature and its interrelationship with humankind. Though von Humboldt did not have firsthand experience of the Pacific beyond the shores of South America, he incorporated into this remarkable scientific *summa* ethnographical accounts of the South Seas, which he saw as allowing European scientific observers to arrive at more securely based generalizations, since they had previously been restricted to "a small portion of the earth." Like his master, George Forster, he was skeptical of the systems of racial classification—whether that of Blumenbach, with its five races, or the more recent one of the English anthropologist, James Pritchard, with seven races: of both he wrote "we fail to recognise any typical sharpness of definition, of any general or well-established principle, in the division of these groups."[141] As with Forster's querying of the Kantian racial divisions, it was a position that reflected an emphasis on the interconnectedness of all of nature together with the wariness of generalizations that follows naturally from the accumulation of empirical data through extensive travel.

Thanks to George Forster's influence, then, Humboldt in some ways carried into the nineteenth century some of the approaches and concerns of the German Enlightenment with its belief in the integration of all knowledge and its preoccupation with the universal history of humankind and the stages of its development. For eighteenth-century Europe, the "new world of the Pacific" was fertile ground for such conjectures, and Germany participated in this European-wide reflection on the significance of the data brought back by Pacific explorers. It did so both by the involvement of some Germans (notably the Forsters) in British and, to a lesser extent, Russian expeditions and through the focused research made possible by its universities (notably Göttingen). While imperial powers like Britain, France, Spain, and Russia looked to Pacific expeditions to promote both the Enlightenment and empire, the German involvement in the Pacific had, perforce, given Germany's fragmentation, to be much more restricted to the rewards brought by the pursuit of Enlightenment science and anthropology. When Germany again focused closely on the Pacific, it was in the Wilhelmite period, when a newly united nation joined other imperial powers in an expansion into the Pacific, in which the relative detachment of the Enlightenment was replaced by the more familiar goals of empire.

NOTES

1. Alan Frost, "The Pacific Ocean: The Eighteenth Century's 'New World,' " in *Captain James Cook: Image and Impact*, ed. Walter Veit, 2 vols. (Melbourne: Hawthorn Press, 1972 and 1979), II, pp. 5–49.

2. John Locke, *Two Treatises of Government*, ed. Peter Laslett (New York: Mentor, 1965), II, p. 343, sec. 49.

3. Peter Hulme and Ludmilla Jordanova (eds.), *The Enlightenment and Its Shadows* (London: Routledge, 1990), p. 10.

4. James Sheehan, "Germany," in *Encyclopedia of the Enlightenment*, ed. Alan Kors, 4 vols. (New York: Oxford University Press, 2003), II, pp. 119–25.

5. Peter Reill, *The German Enlightenment and the Rise of Historicism* (Berkeley: University of California Press, 1975), p. 216; Jonathan Knudsen, *Justus Möser and the German Enlightenment* (Cambridge: Cambridge University Press, 1986), p. 186.

6. Joachim Whaley, "The Protestant Enlightenment in Germany," in *The Enlightenment in National Context*, ed. R. S. Porter and M. Teich (Cambridge: Cambridge University Press, 1981), pp. 106–17.

7. Reill, *German Enlightenment*, p. 133.

8. Nickolaus Jacquin, *Plantarum Rariorum Hortii Caesearei Schoenbrunnensis Descriptiones et Icones*, 4 vols. (Vienna: C. F. Wappler, 1797–1804), vol. 1, *Praefatio*, pp. iii–v. (I owe this reference to Robert King.)

9. Morris Berman, " 'Hegemony' and the Amateur Tradition in British Science," *Journal of Social History* 8 (1974–75): 30–43.

10. Charles McClelland, *State, Society, and University in Germany 1700–1914* (Cambridge: Cambridge University Press, 1980), pp. 39–45.

11. Mareta Linden, *Untersuchungen zum Anthropologiebegriff des 18, Jahrhunderts* (Frankfurt: Peter Lang, 1976), pp. 41, 62; John Zammito, *Kant, Herder, and the Birth of Anthropology* (Chicago: University of Chicago Press, 2002), p. 28.

12. Manfred Urban, "Cook's Voyages and the European Discovery of the South Seas," in *James Cook, Gifts and Treasures from the South Seas*, ed. Brigitta Hauser-Schaüblin and Gundolf Krüger (Munich: Prestel, 1998), pp. 30–55 (esp. 51).

13. Han Vermeulen, "Origins and Institutionalisation of Ethnography and Ethnology in Europe and the USA," in *Fieldwork and Footnotes, Studies in the History of European Anthropology*, ed. Han Vermeulen and Arturo Alvarez Roldan (London: Routledge, 1995), pp. 39–59 (esp. 25).

14. Urban, "European Discovery," pp. 51–52.

15. Vermeulen, "Origins," p. 42.

16. Michael Harbsmeier, "Towards a Prehistory of Ethnography: Early Modern German Travel Writing as Traditions of Knowledge," in *Fieldwork and Footnotes*, pp. 19–38 (esp. 22–23).

17. Lisbet Koerner, "Daedalus Hyperboreus: Baltic Natural History and Mineralogy in the Enlightenment," in *The Sciences in Enlightened Europe*, ed. William Clark, Jan Golinski, and Simon Schaffer (Chicago: University of Chicago Press, 1999), pp. 389–422 (esp. 414).

18. Katherine Faull, "Introduction" to *Anthropology and the German Enlightenment: Perspectives on Humanity*, ed. K. Faull (Lewisburg, Pa.: Bucknell University Press, 1995), pp. 11–19 (esp. 12).

19. J. R. Forster, "Specimen Historiae Naturalis Volgensis," *Philosophical Transactions of the Royal Society* 57 (1767): 312–57.

20. Z. Fedeorowicz, "Zoology in Danzig in the XVII and XVIII Centuries," *Memorabilia Zoologica* 19 (1968): 187–93 (English translation from the Polish at Alexander Turnbull Library, Wellington, MS 1486).

21. Ruth Dawson, "Collecting with Cook: The Forsters and their Artifact Sales," *Hawaiian Journal of History* 13 (1979): 5–16.

22. Ilse Jahn, "Forster (1729–1798) und die Konzeption einer 'Allgemeinen Naturgeschichte'—Zum Gendenken an seiner 200. Todestag," *Sitzungsberichte der Gesellschafte Naturforschender Freunde zu Berlin* 37 (1998): 1–12 (esp. 2).

23. Michael Hoare (ed.), *The* Resolution *Journal of Forster*, 4 vols. (London: Hakluyt Society, 1982), I, p. 107.

24. G. Forster, *Werke; Sämtliche Schriften, Tagebücher, Briefe*, ed. Akademie der Wissenschaften, 18 vols. (Berlin: Akademie Verlag, 1958–2003), XIV, p. 299; G. Forster to [Countess] Maria Wilhelmine von Thun (French), March 6, 1785.

25. G. Forster, *A Voyage Round the World in His Britannic Majesty's Sloop, Resolution*, ed. Nicholas Thomas and Oliver Berghof, 2 vols. (Honolulu: University of Hawai'i Press, 2000), I, p. 9.

26. J. R. Forster, *Observations Made During a Voyage Round the World*, ed. Nicholas Thomas, Harriet Guest, and Michael Dettelbach (Honolulu: University of Hawai'i Press, 1996), p. 9.

27. In his "Der Brodbaum" ("The Breadfruit Tree") of 1784. Thomas Saine, *Georg Forster* (New York: Twayne Publishers, 1972), p. 30.

28. Erwin H. Ackerknecht, "George Forster, Alexander von Humboldt, and Ethnology," *Isis* 46 (1955): 83–95 (esp. 85).

29. John Lyon and Phillip Sloan, *From Natural History to the History of Nature: Readings from Buffon and His Critics* (Notre Dame, Ind.: Notre Dame University Press, 1981), p. 121.

30. Muséum National d'Histoire Naturelle, MS 189, "Catalogue d'un Herbier rapporté des Isles de la Mer Australe par M. Forster et cédé à M. le Comte de Buffon pour le Cabinet du Roy, en November 1777."

31. L. C. Rookmaker, "The Zoological Notes by Johann Reinhold and Georg Forster Included in Buffon's *Histoire Naturelle* (1782)," *Archives of Natural History* 12 (1985): 203–12.

32. Robert Wokler, "From L'Homme Physique to L'Homme Morale and Back: Towards a History of Enlightenment Anthropology," *History of the Human Sciences* 6 (1993): 121–38.

33. J. Forster, "Lectures on Natural History and Especially on Mineralogy . . . Begun to Be Read and Composed 1767 and 1768" (Berlin: Staatsbibliothek), MS germ. oct. 22a, folios 3–4, 9–12, 18.

34. J. R. Forster, "Lectures on Entomology: Drawn Up 1786 and Delivered from Jan.–June 1769" (Berlin: Staatsbibliothek), MS germ. oct. 21, folios 3–4.

35. J. R. Forster to D. Barrington, Peabody Museum, Salem, Mass., December 12, 1768.

36. Charles Withers, "Geography, Natural History and the 18th Century Enlightenment: Putting the World in Place," *History Workshop Journal* 39 (1995): 137–63.

37. J. R. Forster, "Nota rélativement aux Curiosités Artificielles, qu'on a rapports de la Mer du Sud" (Wellington, MS: Alexander Turnbull Library), 3497 (printed in M. Hoare, *The* Resolution *Journal*, IV, pp. 780–82).

38. J. R. Forster, "Anmerkungen über die beste Methode mit Nutzen zu Reisen" (Berlin: Staatsbibliothek), MS germ. quart. 246.

39. Michael Hoare, *The Tactless Philosopher: Johann Reinhold Forster (1729–98)* (Melbourne: Hawthorn Press, 1976), p. 76.

40. J. Forster, *Observations*, p. 185.

41. For example, the "surprisingly accurate" vocabulary list recorded in Balad. J. Hollyman and A. G. Huadricourt, "The New Caledonian Vocabularies of Cook and the Forsters," *Journal of the Polynesian Society* 69 (1980): 215–27 (esp. 226).

42. Karl Rensch, *The Language of the Noble Savage: The Linguistic Fieldwork of Reinhold and George Forster in Polynesia on Cook's Second Voyage to the Pacific 1772–1775* (Canberra: Archipelago Press, 2000), pp. v, 23, 77–80.

43. J. R. Forster, "Vocabularies of the Languages Spoken in the Isles of the South Sea . . . 1774" (Berlin: Staatsbibliothek), MS or. oct. 62, folio 4.

44. Janet Browne, *The Secular Ark: Studies in the History of Biogeography* (New Haven: Yale, 1983), p. 38.

45. Britta Rupp-Eisenreich, "Aux 'Origines' de la 'Völkerkunde' Allemande: De la "Statistik" à l' 'Anthropologie' de Georg Forster," in *Histoires de L'Anthropologie (XVIe–XIXe Siècles)*, ed. Britta Rupp-Eisenreich (Paris: Klincksieck, 1984), pp. 89–115 (esp. 104). On the extent to which Forster's concern with a "total" local history embodied an interest in the forces that shaped history as a whole, see Jörn Garber, "Anthropologie und Geschichte," in *Georg Forster in Interdisziplinärer Perspektive*, ed. Claus-Volker Klenke (Berlin: Akademie Verlag, 1994), pp. 193–210 (esp. 203).

46. Hoare, *The Tactless Philosopher*, p. 143. See, for example, George Forster's critical remarks on those who would explain the character of a society solely in terms of climate in his description of New Caledonia. G. Forster, *Observations*, II, p. 592.

47. Hans Böedeker, "Aufklärerische Ethnolologische Praxis: Johann Reinhold und Georg Forster," in *Wissenschaft als Kulturelle Praxis 1750–1900*, ed. Hans Böedeker (Göttingen: Vandenhoeck and Ruprecht, 1999), pp. 227–53 (esp. 236).

48. Adrienne Kaeppler, "Die Ethnographischer Sammlungen der Fosters aus dem Südpazifik. Klassische Empirie im Dienste der Modernen Ethnologie," in *Georg Forster in Interdisziplinärer*, pp. 59–75.

49. J. R. Forster, *Observations*, pp. 9–10.

50. G. Forster, *Voyage*, p. 6.

51. "Noch Etwas," in G. Forster, *Werke*, VIII, pp. 130, 131, 137.

52. Ibid., p. 141; Thomas Strack, "Philosophical Anthropology on the Eve of Biological Determinism: Immanuel Kant and Georg Forster on the Moral Qualities and Biological Characteristics of the Human Race," *Central European History* 29 (1996): 285–308 (esp. 302).

53. Second edition of 1785, entitled *Über die körperliche Verschiedenheit des Negers vom Europäer (On the Bodily Difference Between Negroes and Europeans)*.

54. "Noch Etwas," in G. Forster, *Werke*, VIII, pp. 143, 153.

55. Ibid., pp. 153–54.

56. Rupp-Eisenreich, "Aux 'Origines,'" p. 103.

57. Saine, *Georg Forster*, p. 35.

58. J. R. Forster, *Observations*, p. 172.

59. Ruth Dawson, *Georg Forster's Reise um die Welt: A Travelogue in Its Eighteenth-Century Context*, University of Michigan PhD, 1973, p. 86.

60. G. Forster, *Voyage*, I, p. 171.

61. Ibid., I, p. 256.

62. Ibid., I, pp. 400–401.

63. J. R. Forster, *Observations*, p. 329.

64. Hoare, *The Tactless Philosopher*, p. 14.

65. Hoare, Resolution *Journal*, IV, p. 613.

66. J. R. Forster, *Observations*, p. 339.

67. Hoare, Resolution *Journal*, III, p. 395.

68. Paul Hazard, *European Thought in the Eighteenth Century* (Harmondsworth: Penguin, 1965), pp. 51–52, 80–81.

69. Hoare, Resolution *Journal*, I, p. 278.

70. J. R. Forster, "On the Human Species in the Isles of the South Seas" (Berlin: Staatsbibliothek), MS germ. oct. 79, folio 3.

71. G. Forster, *A Voyage*, I, p. 200.

72. Ulrich Kronauer, "Rousseaus Kulturkritik aus der Sicht Georg Forster," in *Georg Forster in Interdisziplinärer*, pp. 147–56; Rupp-Eisenreich, "Aux 'Origines,'" p. 103. On George Forster and Rousseau, see also Eberhard Berg, *Zwischen den Welten. Anthropologie der Aufklärung und das Werk Georg Forsters* (Berlin: Reiner Verlag, 1982).

73. G. Forster, *Voyage*, II, p. 481.

74. Muséum National d'Histoire Naturelle, MS 186 (Foster MSS, section on "sago").

75. H. West, "The Limits of Enlightenment Anthropology: Georg Forster and the Tahitian," *History of European Ideas* 10 (1989): 147–60.

76. J. Forster, *Observations*, pp. 205, 195–96.

77. Ibid., pp. lxxviii, 196, 235.

364

78. G. Forster, *Voyage*, II, p. 684.

79. G. Forster, *Werke*, VIII, pp. 65–68 (French).

80. Ackerknecht, "Forster," p. 84.

81. G. Forster, *Werke*, XIV, p. 599; G. Forster to Sömmering, Dec. 1786.

82. I. Jahn, "Scientia Naturae—Naturbetrachtung oder Naturwissenschaft? Georg Forsters Erkenntnisfragen zu biologischen Phänomenon in Vorlesungs-Manuskripten aus Wilna und Mainz," in *Georg Forster in Interdisziplinärer*, pp. 159–77 (esp. 166).

83. Muséum National d'Histoire Naturelle, MS 188, *Rudimenta Scientiae Naturalis* (1786), fol. 7.

84. Timothy Lenoir, "Kant, Blumenbach, and Vital Materialism in German Biology," *Isis* 71 (1980): 77–108 (esp. 86).

85. Stefano Bertoletti, "The Anthropological Theory of Johann Blumenbach," in *Romanticism in Science: Science in Europe, 1790–1840*, ed. Stefano Poggi and Maurizio Bossi (Dordrecht: Kluwer, 1994), pp. 103–25 (esp. 110).

86. Niedersächisische Staats—und Universitätsbibliothek [Göttingen], Blumenbach MSS, IV.

87. J. Forster, *Observations*, p. 1.

88. Jahn, "Forster," p. 6.

89. Hoare, *Tactless Philosopher*, p. 311.

90. Frost, "The Pacific Ocean," pp. 35–36.

91. H. Plischke, "Die Malayaische Varietät Bluemenbachs," *Zeitschrift für Rassenkunde* 8 (1938): 225–31 (esp. 229).

92. Hoare, *Tactless Philosopher*, pp. 317–18, 327.

93. J. Forster, *History of the Voyages and Discoveries in the North*, translated into English (London: G. G. J. Robinson & J. Robinson, 1786), p. xiv.

94. G. Forster, "Neuholland und die brittishche Colonie in Botany-Bay," in his *Werke*, V, pp. 163, 174, 176, 161–62. Cited from the English translation by Robert King, Australian National Library, SR 909.7/S768, pp. 1, 7–8, 10–11.

95. G. Forster, "Cook der Entdecker," in Forster, *Werke*, V, pp. 191–302. Cited from the English translation by P. Klarwill, Alexander Turbull Library, Wellington, MS—Papers—1485, pp. 1, 5, 8, 20, 83, 86–89. M. E. Hoare, "Cook the Discoverer": An Essay by Georg Forster," *Records of the Australian Academy of Science* 1 (1969): 7–16 provides a summary of this important essay.

96. G. Craig, "Engagement and Neutrality in Germany: The Case of Georg Forster, 1754–94," *Journal of Modern History* 41 (1969): 1–16 (esp. 4, 12–13).

97. Saine, *Georg Forster*, p. 134.

98. G. Forster, *A Letter to the Right Honourable, The Earl of Sandwich*, in G. Forster, *Voyage*, II, pp. 790, 799.

99. British Library, Add. MS 8098, fol. 436, Blumenbach to Banks, Jan. 19, 1799.

100. Ibid., fol. 486, Wilhelmina Sprengel to Banks, June 4, 1799.

101. Hoare, *Tactless Philosopher*, p. 310.

102. J. S. Gordan, *Reinhold and Georg Forster in England, 1766–1780*, PhD thesis, Duke University, 1975, p. 62.

103. Forster, *Werke*, XIV, pp. 642–43; G. Forster to Pennant, March 5, 1787.

104. Urban, "European Discovery," p. 55; Manfred Urban, "The Acquisition History of the Göttingen Collection," pp. 60–70 (both in Hauser-Schäublin and Krüger, *James Cook*).

105. Eva Raabe, Eberhard Schlesier, and Manfred Urban (eds.), *Verzeichnis der Völkergundlichen Sammlung des Instituts für Völkerkunde der Georg-August-Universität zu Göttingen,* Teil 1: *Abteilung Ozeanien* (Göttingen: Das Institut, 1988), p. 49.

106. Urban, "Acquisition History," pp. 70–71.

107. Urban, "European Discovery," p. 53.

108. Included in Klaus-Georg Popp (ed.), *Cook der Endecker* [Schriften über James Cook] (Leipzig: Reclam, 1980) along with G. Forster's writings on Cook.

109. Forster, *Werke*, XIII, p. 291; Forster to Banks, May 27, 1780.

110. Robert Leventhal, "The Emergence of Philological Discourse in the German States," *Isis* 77 (1986): 243–60 (esp. 250).

111. Urban, "European Discovery," p. 53.

112. Niedersächisische Staats—und Universitätsbibliothek, Cod. Mich. 327, fol. 229, Pringle to Michaelis, May 6, 1766; folios 239–40, June 1769; fol. 260, Sept. 23, 1771; fol. 314, July 9, 1774; fol. 319, Dec. 30, 1774; fol. 324, July 18, 1775; fol. 335, May 16, 1777.

113. Paul Raabe and Wilhelm Schmidt-Biggemann (eds.), *Enlightenment in Germany* (Bonn: Hohwacht Verlag, 1979), pp. 175–76.

114. John Gascoigne, "Blumenbach, Banks, and the Beginnings of Anthropology at Göttingen," in *Göttingen and the Development of the Natural Sciences,* ed. Nicolaas Rupke (Göttingen: Wallstein Verlag, 2002), pp. 86–98; and *Joseph Banks and the English Enlightenment: Useful Knowledge and Polite Culture* (Cambridge: Cambridge University Press, 1994), pp. 149–59.

115. Niedersächisische Staats—und Universitätsbibliothek, Blumenbach MSS, VI: 12, Andrew Halliday to Dr. Clarke, Jan. 27, 1821.

116. Ibid., V: 20, Isaac Buxton to Blumenbach, n.d. [ca. 1820].

117. Ibid., I, iii: 14, 35; V: 42 (includes John Bigge to Blumenbach, April 16, 1822).

118. Ibid., V: 43, William Davison to Blumenbach, Dec. 6, 1822.

119. Ibid., V: 45, Alex Caldeburgh to Blumenbach, August 30, 1825.

120. Ibid., V: 58, Thomas Traill to Blumenbach, May 12, 1829.

121. Ibid., V: vi, John Crawford to Charles König, May 3, 1832.

122. Ibid., V: 28, Benjamin Barton to Blumenbach August 12, 1796. An undated letter (V: 15 from Michaelis to Mrs. Blumenbach [French]) refers to having "received the head of an Indian chief from a very distinguished anatomist from Philadelphia."

123. Ibid., V: 40, John Sullivan to Blumenbach, October 4, 1824.

124. Ibid., VI, Edward Everett to Blumenbach, November 9, 1835.

125. Georgette Legée, "Johann Reinhold Friedrich Blumenbach, 1752–1840. La Naissance de L'Anthropologie à L'époque de la Révolution Française," *Histoire et Nature* 28–29 (1987–88): 23–45 (esp. 30).

126. British Library, Add. MS 8098, Blumenbach to Banks, May 8, 1792.

127. Niedersächisische Staats—und Universitätsbibliothek, Blumenbach MSS, V: 34; F. P. to Blumenbach, March 24, 1826.

128. Urban, "Acquisition History," p. 74.

129. Legée, "Blumenbach," p. 33.

130. Urban, "Acquisition History," p. 74.

131. Johann von Herder, *Outlines of a Philosophy of the History of Man*, English trans. T. O. Churchill (London: Printed for J. Johnson by Luke Hansard, 1800), p. 153.

132. Herder, *Reflections*, edited by Frank Manuel using the Churchill translation (Chicago: University of Chicago Press, 1968), pp. 4, 6, 51, 62.

133. Immanuel Kant, *Anthropology from a Pragmatic Point of View*, trans. and ed. Mary Gregor (The Hague: Martinus Nijhoff, 1974), p. 4.

134. Emmanuel Eze, "The Color of Reason: The Idea of 'Race' in Kant's Anthropology," in *Anthropology and the German Enlightenment*, pp. 200–41 (esp. 228).

135. Strack, "Philosophical Anthropology," p. 293.

136. Emmanuel Eze, *Race and the Enlightenment: A Reader* (Oxford: Blackwell, 1997), p. 70.

137. Hoare, *Forster's* Resolution *Journal*, I, p. 114.

138. Michael Dettelbach, "Global Physics and Aesthetic Empire: Humboldt's Physical Portrait of the Tropics," in *Visions of Empire: Voyages, Botany, and Representations of Nature* (Cambridge: Cambridge University Press, 1996), pp. 258–92 (esp. 266).

139. Ackerkneckt, "George Forster," p. 85.

140. Ludwig Uhlig, "Georg Forsters Horizont," in *Georg Forster in Interdisziplinärer Perspektive*, pp. 3–14 (esp. 11).

141. Alexander von Humboldt, *Cosmos: A Sketch of a Physical Description of the Universe*, trans. C. E. Otté, 4 vols. (London: Henry Bohn, 1849–52), I, pp. 362, 365.

For a recent work which includes an illuminating discussion of the German response to the Pacific, see Harry Liebersohn, *The Travelers' World: Europe and the Pacific* (Cambridge, Mass.: Harvard University Press, 2006), which appeared while this chapter was in press.

X

The Royal Society, natural history and the peoples of the 'New World(s)', 1660–1800

Suffused with the high expectations that accompanied the foundation of the Royal Society, its second charter of April 1663 proclaimed in the name of the king that: 'We have long and fully resolved with Ourself to extend not only the boundaries of Empire, but also the very arts and sciences.'[1] For the English, as for other European imperial powers, the widening sway of seaborne power did indeed converge with the expansion of the sciences since, properly to possess new territories, one needed to catalogue their products and their peoples.[2] The quest for suitable goods for trade in an increasingly globalized world was, as Cook has recently argued, a catalyst for such scientific values as accurate recording of data.[3] Such a preoccupation was the domain of natural history, a form of knowledge to which the Royal Society was particularly committed.

Its commitment to natural history owed much to the eloquent claims made by Francis Bacon, the Royal Society's philosophical mentor, for the possibilities that such a form

Earlier versions of this article were presented at the Sydney–Dunedin Conference on Early Modern Thought, Sydney, June 2008, and the Three Societies' History of Science Conference, Oxford, July 2008. I have benefited considerably from the helpful comments of Prof. Peter Anstey, University of Otago, and the anonymous *BJHS* referees. I am grateful for financial support for this work from the Australian Research Council and the Faculty of Arts and Social Sciences, University of New South Wales.

1 H. Lyons, *The Royal Society 1660–1940: A History of Its Administration under Its Charters*, Cambridge, 1944, 28.

2 P. H. Smith and P. Findlen, 'Commerce and the representation of nature in art and science', in *Merchants and Marvels: Commerce, Science and Art in Early Modern Europe* (ed. P. H. Smith and P. Findlen), New York, 2002, 1–25, 18; and *idem*, 'Local herbs, global medicines: commerce, knowledge, and commodities in Spanish America', in ibid., 163–81, 165.

3 H. J. Cook, *Matters of Exchange: Commerce, Medicine, and Science in the Dutch Golden Age*, New Haven, 2007, 46.

of knowledge opened up. For Bacon saw natural history as the bedrock for a new form of natural philosophy which would undermine the speculations of the Schools. As Bacon acknowledged, natural history was one form of the broader category of history which he linked with the faculty of memory. History in this sense was simply a form of description and, as Bacon wrote in his *Description of the Intellectual Globe*, 'History is either *Natural* or *Civil*. Natural history relates the deeds and actions of nature; civil history those of men.'[4] But even Bacon's own practice indicated that the boundary between natural and civil history was a wavy and uncertain one. The view that the human realm should be confined to civil history ran in the face of the fact that, increasingly, one of the major forms that natural history took was that of travellers' accounts or works by those such as the Spanish who had systematically studied the flora, fauna and human populations of lands that European expansion had brought under their view.[5]

One of the most notable of such post-Columbian works was *The Naturall and Morall Historie of the East and West Indies* (1590; English translation 1604) by the Spanish Jesuit Joseph Acosta, which Bacon cited in both the *New Organon* and the *History of Winds*, two works which formed parts of his *Great Instauration* of 1620.[6] As his title suggests, Acosta's work merged the realm of nature and of humankind and Bacon himself went some way towards doing the same in his attempt at a model natural history. In his *Sylva Sylvarum*, among the many tedious details about the behaviour of nature, he turns to some lurid accounts of the way in which 'the cannibals in the West Indies eat man's flesh; and the West Indies were full of the pocks when they were first discovered'.[7] This merging of the human and the natural worlds accorded with Bacon's view in the *Advancement of Learning* that there were forms of history, notably cosmography, which were 'manifoldly mixt … being compounded of Naturall history, in respect of the Regions themselves, of History civill, in respect of the Habitations, Regiments and Manners of the people'. Tellingly, this comment was followed by an acknowledgement of the significance of recent global exploration for 'the furder proficience, and augmentation of all Scyences, because it may seeme they are ordained by God to be *Coevalls*, that is, to meete in one Age'.[8] This was an indication of the extent to which travel and natural history were closely intertwined in what Bacon saw as this providential conjunction of both science and empire. There was, however, less coincidence than Bacon allowed. As Richard Drayton has argued, both European global expansion and the quest to understand nature better drew on common theological roots

4 G. Rees (ed.), *The Oxford Francis Bacon*, Vol. VI: *Philosophical Studies c.1611–c.1619*, Oxford, 1996, 99.

5 B. Shapiro, 'History and natural history in sixteenth- and seventeenth-century England: an essay on the relationship between humanism and science', in *English Scientific Virtuosi in the 16th and 17th Centuries* (ed. B. Shapiro and R. Frank), Los Angeles, 1979, 1–55, 7, 18.

6 J. Spedding, R. L. Ellis and D. D. Heath (eds), *The Works of Francis Bacon*, 14 vols., London, 1857–74, v, 152; G. Rees (ed.), *The Oxford Francis Bacon*, Vol. XI: *The Instauratio Magna, Part II, Novum Organum and Associated Texts*, Oxford, 2004, 321.

7 Spedding, Ellis and Heath, op. cit. (6), ii, 348.

8 M. Kiernan (ed.), *The Oxford Francis Bacon*, Vol. IV: *The Advancement of Learning*, Oxford, 2000, 70, 71.

and a belief shared by Bacon in the directing hand of Providence.[9] For Bacon, the study of nature included the study of man. Hence, when outlining the full extent of natural history in his *Parasceve* (*The Preparative towards a Natural and Experimental History*) which formed a part of his *Great Instauration*, Bacon at least gestured towards the need to include the world of humankind. In sketching in rather summary form the task of the natural historian, he included attention to the physical characteristics of human beings along with 'the way these things vary with race and climate'. The programme extended to the more cognitive aspects of humanity with histories of 'the intellectual faculties'.[10]

The early Royal Society went much further in securely including the human world within the remit of the natural historian. It formed, for example, part of the agenda outlined in Robert Boyle's 1666 'General Heads for a Natural History of a Country, Great or Small' – a work that was based on part of Bacon's *Parasceve* but which accorded the study of humankind a much more explicit and conspicuous place than had Bacon.[11] '*Secondly*', wrote Boyle, 'above the ignoble *Productions* of the Earth, there must be a careful account given of the *Inhabitants* themselves, both *Natives* and *Strangers*'.[12] An enthusiastic though critical reader of travellers' accounts, John Locke saw such material as contributing to the study of human nature, which he regarded as part of natural history.[13] As Carey points out, Henry Oldenburg wavered somewhat in the encouragement he gave to correspondents reporting on the natural history of humanity. But, as he put it in the preface to the eleventh volume of the *Philosophical Transactions*, he did aspire to 'making the fullest discovery of Mankind, as Man is the Microcosme'. Hence he urged the need to bring 'under one view, the shapes, features, statures, and all outward appearances, and also the intrinsick mentals or intellectuals of Mankind'.[14]

Such an impulse to study the human world could take either local or global forms, both of which promoted habits of empirical investigation and accurate recording.[15] The Baconian concern for precise description prompted the close study of British localities and their antiquities for which Robert Plot and other Royal Society practitioners of chorography were renowned, but it also helped stimulate the study of more distant societies.[16] This interest in the peoples of the new worlds helps account for the early Royal Society's interest in accounts of voyages to little-known quarters of the earth. Robert Hooke contributed an enthusiastic preface to the *Historical Relation of the Island Ceylon* (1681) by the East India Company captain Robert Knox, a work that was

9 R. Drayton, 'Knowledge and empire', in *The Oxford History of the British Empire*, Vol. II: *The Eighteenth Century* (ed. P. Marshall), Oxford, 1998, 231–52, 233.

10 Rees, op. cit. (6), 479, 481.

11 D. Carey 'Compiling nature's history: travellers and travel narratives in the early Royal Society', *Annals of Science* (1997), 54, 269–92, 273. On the background to this document see M. Hunter, 'Robert Boyle and the early Royal Society: a reciprocal exchange in the making of Baconian science', *BJHS* (2007), 40, 1–23.

12 R. Boyle, 'General heads ... ', *Philosophical Transactions* (1665–6), 1, 186–9, 188.

13 D. Carey, 'Locke, travel literature, and the natural history of man', *Seventeenth Century* (1986), 7, 259–80.

14 Carey, op. cit. (13), 269.

15 C. Withers, *Geography, Science and Natural Identity: Scotland since 1520*, Cambridge, 2001, 37–9.

16 S. Mendyk, '*Speculum Britanniae*': *Regional Study, Antiquarianism, and Science in Britain to 1700*, Toronto, 1989, 165.

also endorsed by Christopher Wren. As the title suggests, this was a text very much devoted to the human as well as the natural history of Ceylon, even though much of its appeal lay in its being something of an adventure story. Hooke nonetheless saw this book as an example of that programme of publication of seamen's accounts which the Royal Society sought to promote. By so doing, the extent of natural history would be widened, for, as Hooke wrote, 'How much of the present Knowledge of the Parts of the World is owing to late Discoveries may be judged by comparing the Modern with the Ancients' Account thereof.'[17]

Robert Southwell, president of the Royal Society from 1690 to 1695, was a close observer of the voyages of William Dampier's epic circumnavigatory voyage of 1679–91 and it was probably Southwell who encouraged John Woodward, a prominent member of the society, to produce in 1696 his *Brief Instructions for Making Observations and Collecting in All Parts of the World in Order to Promote Natural History ... presented to the Royal Society*.[18] Interestingly, the full title refers to the 'Advancement of Knowledg both Natural and Civil', an allusion to Bacon's great work and to the need to draw together human and natural history. Its appendix provides a template for the description of the indigenous peoples across much of the globe, including Africa, the East and West Indies and '*other remote, and uncivilized, or Pagan Countries*'. Woodward urged detailed description of these peoples' anatomy, along with 'their *Tempers, Genius's, Inclinations, Virtues*, and *Vices*'. Religion loomed large with the stipulation that the traveller should 'Enquire into their *Traditions* concerning the *Creation of the World*, the *universal Deluge*' and ascertain 'their *Notions* touching the *Supreme God, Angels*, or other inferior Ministers'. Such concerns were later to be reflected in Woodward's correspondence with the New England divine Cotton Mather on the ways in which the study of the natural history of the North American continent confirmed the biblical narrative. Archaeological evidence of what appeared to be huge antediluvian humans, for example, was used to confirm the references to giants in Genesis.[19] Woodward's attempt to understand the place of America in the history of the globe and, in particular, in the Noachian Flood extended to the attempt to obtain 'curiosities' of artefacts from the country, a quest which was in keeping with his antiquarian and archaeological pursuits within the British Isles.[20] Woodward's *Brief Instructions* also urged the need for a detailed description of different peoples' customs, such as the basic rites of passage, their mode of computing time and their forms of government and law.[21] Overall, it was a work that took a very compendious view of the

17 R. Knox, *Historical Relation of the Island Ceylon, in the East Indies ...*, London, 1681, p. [i].

18 D. Preston and M. Preston, *A Pirate of Exquisite Mind: The Life of William Dampier, Explorer, Naturalist and Buccaneer*, London, 2005, 325. On Woodward's work see C. Withers, 'Geography, natural history and the eighteenth-century enlightenment: putting the world in place', *History Workshop Journal* (1995), **39**, 137–63, 144–5.

19 Royal Society Archives, EL/M2/34, summaries of several letters from Cotton Mather to John Woodward and to Richard Waller, 1712–. A *Philosophical Transactions* article based on this correspondence ((1714–16), **29**, 62–71) includes material by Mather on American Indian ethnology.

20 Joseph Levine, *Dr Woodward's Shield: History, Science and Satire in Augustan England*, Berkeley, 1997, 99; see also 73 on the correspondence with Mather.

21 John Woodward, *Brief Instructions ...*, London, 1698, 8–9.

domain of natural history and accorded the study of humankind a prominent place within it.

The travels of Dampier and others were also accorded reviews within the pages of the *Philosophical Transactions*. This again illustrated the extent to which the Royal Society saw the human world as forming part of its terrain. Hooke's expansive seven-page review of Dampier's *An Account of a New Voyage round the World* (1697) praised the information contained therein, including the description of 'the Natives, their Shapes, Manners, Customs, Clothing, Diet, Art, & c.', as being 'very Curious, Remarkable and New'.[22] The previous volume of the *Philosophical Transactions* had contained a review of an account of the recent voyages by John Narborough into the South Seas in 1669–71, along with earlier voyages of figures such as the seventeenth-century Dutch captain Abel Tasman. The material such voyage accounts provided was described as 'contribut[ing] to the enlarging of the Mind and Empire of Man, too much confin'd to the narrow *Spheres* of particular *Countries*'. For such works, continued the review, with an allusion to the importance of what the following century would term 'the Science of Man', helped provide 'a large Prospect of *Nature* and *Custom*'.[23] A 1698 review of a nine-year voyage to the East Indies and Persia by John Fryer FRS similarly valued not only the work's 'Account of the Nature and Products of the Countries themselves' but also that 'of the Men that inhabit each, their Shape, their Genius, Manners, Customs, Laws'.[24]

Bacon had envisaged that the acquisition of the materials for a true natural history would, as he wrote in his *Parasceve*, mean 'that they ought to be sought out and gathered in (as if by agents and merchants) from all sides'.[25] In this, as in much else, the early Royal Society followed the Baconian lead, viewing travel as central to the promotion of its enterprise: hence Oldenburg's goal of 'appoint[ing] philosophical ambassadors to travel throughout the world to search and report on the works and productions of nature and art' on which it would be possible 'to compose in Time a Natural and Artificial History which will be perfect'.[26] Boyle quite explicitly drew on 'Navigators, Travellers & c.' in compiling 'the Particulars admitted into the Natural History'.[27]

Travellers' accounts thus naturally merged with the promotion of natural history as the early Royal Society understood it. The prominence of such accounts in the early proceedings of the society is one of the reasons why a considerable amount of attention was devoted to the study of the human world. But though travellers' tales might be diverting and were read in quantity by those, such as Locke, engaged in the development of what in the age of the Enlightenment was termed the 'Science of Man', they

22 [Review of] *An Account ...*, *Philosophical Transactions* (1695–7), **19**, 426–433, 428–9. Preston and Preston, op. cit. (18), 330, attribute this review to Hooke.

23 [Review of] *An Account ...*, *Philosophical Transactions* (1694), **18**, 166–8, 167.

24 John Fryer, 'An abstract ... ', *Philosophical Transactions* (1698), **20**, 338–48, 338.

25 Rees, op. cit. (6), 451.

26 Oldenburg to Sorbière, 3 January 1663/4, in A. Rupert Hall and M. Boas Hall (eds.), *The Correspondence of Henry Oldenburg*, 13 vols., Madison, WI and London, 1965–86, ii, 144.

27 Boyle to Oldenburg, 13 June 1666, in M. Hunter, A. Clericuzio and L. Principe (eds.), *The Correspondence of Robert Boyle, 1636–1691*, 6 vols., London, 1999–2000, iii, 171.

were also notoriously unreliable.[28] Over time, with increasing emphasis on experiment, the methods of the Royal Society reduced the attention paid to travel accounts.[29] The growing focus on experimentation also brought with it an increasing stress on the need to witness the process by which scientific information was produced.[30] This in turn also prompted uneasiness about material which was not based on first-hand reporting by a figure with professional competence and, preferably, one who was known to and could be questioned by members of the Royal Society. Accordingly, too, the study of the human world also declined, though the pages of the *Philosophical Transactions* continued to carry descriptions of non-European societies throughout the eighteenth century.

For the early Royal Society, however, the need to embark on a form of natural history that would achieve the goals laid out by Bacon as a form of knowledge which broke with scholastics' logic-chopping made travellers' accounts too valuable a source to be lightly dismissed.[31] The methods of the Baconian natural historian and those of the traveller were too close to be able to discount the information received from travellers around the globe in too drastic a manner.[32] For the promotion of natural history was one of the things that bound the early and rather embattled society together, even if, as Paul Wood has pointed out, Fellows could differ about what they meant by the Baconian programme to which they all outwardly subscribed.[33] As Boyle mused, natural history was so dependent on travel accounts that one had to be willing to accept that 'many things must be taken upon trust in the History of Nature, as matters of fact Extraordinary ... such as are not to be examin'd but in remote Countrys'.[34]

The early society did, however, do what it could to check the veracity of some of the more far-fetched reports from foreign lands. In 1671 it set out to determine the truth of earlier reports about Brazil by sending a lengthy set of queries to a Jesuit there with questions such as whether 'fiery flying dragons appear frequently' and, more plausibly, 'Are the older Brazilians excellent botanists, able with ease to prepare every kind of medicine from materials gathered in all places?' The abiding Western preoccupation with cannibalism was evident in the query 'Is it true that, moved by affection, they seize

28 Carey, op. cit. (13), 259–80.

29 M. Boas Hall, *Promoting Experimental Learning: Experiment and the Royal Society 1660–1727*, Cambridge, 1991; and P. Anstey, 'Experimental versus speculative natural philosophy', in *The Science of Nature in the Seventeenth Century: Patterns of Change in Early Modern Natural Philosophy* (ed. P. Anstey and J. Schuster), Dordrecht, 2005, 215–42, 220.

30 P. Fontes da Costa, 'The making of extraordinary facts: authentication of singularities of nature at the Royal Society of London in the first half of the eighteenth century', *Studies in the History and Philosophy of Science* (2002), 33, 265–88, 282.

31 For a recent affirmation of the importance of the Baconian programme in shaping the early Royal Society see W. T. Lynch, *Solomon's Child: Method in the Early Royal Society of London*, Stanford, 2001, which argues, at 233, that the 'Royal Society was a Baconian institution'.

32 P. Anstey, 'Locke, Bacon and natural history', *Early Science and Medicine* (2002), 7, 65–92, 85.

33 P. B. Wood, 'Methodology and apologetics: Thomas Sprat's *History of the Royal Society*', *BJHS* (1980), 13, 1–26.

34 Boyle to Oldenburg, 13 June 1666, in Hunter, Clericuzio and Principe, op. cit. (27), iii, 174.

the bodies of parents not killed by poison and having dismembered them, bury them inside themselves'.[35]

This list of queries for Brazil also attempted to secure information to test the widespread belief that other peoples often lived longer than Europeans. Hence the item 'Do many Brazilians enjoy a green old age beyond one hundred years?'[36] The Bahamas, too, were reported to be a haven of longevity. In 1668 the *Philosophical Transactions* published a report that there 'some do live to an hundred years and something upwards'.[37] The Royal Society seems to have accepted this, though it did seek further information. Hence in the following year it sent a list of queries to the Bermudas in which it was asked, 'What is conceived to be ye cause of the Longevity of ye inhabitants.'[38] Such reports of longevity, which probably reflect uncertainties in other societies about determining age, flooded in from around the world, and in the absence of any definite counterevidence they appear to have been believed in the early Royal Society. In 1682 John Evelyn repeated Walter Raleigh's report of a Virginian king who lived for over three hundred years. This prompted Hooke to add some credibility to the tale by taking 'notice of what Sir Christopher Wren had formerly acquainted the Society, that the people at Hudson's Bay commonly live up to 120 or 130 years of age; and till that age are very lusty'.[39] Boyle was, however, more sceptical, since his questioning of a Hudson's Bay Company employee elicited the information that the 'Natives never live long by reason of the badness of the waters in the country'.[40]

Establishing the worth of the accounts of travellers whose information from distant lands could not be readily checked was, then, a continuing issue for the Royal Society. As Peter Dear has argued, one of the great issues faced by the Royal Society was that of establishing criteria for what constituted credible data on which the philosopher could build.[41] In taking further Bacon's critique of the methods employed by the scholastics to arrive at accepted knowledge, the early Royal Society had to put others in place. In the new Baconian dispensation knowledge needed to be based on events and empirical findings which, in principle, could be repeated. In establishing the validity of such events and findings, much depended on the corporate response of the Royal Society to the papers it considered. Though publication in the *Philosophical Transactions* did not necessarily amount to full assent, it was at least an indication that the information should be seriously considered by the world of learning. Such papers therefore had to be considered in terms of what members of the Royal Society regarded as plausible. Steven Shapin has pointed to the importance of social position in establishing the worth of scientific claims: by definition a gentleman did not lie (at least to his peers).[42] Yet the

35 Oldenburg to Hill?, 30 August 1671, with enclosure for an unknown Jesuit in Brazil, in Hall and Hall, op. cit. (26), viii, 236, 244.

36 Oldenburg to Hill?, op. cit. (35), 236.

37 'Extracts of Three Letters … ', *Philosophical Transactions* (1668), 3, 791–6, 794.

38 Oldenburg to Hotham, 7 March 1669–70, in Hall and Hall, op. cit. (26), vi, 535.

39 T. Birch, *History of the Royal Society*, 4 vols., London, 1756–7, iv, 165.

40 Royal Society Archives, RB/1/39/9 (Boyle Papers), fol. 49.

41 P. Dear, '*Totius in verba*: rhetoric and authority in the early Royal Society', *Isis* (1985), 76, 144–61.

42 S. Shapin, *A Social History of Truth: Civility and Science in Seventeenth-Century England*, Chicago, 1994.

Royal Society also valued reports from the likes of common seamen, since they were likely to come unvarnished by philosophical speculation.[43] Over time, too, as the Royal Society came to be settled and more confident in its methods, the competence of the reporter rather than his social position became of increasing importance, as Palmira Fontes da Costa has argued.[44] This was to weaken the credibility of many travellers' tales.

In the early Royal Society, with its high hopes for the Baconian programme of natural history, however, travellers' accounts were viewed as a form of data that at least in theory could comply with scientific canons since its results were repeatable or, at least, could be checked for consistency with other travellers' accounts.[45] Hence the *Philosophical Transactions* published in 1666–7 a list of queries about India even though replies had already been received, since ' 'tis altogether necessary, to have confirmations of the truth of these things from several hands, before they be relied on'.[46] But, of course, it was not always possible to get independent verification, so the fact that the strange but indeed accurate account that in the Congo 'there are Serpents twenty five foot long, which will swallow at once a whole Sheep', reached the society through the Jesuits, and after publication in the *Journal des Sçavans*, may have added to its credibility.[47]

Interestingly, on occasions those in distant lands could be more sceptical about the strange and the novel than were members of the Royal Society. In response to an early enquiry about 'Whether Diamonds and other Precious stones grow again', the Royal Society's correspondent in Batavia curtly replied, 'Never, at least as the memory of man can attain to.'[48] Royal Society interest in the novel and the bizarre reflected Bacon's preoccupation with '*Deviating* instances ... and monstrous objects'.[49] This interest extended to the human population of foreign lands with the publication in 1668 of a report claiming that Indians of Virginia and Florida were of 'Gigantick Stature'.[50] This was a tradition which continued well into the eighteenth century. In 1767 the *Philosophical Transactions* solemnly published an account of the alleged giants of Patagonia by Charles Clarke, an officer on the Pacific expedition of John Byron.[51] On the other hand, though travellers' reports figured less in the deliberations of the

43 Dear, op. cit. (41), 156.

44 Fontes da Costa, op. cit. (30).

45 Carey, op. cit. (11), 286.

46 'Inquiries for Suratte ... ', *Philosophical Transactions* (1666–7), 2, 415–22, 419.

47 M. Angelo De Guattini, 'Observations of some animals ... ', *Philosophical Transactions* (1677–8), 12, 977–8, 978.

48 T. Sprat, *History of the Royal Society* (ed. J. I. Cope and H. Whitmore Jones, St Louis, 1959), 158.

49 L. Daston, 'Marvelous facts and miraculous evidence in early modern Europe', in J. Chandler, A. I. Davidson and H. Harootunion, *Questions of Evidence: Proof, Practice and Persuasion across the Discipline*, Chicago, 1994, 243–89, 261.

50 'Extracts of three letters ... ', *Philosophical Transactions* (1668), 3, 791–6, 795.

51 C. Clarke, 'An account of the very tall men ... ', *Philosophical Transactions* (1767), 57, 75–9. The credibility of such claims about the existence of Patagonian giants was much discussed among naturalists, and Buffon devoted a section of his 'Additions à l'article qui a pour titre, Variétés dans l'espèce humaine' to the issue. He was sceptical about the existence of real giants, though he was willing to believe the sort of claim made by Clarke that some Patagonians were nine feet tall. G. Buffon *Histoire Naturelle, Générale et Particulière, Supplément*, Vol. IV, Paris, 1777, 525.

Royal Society over the course of the eighteenth century, the tradition of promoting investigation to test the veracity of remarkable reports did continue. In 1706–7 one of the African informants of the *Philosophical Transactions* went so far as actually to inspect Hottentots to determine the truth of the report that all men had a testicle removed, and concluded that it was true only of married men.[52] In North America another correspondent in 1786 carried out first-hand observation to ascertain whether there was any truth in the reports that the Indians lacked beards and concluded that in this respect they were not different to other men, though they plucked out the hairs.[53]

More proactively, from its beginnings the Royal Society also attempted, where possible, to make the travel accounts on which it drew conform to its canons of truth by directing the information travellers sent back and the form in which it was expounded.[54] The society's determination to shape travellers' accounts into forms appropriate for the stuff of natural history was evident from the first volume of the *Philosophical Transactions*: this included detailed 'Directions for Sea-Men, Bound for Far Voyages' which prescribed a formidable inventory of observations which the Royal Society wished to see carried out. As the preamble stated, this formed part of its Baconian-inspired quest 'to study Nature rather than Books' and hence 'to compose such a History of Her [Nature], as may hereafter serve to build a Solid and Useful Philosophy'. In doing so, the society's Fellows sought to 'increase their *Philosophical* stock by the advantage, which *England* injoyes of making Voyages into all parts of the World'.[55] It was to such aspirations that Sprat referred in his grandiloquent manner when he wrote that the Royal Society's Fellows 'have begun to settle a *correspondence* through all Countreys ... that in short time, there will scarce a Ship come up the Thames, that does not make some return of Experiments, as well as of Merchandize'. In the spirit of the wording of the Second Charter quoted above, Sprat saw such a conjunction of imperial expansion with the growth of knowledge as suited to an England which was 'not only Mistress of the Ocean, but the most proper Seat, for the advancement of Knowledge'.[56]

The Royal Society thus attempted to turn England's growing imperial sway to its advantage.[57] The American colonies offered fertile ground for scientific enquiry both from the point of view of the scientific agenda of the Royal Society and in terms of the not always congruent areas of enquiry of the local population.[58] However, by the late eighteenth century the latter increasingly gave way to the former.[59] From its foundation,

52 J. Maxwell and J. Harris, 'An account of the Cape of Good Hope', *Philosophical Transactions* (1706–7), **25**, 2423–34, 2426.

53 R. McCauseland, 'Particulars relative to the nature and customs of the Indians of North-America', *Philosophical Transactions* (1786), **76**, 229–35, 230.

54 Carey, op. cit. (11), 273.

55 'Directions for Sea-Men ... ', *Philosophical Transactions* (1665–6), **1**, 140–3, 140–1.

56 Sprat, op. cit. (48), 86.

57 R. Illiffe, 'Foreign bodies: travel, empire and the early Royal Society of London, Part II. "The land of experimental knowledge"', *Canadian Journal of History* (1999), **34**, 24–50, 40.

58 R. Stearns, *Science in the British Colonies of America*, Urbana, IL, 1970.

59 S. Parrish, *American Curiosity: Cultures of Natural History in the Colonial British Atlantic World*, Chapel Hill, 2006, 315.

the Royal Society sought to enlist the help of prominent American colonists such as John Winthrop, who, as governor of Connecticut, returned to London in 1661. It was to him that Oldenburg wrote in 1664 by order of the Royal Society's Council to inform him 'that he was invited in a particular manner to take upon him the charge of being the Chief Correspondent of the Royal Society in the West, as Sir Philiberto Vernatti [of the Dutch East India Company] was in the East Indies'.[60] When Winthrop was rather tardy in fulfilling such a role Oldenburg wrote him a gently chiding letter biding him to reflect on the fact that the Royal Society sought knowledge from all around the globe: 'we have taken to taske the whole Universe, and that we were obliged to doe so by the nature of our dessein'.[61] Winthrop took the hint and sent a considerable collection of objects reflecting the natural history of New England, which, tellingly, included such human artefacts as 'some girdles of the Indian mon[e]y'. His description of these objects amounted to an early form of ethnology:

> the white they call wampampeage, the black suckalog ... the black is double the value of the white: six of the white, is a penny and three of the black, according to the Indian account ... that wch is made up in Girdles they use to lay up as their treasure.[62]

Around the same time Oldenburg also attempted to enlist the support of an English colonist in the Bahamas with a request that he provide observations 'concerning natural and artificial things', objects from the world both of nature and of human-kind.[63]

Along with such colonists, the Royal Society had high hopes, only very partially realized, of turning the chartered companies, another major arm of empire, to philo-sophical advantage. After all, from 1662, with the granting of a royal charter, the Royal Society was also a chartered entity.[64] By 1664 there was an attempt to enlist the East India Company into providing 'answeres, as may satisfy the inquiries to be sent to them'.[65] Perhaps such overtures formed part of the background to Oldenburg's enquiry to Boyle in 1666 about how best to use the willingness of the incoming governor of Bombay to be of 'service for Philosophicall purposes'. True to the Royal Society's aspirations to invest such traveller's accounts with an element of scientific rigour and direction, Oldenburg added that the governor would 'performe the more effectually, if he may receave some Instructions from you'. Boyle had to admit, however, that he knew so little about Bombay that, unusually, he could not compile the sort of list of queries of which the Royal Society was so fond.[66]

During the following century the Royal Society continued to seek the somewhat intermittent scientific cooperation of the East India Company. Johann Reinhold Forster

60 Lyons, op. cit. (1), 28.
61 Oldenburg to Winthrop, 13 October 1667, in Hall and Hall, op. cit. (26), iii, 525.
62 Winthrop to Oldenburg, 4 October 1669, in Hall and Hall, op. cit. (26), vi, 255.
63 Oldenburg to Norwood, 24 October 1666, in Hall and Hall, op. cit. (26), iii, 276.
64 M. Hunter, *Science and Society in Restoration England*, Cambridge, 1981, 36.
65 Birch, op. cit. (39), i, 457, 3 August 1664.
66 Oldenburg to Boyle, 13 March 1666; Boyle to Oldenburg, 19 March 1666, in Hunter, Clericuzio and Principe, op. cit. (27), iii, 109, 117.

FRS reflected such hopes in his preface to his 1772 translation of Bougainville's account of his voyage around the world, a translation dedicated to the president of the Royal Society. Forster urged the need for the company to send out on its ships 'men properly acquainted with mathematics, natural history, physic' who should take care 'to observe the manners, customs, learning, and religion of the various nations of the East'.[67] But such aspirations were only very partially realized. The Royal Society did, for example, secure the right in 1778 to allow Dr James Lind FRS to take scientific instruments in his voyage on a company ship to India and China. This the company welcomed as 'an opportunity of forwarding the views of the Royal Society', but it was more circumspect about acceding to the accompanying wish that Lind could request the services of company employees in India.[68]

While the Royal Society had only limited success with the vast East India Company, it had a long and fruitful relationship lasting well into the late eighteenth century with the comparatively small Hudson's Bay Company. This was perhaps in part because it was founded in 1670, not long after the Royal Society. The East India Company, by contrast, was founded in 1600 and was thus less open to new influences.[69] In 1672 Oldenburg could present to the society the responses to a series of queries entrusted to a captain in the service of the Hudson's Bay Company. Interestingly, it included considerable attention to the indigenous human population, with detailed responses to such queries as 'What kind a people the Natives are, where they are winterd?' and 'What governmt and religion they have amongst themselves?'[70] Evidently, the society gained further information about the indigenous peoples of this area, for in 1681 Christopher Wren gave an exposition to the society of their customs and some of the ways in which these had begun to change with the coming of the Europeans: 'These people used to strike fire readily with flint against flint; but have been since furnished by the English with steel, which they use with flints.'[71]

The relationship with the Hudson's Bay Company continued to bear fruit well into the late eighteenth century. One very tangible outcome was the annual arrival at the Royal Society between 1771 and 1774 of large collections of natural-history specimens from the Hudson Bay area. This so impressed the society that it proposed (apparently in vain) that application be made to 'the Directors of the East India, Turkey, Russia and Africa Companies, for the same sort of collections to be transmitted annually'.[72] The task of compiling a catalogue of the first contingent of these objects was entrusted by the Royal Society to the German polymath Johann Reinhold Forster, soon to achieve fame and some notoriety as the naturalist on board Cook's second great Pacific voyage

67 L. de Bougainville, *A Voyage Round the World. Performed by Order of the His Most Christian Majesty, in the Years 1766, 1767, 1768, and 1769* (tr. from the French by J. Reinhold Forster), London, 1772, p. viii.

68 Royal Society Archives, Misc MS3/20, East India Court of Directors to the Royal Society, 19 November 1778.

69 R. I. Ruggles, 'Governor Samuel Wegg, intelligent layman of the Royal Society, 1753–1802', *Notes and Records of the Royal Society* (1978), 32, 181–99.

70 Stearns, op. cit. (58), 705, 707.

71 Birch, op. cit. (39), ii, 92.

72 J. McClelland, *Science Reorganized: Scientific Societies in the Eighteenth Century*, New York, 1985, 303–4.

of 1772–5. Forster singled out as the basis for a paper in the *Philosophical Transactions* the Indians' practice of dyeing porcupine quills with a root which he determined to be hellebore. With a view to advantaging the company as a return for its scientific largesse, Forster recommended that 'the directors of the Hudson's-bay Company ... order larger quantities of this root from their settlements, as it will no doubt become an useful article of commerce'.[73]

If possible, then, the Royal Society attempted to turn England's imperial or commercial sway to its advantage, but, where necessary, it also turned to foreigners if they were willing to cooperate. One of its most enthusiastic early informants was, as we have seen, Sir Philiberto Vernatti, governor of Batavia, the great entrepôt of the Dutch East India Company. It was thanks to Vernatti that in 1668 the society received diverse curiosities including a poison produced by the Macassan people.[74] Surprisingly, too, the Royal Society sought an alliance with the Jesuits. The society realized that in spite of all the religious prejudices that stood between that order and Protestant England it could draw on a unique and far-flung network in providing reliable scientific information from otherwise inaccessible regions of the globe.[75] As Sprat put it, the Jesuits were permitted to 'bestow some labours about Natural Observations, for which ye have great advantages by their Travails'.[76] Much of the Royal Society's knowledge of the ancient civilizations of China and India, for example, derived from the Jesuits, along with information on little-known areas of the globe such as the Philippines.[77]

From such scattered sources the Royal Society added to its store of natural history, including information on the human world. The society did what it could to ensure accuracy and, thanks to its queries and questionnaires, some element of consistency. Not only did its informants add to its growing paper archives of information, but they also contributed to its Repository or museum, an institution which embodied the quest for the accumulation of material on which a secure natural history could be erected.[78] Museums such as the Repository that allowed for both observation and even experiment, as Ken Arnold has argued, acted as important early sites of the scientific movement.[79] The natural-history contributions to the *Philosophical Transactions* looked both inwards to the heart of England and outwards to a globe increasingly coming under European sway. So, too, the Repository included objects reflecting both local history and the larger world. This two-edged character of the Baconian collecting

---

73 J. Reinhold Forster, 'A Letter ... ', *Philosophical Transactions* (1772), **62**, 54–9, 56.

74 Birch, op. cit. (39), ii, 314.

75 C. Reilly, 'A catalogue of Jesuitica in the *Philosophical Transactions of the Royal Society of London*, 1665–1715', *Archivum Historicum Societatis Jesu* (1958), **27**, 339–62.

76 Sprat, op. cit. (48), 373.

77 P. Clain, Father Le Gobien, 'An extract of two letters from the missionary Jesuits ... ', *Philosophical Transactions* (1708–9), **22**, 189–99.

78 M. Hunter, 'The cabinet institutionalised: the Royal Society's "Repository" and its background', in *The Origins of Museums: The Cabinet of Curiosities in Sixteenth- and Seventeenth-Century Europe* (ed. O. Impey and A. MacGregor), Oxford, 1985, 159–68, 159.

79 K. Arnold, *Cabinets for the Curious: Looking back at Early English Museums*, Aldershot, 2006, 2.

impulse was reflected in John Norris's poetic tribute to the eminent Royal Society chorographer, Robert Plot:

'Tis all one
New Worlds to *find*, or nicely to describe the *Known*.[80]

Within the Repository there was a natural tendency to attract attention by high-lighting the rare and exotic, including specimens which related to the natural history of human society. The 1694 edition of Nathaniel Grew's catalogue of the society's 'Natural and Artificial [i.e. man-made] Rarities' included a very diverse range of eth-nological materials: along with 'A pot of Macassar Poyson Given by Sir Phil. Vernatti' there were, inter alia, a canoe from Greenland, 'An Indian Poyson'd Dagger', 'A Tomahauke, or Brazilian Fighting-Club', 'A West Indian Bow, Arrows and Quiver' and 'A Japan Wooden-Cup; cover'd with a Red Varnish within'. When describing the items of flora or fauna on display, Grew often also alluded to the uses to which they were put by the indigenous populations of the parts of the world whence they came. The plumes of a bird of paradise prompted the remark that such birds were worshipped as gods by the natives of the Molucca Islands. Cacaw-nuts (coconuts), he noted, on the authority of Acosta, were used as money, and the husks of Indian maize were woven by the women into 'Baskets of several fashions'. Tellingly, however, the 'Artificial Rarities' did not fit neatly into Grew's overall introductory system of classification and were relegated to the rather miscellaneous last category 'Of Coyns, and other matters relating to Antiquity'.[81] Such problems of ordering ethnological materials were to grow in the eighteenth century as systems of classification gained greater ascendancy. Since classificatory order increasingly became the hallmark of scientific standing, this tended to weaken the status of the study of humankind.

In some ways, then, ethnological items helped to make the Repository more like a traditional virtuoso cabinet of curiosities than the more chaste and considered collec-tion which greater attention to classificatory neatness would have dictated.[82] Such a highlighting of the exotic helps explain why the Royal Society's Repository became one of the late seventeenth-century London tourist sights.[83] Nonetheless, the society did try to exercise some control over the items added to the collection so that it would, at least to some degree, be in keeping with its larger goals of building up a reliable natural history. Those who were not Fellows were, as Oldenburg wrote in 1667/8, obliged to show intended donations to the president 'for fear of lodging unknowingly ballads and buffooneries in these scorching times'.[84] Censorship of the gifts of the Fellows them-selves was presumably considered too delicate a matter, even in the face of the satire of

---

80 Arnold, op. cit. (79), 110.

81 N. Grew, *Musaeum Regalis Societatis, or a Catalogue and Description of the Natural and Artificial Rarities Belonging to the Royal Society and Preserved at Gresham College ...*, London, 1694, 365–72, 56, 205, 222.

82 E. Hooper-Greenhill, *Museums and the Shaping of Knowledge*, London, 1992, 158.

83 D. Collett, 'An empire of things – exotische Objekte in "Musaeum" der Royal Society', *WerkstattGeschichte* (2006), **43**, 5–21, 20.

84 Lyons, op. cit. (1), 49.

the Royal Society to which Oldenburg here alludes. But the collection of foreign arte-facts continued to loom large in the museum, perhaps because of the public interest they inspired, and when a new Repository building was erected in 1712 a committee was formed 'to take due placing of the Curiosities'.[85] Eventually, in 1781, the society do-nated the Repository to the British Museum: a symptom of the by then increasingly half-hearted commitment to the Baconian ideal of a natural history based on descrip-tion and collection of data.[86]

It was to this Repository that Winthrop's collection of 'American Curiosities' was sent after being viewed by the king. Thus, as Oldenburg wrote to him in 1670, your 'rich Philosophical present' has 'increased the stock of their [the society's] reposi-tory'.[87] The Winthrop family tradition of largesse to the Royal Society later continued with the donation by Winthrop's grandson and namesake of 364 items from New England in the 1730s.[88] In 1671, shortly after the original Winthrop donation to the Royal Society Repository, there was also a donation by Lord Willughby 'of several curiosities from Barbadoes and other American islands'.[89] Another part of the globe was represented in the Repository when, in 1683, Captain Robert Knox, the historian of Ceylon with Royal Society connections, donated a collection of objects from the Gulf of Tonkin, Vietnam, a collection which included such examples of indigenous customs as material for consuming betel nut.[90]

The early Royal Society's project for compiling what Oldenburg referred to as a 'universal history of nature' drew into its archives and collections material relating to the human as well as the natural worlds.[91] This was prompted in part by simple curi-osity of the sort evident in descriptions of African music in 1686.[92] It was evident later, in 1775, in discussions of the nature of Tahitian musical instruments.[93] There was also a strong element of admiration for some of the achievements of the ancient cultures of China and India, such as that which prompted a paper in the *Philosophical Transactions* on India in the course of which tribute was paid to the 'most ingenious Invention of Figures by the Sagacious *Indians*' which 'can never be sufficiently enough admired'.[94] The early Royal Society was also much preoccupied with the nature and origin of Chinese character script, prompting John Wallis to make a detailed study of Chinese sources in the Bodleian Library.[95]

85 L. Jardine, 'Paper monuments and learned societies: Hooke's Royal Society Repository', in *Enlightening the British: Knowledge, Discovery and the Museum in the Eighteenth Century* (ed. R. G. W. Anderson *et al.*), London, 2004, 49–54, 50.

86 D. Miller, '"Into the valley of darkness": reflections on the Royal Society in the eighteenth century', *History of Science* (1989), 27, 155–66, 161.

87 Oldenburg to Winthrop, 26 March 1670, in Hall and Hall, op. cit. (26), vi, 594. Grew's catalogue, at 370, duly records 'Several sorts of Indian Money, called Wampampeage'.

88 D. Stimson, *Scientists and Amateurs: A History of the Royal Society*, New York, 1968, 154.

89 Birch, op. cit. (39), ii, 495.

90 Birch, op. cit. (39), iv, 226–7.

91 Oldenburg to Hevelius, 13 June 1677, in Hall and Hall, op. cit. (26), xiii, 299.

92 Birch, op. cit. (39), iv, 493.

93 J. Steele, 'Account of a musical instrument ... ', *Philosophical Transactions* (1775), 65, 67–71, 72–8.

94 J. Cope, 'Some considerations ... ', *Philosophical Transactions* (1735–6), 131–5, 131.

95 Birch, op. cit. (39), iv, 504.

In the Baconian spirit there was, however, also the strong hope that study of these novel areas of natural history would yield useful results that could contribute to 'the relief of [European] man's estate'. A continuing theme in the reports on the customs of indigenous peoples is how their skills might be emulated to the advantage of others. In doing so, the Royal Society threw a wide geographical net. Its 1666–7 list of inquiries for the East Indies included information about iron-making in Japan, along with queries about how the Chinese made and coloured their dishes and how the Chinese and Japanese made black varnish, while those for Persia in the same year sought information on 'What other Trades or Practices, besides Silk- and *Tapistry* making, they are skilled in.'[96] A report from Canada in 1685 described the method the Indians used to make maple syrup, a process again described in a report from New England in 1720–1.[97] From Canada, too, came a 1686–7 account of how to make 'several impressions on folds of a very thick bark of birch'.[98] At much the same time there was further attention to the potentially useful customs of the Americas with the publication of a paper by Sir Robert Southwell, president of the Royal Society, of 'The Method the Indians in Virginia and Carolina use to Dress Buck and Doe Skins'.[99] The benefits of studying and possibly learning from the American Indians were urged again much later in the Royal Society when Forster compiled his 1772 report on the curiosities from Hudson Bay. In this he expounded on the way in which the

> wild inhabitants of North America are certainly possessed of many important arts; which, thoroughly known, would enable the Europeans to make a better, and more extensive use of many unnoticed plants, and productions of this vast continent, both in physic, and in improving our manufactures, and erecting new branches of commerce.[100]

This willingness to learn (and also to profit) from other cultures is a marked feature of these reports from foreign lands, and contrasts with the contempt for indigenous cultures which was to be more of a feature of the high imperialism of the nineteenth century. Drawing on its Jesuit contacts, the Royal Society published in 1713 an account of India, which glowingly wrote that 'This Country furnishes Materials for Mechanic Arts and Sciences more than any Country that I know of. The Artifacts here have wonderful Skill and Dexterity'.[101] It was to the Jesuits, too, that the society turned for a detailed account of varnish-making in China.[102] Methods of dyeing attracted particular interest and admiration, as instanced by Hooke's 1686–7 account of the techniques

---

96 *Philosophical Transactions* (1666–7), **2**, 415–22, 417, 420.

97 'An account of a sort of sugar … ', *Philosophical Transactions* (1685), **15**, 988; P. Dudley, 'An account of the method of making sugar … ', *Philosophical Transactions* (1720–1), **31**, 27–8.

98 Birch, op. cit. (39), iv, 520.

99 R. Southwell, 'The method the Indians in Virginia and Carolina use … ', *Philosophical Transactions* (1686), **16**, 532–3.

100 J. Reinhold Forster, 'A letter … ', *Philosophical Transactions* (1772), **62**, 54–9, 57.

101 Father Papin, 'A letter … upon the mechanic arts and physick of the Indians', *Philosophical Transactions* (1713), **28**, 225–30, 225.

102 W. Sherard, 'The way of making several China varnishes', *Philosophical Transactions* (1700–1), **22**, 525–6.

used by the Indians, or, in a much later period, the interest in a new 'colouring substance' brought back from Tonga on Cook's second voyage.[103]

This same openness and willingness to concede that other cultures had much to teach was also evident in the Royal Society's eagerness to learn about the forms of medicine employed in distant lands. The tone for such enquiries was set by Boyle in the first part of his *Usefulnesse of Experimental Naturall Philosophy* (1663), in which he argued, 'Nor should we onely expect some improvements to the *Therapeutical* part of Physick, from the writings of so ingenious a People as the *Chineses*' but rather should also 'take notice of the Observations and Experiments' even of the '*Indians* and other barbarous Nations, without excepting the People of such part of Europe it selfe'.[104] The early Royal Society did indeed make considerable enquiries about the forms of medicine employed by the North American Indians. In 1666 Oldenburg presented a report from a long-standing resident of Virginia on the types of plant employed there to cure disease.[105] Ten years later, in the course of a generalized description of the customs of the Indian peoples of Virginia, an article in the *Philosophical Transactions* of 1676 stated that the '*Indians* being a rude sort of people use no Curiosity in preparing their Physick; yet are they not ignorant of the nature and uses of their plants'. It also detailed the way in which dried substances from a particular fish 'procureth speedy delivery to women in labour'.[106] Such widespread colonial interest in American Indian pharmacology, however, was, as Joyce Chaplin notes, often accompanied by a dismissal of the cultural and religious beliefs with which it was associated.[107] The Royal Society also followed Boyle's recommendation that particular attention be paid to Chinese medicine. This is evidenced by articles such as that which appeared in 1698, thanks to an East India surgeon, on 'a *China* Cabinet, full of the Instruments and Simples used by their Surgeons'. It added pointedly that among them were several that were 'new and different Shapes from the same used in *Europe*'.[108] Such admiration continued with a paper of 1733 paying tribute to the skill of Chinese physicians 'in judging of Distempers by the Pulse'.[109]

Interest and even admiration for indigenous cultures did not, however, preclude some instances of reporting on local customs with a view to the better exploitation of native peoples for European advantage. A 1669 list of queries from the society for an informant in Virginia included a request for information on how 'long any Savages may be train'd to endure ye water Diving or Swimming; ye use of such men being very great for Merchants, Ship-Masters and others'.[110] A report of instructions given to a Dutch East India Company captain exploring the lands near Japan published in the

103 Birch, op. cit. (39), iv, 520; *Philosophical Transactions* (1775), 65, 91–3, 93.
104 Cook, op. cit. (3), 414.
105 Birch, op. cit. (39), ii, 75.
106 T. Glover, 'An account of Virginia … ', *Philosophical Transactions* (1676), 11, 623–36, 634, 624.
107 J. Chaplin, *Subject Matter: Technology, the Body, and Science on the Anglo-American Frontier, 1500–1676*, Cambridge, MA, 2001, 197–8.
108 H. Sloane, 'An Account … ', *Philosophical Transactions* (1698), 20, 390–2, 390.
109 Royal Society Archives, RBO/18/45, Doctor Mortimer's account of the … proposals for printing a geographical and historical account of China, Read 7 June 1733, fol. 278.
110 Stearns, op. cit. (58), 697.

Philosophical Transactions in 1674 stipulated close attention to 'the nature and condition of the people' with a view to ascertaining whether they might be appropriate customers. If gold or silver were sighted, special care was to be taken to avoid giving the impression that these metals were particularly valued by the Europeans.[111]

Such scattered and miscellaneous information about different societies around the globe poured into the early Royal Society as grist to the mill of the natural historians. It illustrated the extent to which the study of human societies formed part of what the early Royal Society understood by that very compendious term 'natural history'. The territory of natural history was vast indeed. Like Bacon before them, the major figures of the early Royal Society thought of natural history as an activity involving the collection of data on all aspects of nature. Such a massive collection of information was to provide the foundation for a true system of causes, a natural philosophy that superseded the airy speculations and logic-chopping of the scholastics. Bacon himself certainly gave the impression that he regarded as distinct the compilation of natural history, with its vast storehouse of facts, and the erection of a system of natural philosophy. In his *Advancement of Learning* he had argued that natural history was the essential, but in some senses subordinate, prerequisite for 'physics', in its seventeenth-century sense as a synonym for natural philosophy. For Bacon, the appropriate division of labour was that 'NATVRAL HISTORY describeth the *varietie of things*; PHYSICKE [physics], the CAVSES'.[112] Such views were further developed in the *Description of the Intellectual Globe* (written in 1612), in which he viewed natural history as 'the primary material of philosophy', contending that 'the noblest end of natural history is this; to be the basic stuff and raw material of the true and lawful induction'.[113] In the *New Organon* he again presented a crucial but nonetheless subordinate role for the natural historian in promoting the endeavours of the natural philosopher: 'So we should have good hopes of natural philosophy once natural history (which is its basis and foundation) has been better organized, but none at all before.'[114]

Yet whether Bacon meant that the two activities should be kept in separate compartments can be questioned. He made much of the fact that his kind of natural history should be informed by the need to develop a true natural philosophy and that it should not be characterized by the random, curiosity-driven pursuits of the virtuosi. His goal for natural history, so he insisted in the *Great Instauration*, was 'not so much to give pleasure by displaying the variety of things ... as to illuminate the discovery of causes and nourish philosophy with its mother's milk'.[115] In the same spirit, he defined the purpose of natural history in his *Parasceve* as being 'to seek out and collect the abundance and variety of things which alone will do for constructing true axioms'.[116] In short, Bacon envisaged the ideal natural history as having an element of rationale and purpose which would enable it to promote the theorizing of the natural philosopher.

111 'A narrative ... ', *Philosophical Transactions* (1674), 9, 197–208, 200.
112 Kiernan, op. cit. (8), 82.
113 Rees, op. cit. (4), 105.
114 Rees, op. cit. (6), 157.
115 Rees, op. cit. (6), 39.
116 Rees, op. cit. (6), 457.

Purver goes so far as to argue that Bacon envisaged natural history and natural philosophy as working together in a form of syncopation with a continuous process, where one was adapted better to suit the goals of the other.[117]

In the early Royal Society, this uncertain boundary between the work of the natural historian and that of the natural philosopher was even more pronounced. In the manner of Bacon, notable figures within the early Royal Society seemed again to portray as separate activities the collecting linked with natural history and the more theoretical demands of natural philosophy. For example, Oldenburg adopted the familiar Baconian division between natural history and philosophy when seeking information from around the globe. His attempt to enlist the services of an informant in Syria prompted a declaration that it was the aim of the Royal Society 'to put together such a Natural History as our illustrious Bacon designed'.[118] Similarly, a request for scientific information from Florence was accompanied by the assertion that the Royal Society sought 'to compose a good Nat. History, to superstruct, in time, a solid and usefull Philosophy upon' – a phrase he also used in his introduction to Boyle's 1666 paper on the 'General Heads for a Natural History'.[119]

But the practice of Oldenburg's close ally Robert Boyle, whose work helped to set the tone for the early Royal Society more generally, indicates that there was some elision between the realm of the natural historian and that of the natural philosopher. When Boyle wrote to Oldenburg in 1666, for example, his discussion of the role of natural history indicated that he saw it as needing to be at least theoretically informed, as at times Bacon himself appears to have implied. Boyle rejected the idea that a system of natural philosophy (or what he called 'whole Body of Physicks') should 'be propos'd as the *Basis* of our Natural History', but nonetheless viewed the role of natural history as being to 'amplify & correct' such systems. Boyle argued that by thus linking natural history and natural philosophy it was possible to conduct experiments in a more fruitful manner:

> the knowledg of differing Theorys, may admonish a man to observe divers such Circumstances in an Experiment as otherwise 'tis like he would not heed; and sometimes too may prompt him to stretch the Experiment farther than else he would (and so make it produce new *Phaenomena*) ... [which] will conduce to make the History *both* more exact and compleat in it self.[120]

Such considerations have prompted Anstey and Hunter to argue in a recent paper for a major reappraisal of Boyle's methodology as being 'a two-stage reciprocal enterprise in which theory informs experiment with a view to constructing a natural history, which in turn informs theory'.[121]

---

117 M. Purver, *The Royal Society: Concept and Creation*, London, 1967, 35.

118 Oldenburg to Harpur, 22 May 1668, in Hall and Hall, op. cit. (26), iv, 422.

119 Oldenburg to Sir John Finch, 10 April 1666, in Hall and Hall, op. cit. (26), iii, 86; *Philosophical Transactions* (1666), 1, 186.

120 Boyle to Oldenburg, 13 June 1666, in Hunter, Clericuzio and Principe, op. cit. (27), iii, 171.

121 P. Anstey and M. Hunter, 'Robert Boyle's "Designe about natural history"', *Early Science and Medicine* (2008), 13, 83–126, 107.

Boyle might elsewhere fall back more unequivocally on the Baconian division between the natural historian and the natural philosopher. He described himself as an 'under-builder' whose humble role it was to conduct experiments so that 'men may in time be furnished with a sufficient stock of experiments to ground hypotheses and theories on'.[122] Yet in practice Boyle's example and the more general growth and success of experimental philosophy brought with them a transformation in natural history that over time was to lead to a narrowing of its domain to areas that did not lend themselves to experiment, especially living things. The Baconian division continued to throw a long shadow over the eighteenth century, particularly in Scotland, where his work was studied so avidly. Late in the century Lord Kames, for example, remarked that 'Natural History is confined to effects, leaving causes to Natural Philosophy'.[123] However, the expansion of the territory of experimental philosophy meant that natural history, which for Bacon had encompassed all aspects of the study of nature, became increasingly confined to areas which were reliant on description rather than experiment. As the eighteenth century wore on this meant that natural history increasingly referred to the study of the terraqueous globe and the classic kingdoms of animal (generally including humankind), vegetable and mineral. It was in this sense that William Wales, future astronomer on Cook's second and third voyages, reported back to the Royal Society from a 1768–9 expedition in the Hudson Bay area that he sought to provide some account relative to 'the natural history of the country ... first with respect to the inhabitants'.[124]

There were, however, a growing number in the eighteenth-century Royal Society who considered description alone, without recourse to experiment or mathematical analysis, not altogether scientifically worthy. In particular, throughout the eighteenth century there was tension between the mathematical practitioners and the natural historians that occasionally erupted within the society. The succession crisis for the presidency that followed the death of Newton in 1727 led James Jurin, the then editor of the *Philosophical Transactions*, to insert a mathematicians' manifesto. He contended that Newton 'was sensible, that something more than knowing the Name, the Shape and obvious Qualities of an Insect, a Pebble, a Plant or a Shell, was requisite to form a Philosopher, even of the lowest rank'. He also reiterated Newton's oft-repeated Baconian-based view '*That Natural History might indeed furnish Material for Natural Philosophy; but, however, Natural History was not Natural Philosophy*'.[125]

This was a passage which again surfaced in the politicking associated with a much later presidential election, that of 1772.[126] This was an indication that the issue still simmered. Finally, it led to open conflict in 1783–4 in the early presidency of the gentleman–collector Joseph Banks. This was a battle that Banks's lieutenant, Charles

---

122 R. Sargent, 'Learning from experience', in *Robert Boyle Reconsidered* (ed. M. Hunter), Cambridge, 1994, 57–78, 58.

123 H. Kames, *The Gentleman Farmer*, 4th edn, Edinburgh, 1798, 1 (first published 1776).

124 W. Wales, 'Journal of a voyage ... ', *Philosophical Transactions*, 1770, 100–36, 127.

125 See M. Feingold, 'Mathematicians and naturalists: Sir Isaac Newton and the Royal Society', in *Isaac Newton's Natural Philosophy* (ed. J. Z. Buchwald and I. Bernard Cohen), Cambridge, MA, 2001, 77–102, 77.

126 Feingold, op. cit. (125), 97.

Blagden, characterized to him as being described by his opponents as 'a struggle of the men of science against the Macaronis [virtuosi] of the Society, dignifying your friends by the latter title'.[127] The fact that Banks remained president of the Royal Society until his death in 1820 indicates that collecting still retained a place in the society. However, for Banks, as for Bacon, collecting should be linked with utility. Banks and his fellow natural historians could also draw confidence from the rising prestige of natural historians that derived from the increasing sway of classificatory systems, of which the most notable was that of Linnaeus.[128] When it came to the study of human society, however, such developments did little to promote a sense of its importance and scientific standing. Studies of human society appeared to have only limited direct utility: hence Johann Reinhold Forster's complaint in his 1772 study of the artefacts from Hudson Bay about 'the vulgar opinion, that it [natural history] is merely speculative, and incapable of being of the least utility in common life'.[129]

Nor did the study of human society combine readily with the classificatory impulse that was sweeping natural history. Ethnological descriptions of human customs were not readily neatly categorized, nor yet were the growing piles of artefacts, so prone to the ravages of time and insects, which had once been a feature of the Repository. Stadial views of the development of society helped to provide French and Scottish moral philosophers with the beginnings of a system of social science which drew on the Baconian goal of a natural history of humanity.[130] But such systems did not neatly link with the Linnaean quest to place all of nature (including the human world) in a well-ordered hierarchical system of classification. Such difficulties were compounded by the fact that, for all the early attempts to produce reports based on prescribed guidelines, what information it received on foreign peoples largely came from correspondents who were not very directly under the authority of the society. As in other areas of natural history, such correspondents tended to be more interested in description than in theory.[131]

The result then was a decline in the eighteenth century in the number of papers in the *Philosophical Transactions* largely devoted to ethnological subjects. Drawing on the table below, there were in the period between 1665 and 1699 a total of thirty-six ethnological papers on extra-European subjects (2.1 per cent of the total papers published in this period), while from 1700 to 1750 there were a mere sixteen (0.7 per cent of the total). There was a small revival in the second half of the century. The figure for the period between 1750 and 1799 was thirty (1.5 per cent of the total), though this was still

127 Fitzwilliam Museum, Cambridge, Perceval Collection, MS 215, Charles Blagden to Joseph Banks (27 December 1783). On this conflict see J. Gascoigne, *Joseph Banks and the English Enlightenment: Useful Knowledge and Polite Culture*, Cambridge, 1994, 62–6; J. L. Heilbron, 'A mathematicians' mutiny, with morals', in *World Changes: Thomas Kuhn and the Nature of Science* (ed. Paul Horwich), Cambridge, MA, 1993, 81–129.

128 On the rising prestige of natural history in the late eighteenth century see W. P. Jones, 'The vogue of natural history in England, 1750–1770', *Annals of Science* (1937), **2**, 345–52; T. L. Hankins, *Science and the Enlightenment*, Cambridge, 1985, 169; Mendyk, op. cit. (16), 243.

129 J. Reinhold Forster, 'A letter ... ', *Philosophical Transactions* (1772), **62**, 58.

130 P. Wood, 'The science of man', in *Cultures of Natural History* (ed. N. Jardine, J. A. Secord and E. C. Spary), Cambridge, 1996, 197–210.

131 A. Rusnock, 'Correspondence networks and the Royal Society', *BJHS* (1999), **32**, 155–69, 157.

Table 1. *Extra-European ethnologically related articles in the* Philosophical Transactions, *1665–1799*

Decade (to nearest *PT* vol.)	1665–9	1670–8	1683–92	1693–9	1700–9	1710–19	1720–30	1731–41	1742–50	1751–60	1761–9	1770–9	1780–9	1790–9
Africa	1	1			1	1					1			
Americas	6	3	3	2	1			1	2	2	2	4	1	
Asia		2	4	9	5	3			1	3		3		6
Middle East	1	1		2				1			1		1	2
South Pacific				1							1	3		
Total ethnological articles	8	7	7	14	7	4		2	3	5	5	10	2	8
Total all articles	358	525	344	459	431	256	416	468	574	530	471	463	343	248

below the proportion for the period up to 1700. This small late eighteenth-century revival reflects the renewed exploration of the South Pacific and other regions which followed the end of the Seven Years War in 1763 and increasing commercial contact with Asia. Interestingly, articles on Asian subjects loomed largest throughout the entire period from 1665 to 1799, both because of trading contacts and because of the number of reports sent back by the Jesuits, particularly on China.

A partial exception to this general trend was the study of race, a topic that became ever more heated and controversial over the course of the eighteenth century, in large part because of the debates about the slave trade.[132] In the first place, race was relevant to classification because it raised the issue of whether the human species was one or many, and what status should be attributed to racial differences. This was not a debate that appears to have loomed large within the Royal Society itself, perhaps because it was too divisive or because it involved dangerous questioning of the traditional Christian view that all human beings were of 'one blood'. But race could be discussed in limited ways with recourse to experimental data to determine the validity of the dominant theoretical construct, namely that race was the outcome of climatic variation. This was an issue frequently addressed within the Royal Society from soon after its foundation. In 1675 Martin Lister drew on observations from Barbados to argue that the colour of the blood of those of black African descent indicated that their skin colour was not a product of the climate, especially as those 'that live in the same Clime and heat with them, have as florid Blood as those that are in a cold Latitude'.[133] That industrious late seventeenth-century reporter on the natural history of Virginia, John Clayton, questioned whether skin colour could be explained in terms of climate, given that American Indians and African Americans lived on similar latitudes.[134] But others took a contrary view. It was argued in 1682 that 'Europeans by continuing to inhabit in Africa have been found to turn black, and that blacks in England after a few generations become white'.[135]

The fullest treatment of the subject was in a 1744–5 paper by John Mitchell MD of Virginia, entitled 'An Essay upon the Causes of Different Colours of People in Different Climates'. This was originally intended for a prize competition offered by the Bordeaux Academy. The paper set out to refute the view, most influentially advanced by Malpighi, that black skin colour could be explained in physiological rather than climatic terms.[136] In so doing, Mitchell argued that both science and Scripture supported the proposition that all human beings were of the same species: 'there is not so great, unnatural, and unaccountable a Difference between Negroes and white people on account of their Colours'.[137] Another American MD, James Bate of Maryland, was more uncertain about how far skin colour was physiologically based, using a case study of an

---

132 R. Bernasconi, *Concepts of Race in the Eighteenth Century*, Vol. I: *Bernier, Linnaeus and Maupertuis*, Bristol, 2001, pp. xi–xii.

133 'An extract of a letter of Mr. Lister's ... ', *Philosophical Transactions* (1675), 10, 399–400, 400.

134 Royal Society Archives, RB/1/39/10, 'An Account of Virginia', fol. 132.

135 M. Govier, 'The Royal Society, slavery and the island of Jamaica: 1660–1700', *Notes and Records of the Royal Society of London* (1999), 53, 203–17, 215.

136 Stearns, op. cit. (58), 545.

137 J. Mitchell, 'An essay ... ', *Philosophical Transactions* (1744–5), 43, 102–50, 131.

African American woman whose skin changed colour. In publishing the case in the *Philosophical Transactions* of 1759–60 he used almost Baconian language to deny that he was 'endeavouring to establish a favourite hypothesis'. He claimed he sought only to 'confine myself to a simple narration of such facts, as may prevent mistakes, or obviate difficulties, arising in the investigation of this difficult piece of physical history'. But in so doing he followed in the wake of Boyle by presenting a theoretically informed natural history, for he questioned both the theories that black colour could be explained by the effects of bile and also the action of heat. He did, however, invite his readers to suggest further experiments which he would 'be glad to execute'.[138]

However, such ventures into physical anthropology were rare in England. The field was largely to be developed in Germany under the tutelage of Johann Blumenbach in the setting of the medical faculty of the innovative University of Göttingen.[139] With its institutional base of gentlemanly amateurs, the Royal Society was a less conducive environment for such studies.[140] The lack of such a theoretically based approach to the study of human society meant that it tended to languish, even by comparison with other branches of natural history. These latter could at least be more readily combined with the great enterprise of neatly docketing away the manifold productions of nature in the systems of classification that loomed so large in the late eighteenth century. The study of humankind that the early Royal Society had considered an integral part of the large enterprise of natural history tended to be overshadowed as natural history itself changed character. With the rise of experimental philosophy, the domain of natural history was gradually reduced.

What remained were largely the more descriptive accounts of the animate and in-animate world. But these were transformed by the rise of classifactory systems and, by the late eighteenth century, claims that natural history could venture into the realm of causes once reserved for natural philosophy by approaching explanation in historical rather than mechanistic terms.[141] Such historically based approaches to the 'Science of Man' began to make an impact in Scotland, France and Germany.[142] However, they did not mesh too readily with the culture of the Royal Society, with its emphasis on ex-periment or classification. There the study of human society remained for some time in the traditional mould of a descriptive 'history'. Its accumulation of colourful detail and

138 J. Bate, 'An account … ', *Philosophical Transactions* (1759–60), 51, 175–8, 177–8.

139 J. Gascoigne, 'Blumenbach, Banks, and the beginnings of anthropology at Göttingen', in *Göttingen and the Development of the Natural Sciences* (ed. N. Rupke), Göttingen, 2002, 86–98.

140 M. Berman, '"Hegemony" and the amateur tradition in British science', *Journal of Social History* (1974–5), 8, 30–43.

141 J. Lyon, and P. R. Sloan, *From Natural History to the History of Nature: Readings from Buffon and His Critics*, Notre Dame, 1981, 121; P. Sloan, 'Natural History, 1670–1802', in *Companion to the History of Science* (ed. R. C. Olby, G. N. Cantor, J. R. R. Christie and M. J. S. Hodge), London, 1996, 295–313; P. Sloan, 'Natural History', in *The Cambridge History of Eighteenth-Century Philosophy* (ed. K. Haakonssen), 2 vols., Cambridge, 2006, ii, 903–38.

142 P. Wood, 'The science of man', in *Cultures of Natural History* (ed. N. Jardine, J. A. Secord and E. C. Spary), Cambridge, 1996, 197–210; D. Outram, 'New spaces in natural history', in ibid., 249–65; G. Stocking, 'French anthropology in 1800', in *Race, Culture and Evolution* (ed. G. Stocking), Chicago, 1982, 13–41; J. Gascoigne, 'The Pacific and the German Enlightenment', in *The Anthropology of the Enlightenment* (ed. L. Wolff and M. Cipolloni), Stanford, 2007, 141–71.

X

cumbersome artefacts seemed too redolent of an earlier form of natural history to be embraced with scientific enthusiasm. The rise of the study of human society as a scientific discipline in the form of anthropology was to be a feature of the nineteenth century, when the specialist societies challenged the Royal Society's traditional oversight over all of nature. Bodies such as the Ethnological Society of London (founded 1843) were to provide a more congenial setting for the evolutionary theoretical frameworks that gave this new discipline the scientific status it had once enjoyed in the older Baconian understanding of 'natural history'.

INDEX